T4-AHW-007

Proceedings of the Lebedev Physics Institute
Academy of Sciences of the USSR
Series Editor: N.G. Basov

Proceedings of the Lebedev Physics Institute
Academy of Sciences of the USSR

Volume 165	**Solitons and Instantons, Operator Quantization,** Edited by V.L. Ginzburg
Volume 166	**The Nonlinear Optics of Semiconductor Lasers,** Edited by N.G. Basov
Volume 167	**Group Theory, Gravitation and Elementary Particle Physics,** Edited by A.A. Komar
Volume 168	**Issues in Intense-Field Quantum Electrodynamics,** Edited by V.L. Ginzburg
Volume 169	**The Physical Effects in the Gravitational Field of Black Holes,** Edited by M.A. Markov
Volume 170	**The Theory of Target Compression by Longwave Laser Emission,** Edited by G.V. Sklizkov
Volume 171	**Research on Laser Theory,** Edited by A.N. Orayevskiy
Volume 172	**Phase Conjugation of Laser Emission,** Edited by N.G. Basov
Volume 173	**Quantum Mechanics and Statistical Methods,** Edited by M.M. Sushchinskiy
Volume 174	**Nonequilibrium Superconductivity,** Edited by V.L. Ginzburg
Volume 174 Supplemental Volume 1	**Thermodynamics and Electrodynamics of Superconductors,** Edited by V.L. Ginzburg
Volume 174 Supplemental Volume 2	**Electron Liquid Theory of Normal Metals,** Edited by V.P. Silin
Volume 175	**Luminescence Centers of Rare Earth Ions in Crystal Phosphors,** Edited by M.D. Galanin

Volume 175 Supplemental Volume	**Luminescence and Anistropy of Zinc Sulfide Crystals,** Edited by M.D. Galanin
Volume 176	**Classical and Quantum Effects in Electrodynamics,** Edited by A.A. Komar
Volume 176 Supplemental Volume	**Quantization, Gravitation and Group Methods in Physics,** Edited by A.A. Komar
Volume 177	**Stoichiometry and Its Influence on the Physical Properties of Crystalline Compounds,** Edited by Yu. M. Popov
Volume 178	**The Delfin Laser Thermonuclear Installation: Operational Complex and Future Directions,** Edited by G.V. Sklizkov
Volume 179	**X-Ray Plasma Spectroscopy and the Properties of Multiply-Charged Ions,** Edited by I.I. Sobel'man
Volume 179 Supplemental Volume 1	**Interaction of Ultrashort Pulses with Matter,** Edited by M.D. Galanin
Volume 179 Supplemental Volume 2	**Electron-Excited Molecules in Nonequilibrium Plasma,** Edited by N.N. Sobolev
Volume 180	**Inelastic Light Scattering in Crystals,** Edited by M.M. Sushchinskiy
Volume 181	**Metal Vapor and Chalcogenide Lasers,** Edited by G.G. Petrash

Proceedings of the Lebedev Physics Institute
Academy of Sciences of the USSR
Series Editor N. G. Basov

Volume 177

STOICHIOMETRY IN CRYSTAL COMPOUNDS AND ITS INFLUENCE ON THEIR PHYSICAL PROPERTIES

Edited by Yu. M. Popov

Translated by Kevin S. Hendzel

NOVA SCIENCE PUBLISHERS
COMMACK

NOVA SCIENCE PUBLISHERS
283 Commack Road
Suite 300
Commack, New York 11725

This book is being published under exclusive English Language rights granted to Nova Science Publishers, Inc. by the All-Union Copyright Agency of the USSR (VAAP).

Library of Congress Cataloging-in-Publication Data

Stekhiometrii͡a v kristallicheskikh soedinenii͡akh i ee vlii͡anie na ikh fizicheskie svoĭstva. English.
Stoichiometry and its influence on the physical properties of crystalline compounds.

(Proceedings of the Lebedev Physics Institute of the Academy of Sciences of the USSR ; v. 177)
Translation of: Stekhiometrii͡a v kristallicheskikh soedinenii͡akh i ee vlii͡anie na ikh fizicheskie svoĭstva.
Bibliography: p.
1. Crystals--Growth. 2. Stoichiometry. I. Popov, I͡U. M. II. Title. III. Series: Trudy Fizicheskogo instituta. English ; v. 177.
QC1.A4114 vol. 177 [QD921] 530 s 88-15146
ISBN 0-941743-21-7 [548'.5]

The original Russian-language version of this book was published by Nauka Publishing House in 1987.

Printed in the United States of America

CONTENTS

THE KINETICS OF VOID NUCLEATION AND GROWTH IN CRYSTALS

F.Kh. Mirzoev, E.P. Fetisov, L.A. Shelepin

FORMATION OF VOID SUPERLATTICES IN SOLIDS

F.Kh. Mirzoev, E.P. Fetisov, L.A. Shelepin

STRUCTURAL RELAXATION IN IMPACTED SOLIDS

S.P. Merkulova, L.A. Shelepin, A.A. Shubin

THE INFLUENCE OF STOICHIOMETRY IN A^2B^6 MONOCRYSTAL COMPOUNDS ON THE CHARACTERISTICS OF A SEMICONDUCTOR ELECTRON-BEAM PUMPED LASER

I.V. Akimova, V.I. Kozlovskiy, Yu.V. Korostelin, A.S. Nasibov, A.N. Pechenov, P.V. Reznikov, V.I. Reshetov, Ya.K. Skasyrskiy, P.V. Shapkin

PIEZOOPTIC EFFECTS AND THE INFLUENCE OF ANISOTROPIC DEFORMATION ON GaInPAs/InP HETEROLASERS

P.G. Eliseev, B.N. Sverdlov, N. Shokhudzhaev

RADIATIVE CHARACTERISTICS OF NONSTOICHIOMETRIC MELT -GROWN AlGaAs/GaAs LASER HETEROSTRUCTURES

P.G. Eliseev, A.A. Zherdev, V.S. Kargapol'tsev, O.N. Talenskiy, G.G. Kharisov

LUMINESCENCE OF CdS AS A FUNCTION OF CRYSTAL POSITION IN THE GROWTH ZONE

N.A. Martovitskaya, S.A. Pendyur, O.N. Talenskiy

FOREWORD

This volume is devoted to an investigation of A^2B^6 and A^4B^6 crystal compounds and certain A^3B^5 compound heterostructures. Materials based on these compounds play an important and increasingly critical role in quantum electronics, semiconductor physics and engineering and luminescence. This volume analyzes the properties of these materials as well as fabrication techniques and areas of application. An important feature of these crystal compounds is the degree of deviation from stoichiometry determined by the chemical formula.

In crystal compounds composition varies while homogeneity is conserved. An excess amount of a specific component determined by the deviation from stoichiometry can have a significant influence on the properties of the material. However the problem of stoichiometry has not received significant attention to date. This is not surprising, since deviations from stoichiometry begin to play a determinant role if the crystalline compound is purified to a sufficient degree to eliminate extraneous impurities so that their concentration is at least an order of magnitude below the concentration of the thermodynamic intrinsic defects. The latter will then determine such characteristics as the type of conductivity, the concentration, mobility and lifetime of the free charge carriers, photosensitivity and luminescence as well as the semiconductor doping processes.

A detailed analysis of intrinsic defects in crystalline compounds and their dependence on deviation from stoichiometry is found in the chapter entitled "The Problem of Stoichiometry in Type A^2B^6 and A^4B^6 Semiconductors of Variable Composition", which also provides a survey of the crytallochemical and physiochemical features of a number of A^2B^6 and A^4B^6 compounds. The composition stability ranges, methods of controlling deviation from stoichiometry and the aspects of the electrical and luminescence properties are investigated experimentally.

An important conclusion of this study is the establishment of the influence of cluster formation processes on material properties: the formation of precipitates due to excess atoms, and the formation of microvoids due to excess vacancies. Cluster formation changes the number of free carriers and influences the luminescence properties and also plays an important role in growing crystal films. Clearly an analysis of these properties can be considered an important theoretical problem for investigating the properties of crystalline compounds.

The next three theoretical articles examine various aspects of cluster - void and precipitate - formation. The chapter entitled "The Kinetics of Void Nucleation and Growth in Crystals" presents a theory of the formation of the new phase that includes all stages of the process within a single approach. This theory will make it possible to determine its primary characteristics including the critical nucleus nucleation time, the duration of the initial stage, the characteristic vacancy condensation time and maximum void dimensions. The influence of stresses on the void formation process is investigated.

The study "The Formation of Void Superlattices in Solids" considers formation processes of defect superlattices. Primary attention is given to an analysis of spatial void ordering and the formation a lattice with crystalline matrix symmetry under irradiation by fast particles. Fabricating materials with an ordered defect structure is of significant interest for a number of applications.

A number of new aspects for fabrication of materials with given properties can be found in the chapters "The Problem of Stoichiometry in Type A^2B^6 and A^4B^6 Semiconductors of Variable Composition" and "External Stimulation Mechanisms for Crystal Growing Processes." The crystal growing processes that employ UV radiation from an ultrahigh pressure xenon lamp and radiation from pulsed lasers emitting in the UV are of special interest. This crystal growing technique has made it possible to improve their quality, reduce the number of defects, reduce the epitaxy temperature and increase the growth rate. The mechanisms of radiation action on epitaxial growth of crystalline compounds are proposed and analyzed.

The chapter "Structural Relaxation in a Solid Under Impact Action" contains results from experimental investigations of the change in the spatial structure of defects under impact current and laser irradiation of thin foil and films. Data are derived on the structural relaxation processes and the characteristic temporal relations of the change in the spatial structure of defects in semiconductor materials exposed to pulsed action.

This volume focuses significant attention on the fabrication and application of semiconductor materials in quantum electronics. The study "The Influence of Stoichiometry of A^2B^6 Monocrystal Compounds on the Characteristics of an Electron-Beam-Pumped Semiconductor Laser" uses results from electron microscope photoluminescence and cathodoluminescence investigations of a number of A^2B^6 compound monocrystals to demonstrate that the degradation in the characteristics of lasers based on these materials can be attributed to cluster formation processes. As the crystal composition deviates from stoichiometry the density of metal microprecipitates increases in the excess case and the vacancy concentration increases in the shortage case. These results are of interest to designers of laser cathode-ray devices.

The chapters entitled "Piezoelectric Effects and the Influence of Anisotropic Deformation in GaInPAs/InP Based Heterolasers" and "Radiative Characteristics of Nonstoichiometric-Melt-Grown

AlGaAs/GaAs Laser Heterostructures" carry out an experimental investigation of laser heterostructures and their fabrication technology. Specifically results from research on the technology of AlGaAs/GaAs heterostructures developed for high-power semiconductor injection lasers are generalized. The experience of fabricating lasers based on these structures that emit in the visible (red) portion of the spectrum by introducing aluminum into the active layer of the heterostructure is also generalized. It is demonstrated that achieving a radiation power of greater than 0.1 w at 700 nm with liquid nitrogen cooling is possible.

Overall this volume contains an extensive collection of new and useful information on the properties of crystalline compounds, the influence of deviation from stoichiometry on these compounds and the use of a number of structures in semiconductor lasers.

Yu.M. Popov

THE PROBLEM OF STOICHIOMETRY IN TYPE A^2B^6 AND A^4B^6 SEMICONDUCTORS OF VARIABLE COMPOSITION

G.A. Kalyuzhnaya, K.V. Kiseleva

ABSTRACT

This survey presents results from experimental investigations of the nature of intrinsic defects and estimates their concentration for two model materials: CdTe and PbTe which are widely used A^2B^6 and A^4B^6 semiconductors. The total volumetric concentration of electrically-charged and neutral intrinsic defects were estimated by using a set of independent experimental techniques and then carrying out a joint analysis of the derived results. Methods of controlling deviation from stoichiometry by selected doping and annealing are examined including the case of electromagnetic irradiation and its influence on the crystallization of these semiconductors.

Introduction

Today the problem of stoichiometry or more accurately the problem of the deviation of a compound from stoichiometry is one of the most important issues in the material processing of semiconductors. This is because without exception semiconductors, including elementary semiconductors, are phases of variable composition. They are differentiated by the size of their composition stability ranges in solids and the methods of achieving a nonstoichiometric composition on the atomic level.

Until very recently the problem of stoichiometry lay outside the field of semiconductor physics. These materials were no longer considered compounds of consistent stoichiometric composition only after investigators ran up against the "unexplainable" nonreproducibility of the physical properties of semiconductors that had been purified to the maximum possible degree which, for example, for A^4B^6 is 10^{16}-10^{17} cm^{-3}.

Further investigations revealed that if it is possible to purify semiconductors to eliminate extraneous impurities so that their concentration is at least an order of magnitude below the concentration of intrinsic thermodynamic-equilibrium defects these defects will themselves determine the primary semiconductor characteristics: type of conductivity, concentration, mobility, and lifetime of the free charge carriers, photosensitivity and luminescence. Moreover the intrinsic defects have a significant influence on the semiconductor doping processes. The physics of actual semiconductor crystals is therefore closely related to the problem of investigating the nature and properties of the intrinsic defects in the crystal lattice. The importance and practical value of such research is clearly demonstrated by the following fact. We know that an increase in the concentration of intrinsic thermodynamic-equilibrium defects increases the practical difficulties of employing semiconductor materials. Indeed the following problems become more complex in a number of classes of A, A^3B^5, A^2B^6 and A^4B^6-type semiconductors with composition stability ranges $<10^{-5}$; $\sim10^{-2}$; 0.1, and 1 at.%, respectively:

a) development of fabrication technology for materials with given specific and controllable properties;

b) control over the doping processes of these materials;

c) assuring operational stability and reliability of semiconductor devices fabricated on the basis of such materials.

The present study contains results from experimental investigations carried out at the Physics Institute of the Academy of Sciences into the nature of intrinsic defects together with estimates of their concentrations for two model materials: CdTe and PbTe; these are standard, widely-employed type A^2B^6 and A^4B^6 semiconductors.

We know that the previous practice for undoped semiconductors was to identify the intrinsic defect concentration, i.e., the dimensions of the composition stability range with the free charge carrier concentration measured by the Hall effect. In this case when estimating the dimensions of the composition stability ranges the authors naturally could not account for the electrically neutral intrinsic defects and consequently could not determine the true deviation of the compound from stoichiometry.

At present there is no direct method in scientific practice for determining the total volumetric concentration of electrically charged and neutral intrinsic defects in solids. This value can be estimated for CdTe and PbTe only by using a set of independent experimental techniques followed by a joint analysis of the derived results (see Section 2).

We employed derived results to develop techniques for controlling the semiconductor properties of cadmium, lead, and tin tellurides. Methods of controlling deviation from stoichiometry were proposed and implemented; these techniques employed special annealing and doping combined with electromagnetic radiation to influence the crystallization process of these semiconductors (see Sections 3, 4).

Before presenting the primary material in this survey we will describe the features of the physiochemical and crystallochemical properties of A^2B^6 and A^4B^6 compounds. General information used to classify crystal lattice defects and their formation energies are given and the phase microdiagrams for PbTe are described together with the crystalline structure of A^2B^6 and A^4B^6 compounds.

1. CRYSTALLOCHEMICAL AND PHYSIOCHEMICAL FEATURES OF CADMIUM AND LEAD TELLURIDES

1.1. Crystal Lattice Defects

The crystal lattice of an ideal crystal consists of an infinite collection of atoms (or groups of atoms) separated by distances called the lattice periods in three directions. However there are always imperfections in ideal periodicity in an ideal crystal; these are called crystal structural defects. Disruptions arising due to atomic (ionic) thermal vibrations near the centers of gravity (the periodic lattice sites) or from the alternation of isotopes of the same element in the lattice will not be considered here, since it is possible to neglect such imperfections in a first approximation when considering the problem of stoichiometry.

Crystal structural defects are normally classified by length and chemical composition. There are two types of defects in the first case: point defects that encompass one-two lattice sites or interstitials, and extended defects in one or more dimensions. Point defects include vacancies and interstitial atoms, while extended defects include dislocations, block boundaries, microprecipitates (precipitates), microcaverns, clusters, and cracks. In turn defects of any size can also be classified among two groups in terms of chemical composition: intrinsic defects consisting of atoms intrinsic to the given compound or their vacancies and extrinsic defects containing atoms that are impurity atoms with respect to the composition of the initial crystal.

In the general case of a binary crystal MN, where M is a metal while N is a more electrically negative element, different defects can coexist, although there is a strict relation between their concentrations as described by the following nine relations (the figures in brackets represent concentration):

$$[M_i][N_i] = K_1, \tag{1}$$

$$[M_N]^{1/2}[N_i] = K_2, \tag{2}$$

$$[V_N][N_i] = K_3 \quad \text{(Frenkel disorder)} \tag{3}$$

$$[M_i][N_M]^{1/2} = K_4 \tag{4}$$

$$[M_N][N_M] = K_5 \quad \text{(antistructural disorder)} \tag{5}$$

$$[V_N][N_M]^{1/2} = K_6 \tag{6}$$

$$[M_i][V_M] = K_7 \quad \text{(Frenkel disorder)} \tag{7}$$

$$[M_N]^{1/2}[V_M] = K_8 \tag{8}$$

$$[V_M][V_N] = K_9 \quad \text{(Shottky disorder)} \tag{9}$$

where K_i are interrelated constants, for example, $K_1K_9 = K_3K_7$ [1]. In principle all possible disorder types can exist simultaneously, although normally one disorder type is prevalent and it is determined by the crystalline structure of the condensed phase and the nature of its constituent atoms. Therefore (as will be demonstrated below) the prevalent defect is Shottky disorder described by relation (9) in undoped cadmium and lead tellurides, which are bilateral phases of variable composition, in the entire composition stability range.

1.2. Certain Thermodynamic Elements of Phases of Variable Composition

In 1930 Wagner and Shottky developed a physiochemical theory of defects in semiconducting compounds [2]. The mathematical apparatus was then developed for a thermodynamical description of the chemistry of defects and experiments were carried out that confirmed the validity of the new theory. Kröger and Vink's graphical method of describing crystal lattice disorder [3, 4] was useful in the successive application of the Wagner-Shottky theory. Chemical and physical methods of investigating solid semiconductor compounds have also been improved significantly in recent decades and in conjunction with theory and the graphical technique for describing the thermodynamics of intrinsic crystal defects it has become possible to establish a number of characteristics of these defects as well as their interrelationship and the temperature dependence of their thermodynamically equilibrium concentration (i.e., the composition stability ranges of various compounds).

In the general case of a multiphase system the number of independent phases and the width of the composition stability range of each phase as well as the number and length of the bilateral phase regions can be determined graphically by a set of graphs plotting the dependence of the Gibbs' free energy G_i on the composition x of each phase of the system (Fig. 1). We then draft the common tangents to the curves $G_i(x)$ the coexisting phases at a given pressure P and temperature T and derive the envelope G-curve for the entire system consisting of the line segments of the common tangents to the neighboring phases and the segments of $G_i(0)$ of the separate phases bounded by the points of tangency. In this case the number of independent phases in the system will be equal to the number of segments of the curves G of the separate phases belonging to the envelope G-curve, while the number of bilateral phase regions is equal to the number of line segments of the envelope G-curve of the system (see Fig. 1).

We will use this approach to determine the region of existence of the binary compound MN where, as before, M is a metal while N is a more electrically negative element. A crystal of such composition at the given pressure P and temperature T will have a structure corresponding to minimum Gibbs free energy

$$G = H - TS, \tag{10}$$

where H is enthalpy, T is absolute temperature and S is entropy. At absolute zero the entropy term of equation (10) vanishes and $G = H$. In this case the MN crystal is completely ordered, i.e., the M and N atoms properly occupy all sites of the corresponding M- and N- sublattices which results in minimum free energy G and enthalpy H. Such a completely ordered and strictly stoichiometric hypothetical crystal is called a perfect crystal. Fig. 2a shows the dependencies of G on the composition x for the MN compound and the elements M and N for $T = 0$ K and at pressure P. Drawing the tangents for the M, MN and N phases we can determine that in this case the exact stoichiometric composition of MN is the only stable composition (see Fig. 2a), i.e., there is no composition stability range in a perfect crystal when $T = 0$ K.

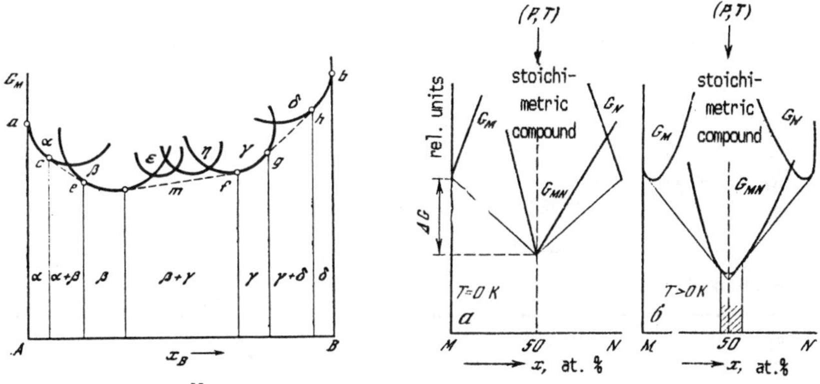

Fig. 1. Envelope G^M curve

Fig. 2. The isothermic Gibbs free energy 1 g·atom of the MN compound and the coexisting M and N phases at a constant pressure plotted as a function of the atomic percent of component N
a – $T = 0°$ K. The MN compound is stable only with a stoichiometric composition; b – $T = 0°$ K. The MN compound is thermodynamically stable in a certain range of compositions

When the temperature rises the entropy term in equation (10) grows and the minimum of the Gibbs free energy G can correspond to a disordered crystal state where, in principle all forms of intrinsic defects can exist. With all the structural varieties of defects their formation processes are identical from the energy viewpoint: these are endothermic processes causing an increase in the internal energy (enthalpy) of the system. Fig. 2b gives the corresponding curves of the Gibbs free energy G of the same M, MN, and N phases for $T > 0$ K accounting for the entropy of the resulting disorder. Using the method discussed above we draw tangents to the free energy curves G_M, G_{MN}, and G_N and then see that a range of stable compositions (the hatched region in Fig. 2b) is formed in place of a single stable composition. Then due to the natural disorder of the crystal lattice the MN phase at $T > 0$ K becomes a phase of variable composition that is stable within a specific range of compositions, called the composition stability range. In the general case the composition stability range of a compound is asymmetric with respect to the stoichiometric composition, i.e., $\Sigma M^* \neq \Sigma N^*$, where M* and N* designate the prevalent disorder type of the M- and N-atoms.

It is easy to understand that the degree of disorder (the concentration of intrinsic defects) depends on the defect formation enthalpy ΔH: if it is large the number of formed defects is small and vice versa. Consequently the formation enthalpy of the prevalent intrinsic defects will itself determine the dimensions of the composition stability range of the compound of variable composition. A comparison of vacancy formation enthalpies ΔH_V obtained from the literature for certain classes of A^2B^6 and A^4B^6 superconductors in which vacancies (or Shottky defects) are the prevalent type of defects to known dimensions of their composition stability ranges confirms the validity of the conclusion derived above (see Table 1).

Table 1

Vacancy formation enthalpies ΔH_V for select type A, A^2B^6 A^4B^6 semiconductors [5].

Semiconductor	ΔH_V^A, eV	ΔH_V^B, eV	Maximum deviation from stoichiometry δ, at.%
Si	2.4	—	$< 10^{-5}$
Ge	2	—	
GaAs	1.6	2	$\sim 10^{-2}$ [6]
ZnSe	1.25	1.85	
InP	1.5	1.3	
GaP	2.5	1.6	
GaSb	1.3	1.7	
InAs	1.4	1.8	
InSb	1.2	1.4	
CdTe	0.8	1.1	$\sim 10^{-1}$ (see this volume)
HgTe	0.7	0.85	
PbSe	0.7	0.95	~ 1 (see this volume)
PbTe	0.69	0.8	
SnTe	0.85	1.35	

1.3. Phase Diagrams of A^4B^6 Compounds

We know that the phase diagrams of semiconductor systems that establish a functional relation between the properties of compounds and their composition make it possible to substantiate selection of the material growing technique. Knowledge of the phase equilibria is no less important for an experimental investigation of deviations from stoichiometry in semiconductors of variable composition. In such materials as lead chalcogenides the electrical conduction and Hall effect vary by several orders of magnitude with a variation of 0.001-01 at.% in the composition. This fact was used to draft microdiagrams of A^4B^6 compounds that are different from the regular diagrams or classical phase diagrams with a significantly expanded scale of the composition axis (by a factor of 10^6) in the vicinity of the compound. The charge carrier concentration has been the only information available on the composition of a solid compound over a long time period for lead chalcogenides with a narrow composition stability range. However this resulted in the mistaken and widely-held view that the deviation from stoichiometry can be identified with the carrier concentration.

This section provides the primary types of microdiagrams of A^4B^6 compounds which have been investigated more extensively than other classes of binary semiconductors. Kröger and Bloem [7], Brebrick and Strauss [8], and Novoselova, et al. [9] made the primary contributions to the investigation of phase diagrams.

***T* - *X* diagrams.** The majority of A^4B^6 compounds have a single congruently-fluid compound (Fig. 3) whose ideal crystal structure corresponds to the formula MN (M is a group IV element, N is a group VI element). The composition of the compound in a regular *T-X* diagram is shown in Fig. 3 by the line with x = 0.5. This type of phase diagram is characteristic of the systems Pb-S, Pb-Se, Pb-Te, Sn-Te [10].

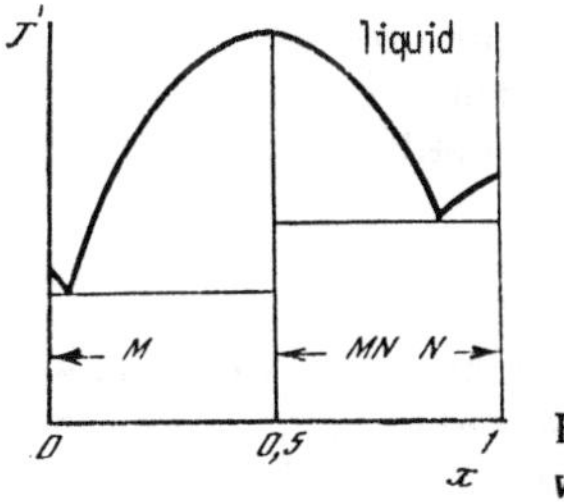

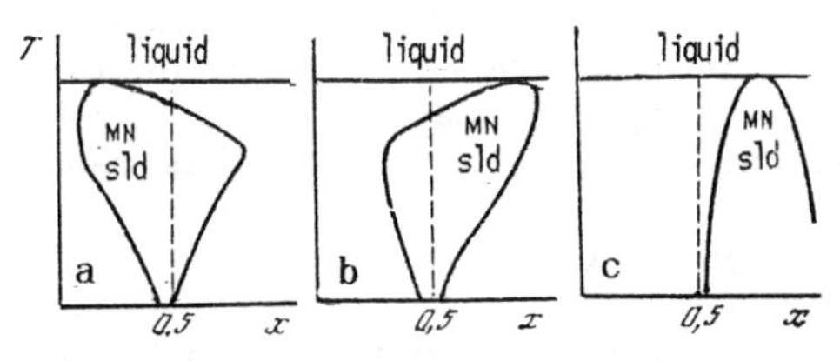

Fig. 4. Types of *T-X*-projections for IV-VI with an expanded X-axis scale in the vicinity of MN (x = 0.5)

Fig. 3. *T-X* projection scheme for the IV-VI binary system consisting of elements M (IV) and N(VI)

In order to investigate the composition stability range of the MN compound the part of the diagram in the vicinity x = 0.5 was investigated in detail. Fig. 4 gives a scheme of this section with a significantly-expanded X-axis scale, illustrating three types of composition stability ranges for the A^4B^6 compounds. The diagrams in Fig. 4a, b show the composition stability ranges that include the stoichiometric composition but only at temperatures below the maximum T_m; in Fig. 4c the composition stability range does not include the stoichiometric composition. In all cases the maximum melting point where solid and liquid phases of identical composition coexist do not correspond to the stoichiometric composition. One other feature is that beginning at certain temperatures the solubility of elements in the compound drops with a drop in temperature; so-called retrograde solubility of the components in the compound exists in this range causing precipitation of microprecipitates upon cooling.

We will consider the cooling of sample MN of composition A saturated by N (Fig. 5). If its composition did not change with cooling, its cooling curve will be represented by line *AB* which lies outside the composition stability range of MN and therefore the sample is supersaturated by N at low temperatures. The excess component N forms microprecipitates or a phase based on N in the crystal bulk. This precipitation process occurs much faster than the release of N to the vapor phase through the crystal surface, since this occurs by diffusion at short distances up to $T = C$ (the effective hardening temperature) at which the diffusion rate begins to be much lower and this will eliminate further precipitation of microprecipitates from cooling a crystal of composition *C*. The M or N microprecipitates will remain dissolved in the lattice, although they will make no contribution to the charge carrier concentration. This is why estimating the composition stability range based on data on free carrier concentration yields an error.

In order to determine the solidi forming the boundaries of the composition stability range of MN it is necessary to prepare samples of "extreme" or saturated composition in equilibrium with a liquid enriched with M or N and to determine the composition of these samples. Qualitative data for determining the composition of PbTe and other A^4B^6 compounds have been obtained by measuring the carrier concentration. Here Brebrick [11] has demonstrated that

$$n - p = [\mathrm{Pb}] - [\mathrm{Te}], \tag{11}$$

where n and p are the electron and hole concentrations, respectively for T = 298 K, while [ ] is the atomic concentration. The solidus of PbTe on the *T-X* microdiagram was calculated based on carrier concentration data in n- and p-PbTe at 25°C as a function of the annealing temperature by the formula

$$(\text{at.\% Te})/100 = 1/2 - (n - p)/4 + 1.485 \cdot 10^{22} \tag{12}$$

(the density of PbTe is 8.25 g/cm^3 with a molecular weight of 334.8 g).

Equation (12) is based on the observation that, first, lead telluride is a degenerate semiconductor at room temperature and, second, there is no proof of multiple ionization of defects associated with deviation from stoichiometry. Here we have ignored the influence on the composition of impurities, inadequate hardening rate, and point defect clusters. According to the Hall effect data all donors and acceptors at room temperature are completely ionized and therefore the carrier concentration is equal to the difference in the defect concentrations.

The *T-X* phase diagram of the Pb-Te system and the composition stability range of the PbTe compound determined in studies [9, 11, 12] are given in Fig. 6 and 7. As indicated by Fig. 7 the composition stability range of PbTe includes the stoichiometric composition, while its largest size estimated by carrier concentration is observed at temperatures 700-800°C and amounts to ~10^{18} cm^{-3} Pb and Te; the maximum melting point of 934°C corresponds to a composition with excess Te. The composition stability range estimated from the partial pressure of telluride vapors was somewhat broader [9]. We find from the experimental data that PbTe samples containing more than 50% Pb have *n*-type conductivity, while samples containing more than 50% Te have *p*-type conductivity. Here the carrier concentration grows monotonically with growth of deviations from stoichiometry. Unlike PbTe the composition stability range of SnTe (Fig. 8) lies wholly on the side of excess Te, does not include the stoichiometric composition and is largest at 600°C. The maximum melting point is 806°C and corresponds to a composition with 50.9 at.% Te [13].

Lead and tin tellurides form a continuous series of $Pb_{1-x}Sn_xTe$ solid solutions following the quasi-binary PbTe-SnTe plan; their composition stability range expands towards Te with growth of *x* (Fig. 9) [14].

The PbTe compound in solid form is a bilateral phase of variable composition, while SnTe in solid form is a unilateral phase. The maximum melting point of the compound $T_{max\ m.}$ does not correspond to the stoichiometric composition but rather is shifted towards Te. The shift of $T_{max\ m.}$ can be related to the difference in the free defect formation energies on different sides of the stoichiometry line. The width of the composition stability range of PbTe and the solid solutions based on PbTe narrows with a reduction in temperature (retrograde solubility). This breaks down the single-phase composition of the material when it is cooled and results in microcondensation of the components.

***P-X* diagrams.** Knowledge of the vapor-solid equilibrium is needed to fabricate materials with a specific and controllable composition by growing crystals from the gas phase and also for subsequent annealing.

The vapor phase in equilibrium with the A^4B^6 compound consists of MN molecules, M atoms, and one or more associations of N atoms. For the 500-900°C range under examination the prevalent N vapor component is the diatomic molecule N_2. In the presence of MN·the relation of the partial pressures of the components is given by the equation

$$P_M P_{(N_2)^{1/2}} = \exp[2\Delta G_f/RT], \quad (13)$$

where P_M and P_{N_2} are the partial pressures, respectively, R is the gas constant, T is temperature in degrees K, ΔG_f is the free energy of the reaction

$$1/2M_{gas,atm} + 1/4N_{2gas,atm} \rightarrow 1/2MN_{cryst}. \quad (14)$$

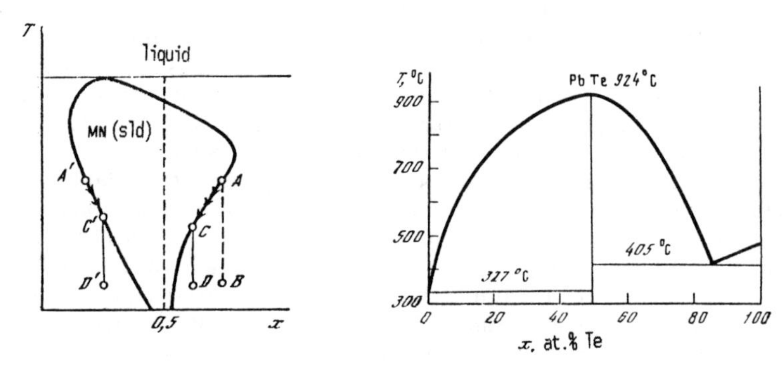

Fig. 5. Configuration of the T-X-projection near $x = 0.5$ illustrating the influence of microprecipitation of the components

Fig. 6. T-X phase diagram of the Pb-Te system

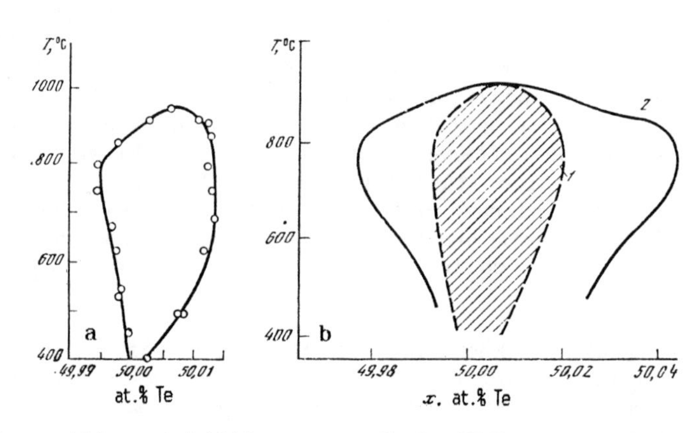

Fig. 7. Composition stability range of the PbTe compound
a - calculated by the carrier concentration; b - calculated by the saturated vapor pressure
1, 2 - based on data from studies [11] and [9], respectively

In a vapor in equilibrium with the stoichiometric composition of solid MN the partial pressure of N_2 is equal to half the partial pressure of M: $P_{N_2} = 1/2P_M$. The total pressure of the vapor phase for T = const is at a maximum for a congruently evaporating composition when vaporization occurs without a change in composition and in the general case will not necessarily correspond to the stoichiometric composition.

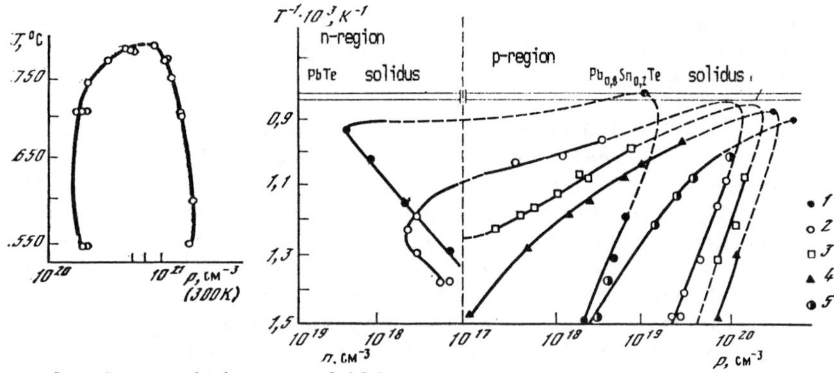

Fig. 8. Composition stability range of the SnTe Compound

Fig. 9. Composition stability range of $Pb_{1-x}Sn_xTe$ determined by the carrier concentration for different solid solution compositions 1 - x = 0.0; 2 - x = 0.13; 3 - x = 0.2; 4 - x = 0.32; 5 - x = 0.5

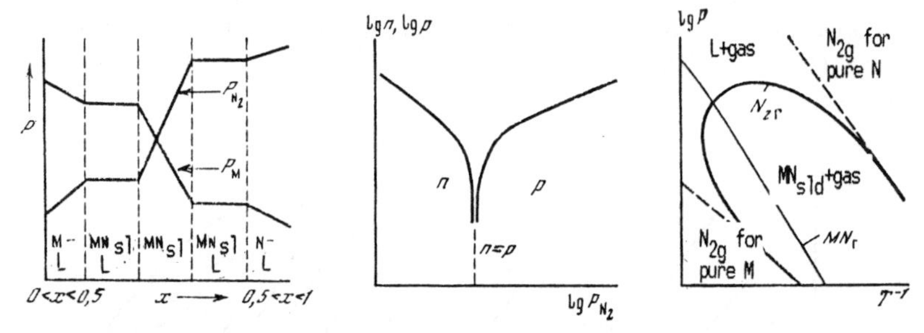

Fig. 10. Configuration of the P-x-projection for the M-N system in the vicinity of solid MN

Fig. 11. Schematic dependence of carrier concentration on the partial pressure of N_2 for MN_s at a constant temperature

Fig. 12. Configuration of the P-T-projection for the M-N system in the vicinity of solid MN

It has been discovered [8] that the partial pressure of MN molecules does not change significantly in the composition stability range of solid MN at the same time that the partial pressure of the components can vary by an order of magnitude. This is characteristic of the highly-ordered solid phase of substances with a narrow composition stability range. Consequently the relation between the solid and gaseous phases at this temperature is best represented as the dependence of the partial pressure of M or N_2 on the composition in the *P-X* coordinate system. Fig. 10 shows a schematic *P-X* diagram for the M-N system. The partial pressures of the components change together with the average composition in the range containing only a single condensed phase and remain constant in the two-phase regions. Since the atomic percent of M(1-x) drops with growth of x the partial pressure of M drops with growth of the partial pressure of N_2.

The type of *P-X* diagram represented schematically in Fig. 11 is of significant practical importance. We see that the type of conductivity and carrier concentration depend on the partial pressure of the component. Significantly simplifying the situation we can assume that in vapor-crystal equilibrium the number of atoms in the vapor $1/2N_2$ corresponds to the number of point defects responsible for the deviation of the MN composition from stoichiometry and there then exists a range of compositions near stoichiometry in which a "thermodynamical M-N-transition" occurs (B.F. Ormont's term). An increase in the percentage x of the N component in the vapor in equilibrium with MN_s will cause a monotonic drop in the electron concentration in the n-type samples and a growth in the hole concentration in the p-type samples.

In 1957 Bloem and Kröger [7] obtained the first experimental data from *P-X* microdiagrams for the Pb-S system. The authors carried out an investigation of the dependence of the charge carrier concentration in lead sulfide on the partial pressure of S_2. The composition stability range of PbS obtained based on Hall effect measurements ran from $p = 10^{18}$ cm^{-3} (excess S atoms) to $n = 10^{19}$ cm^{-3} (excess

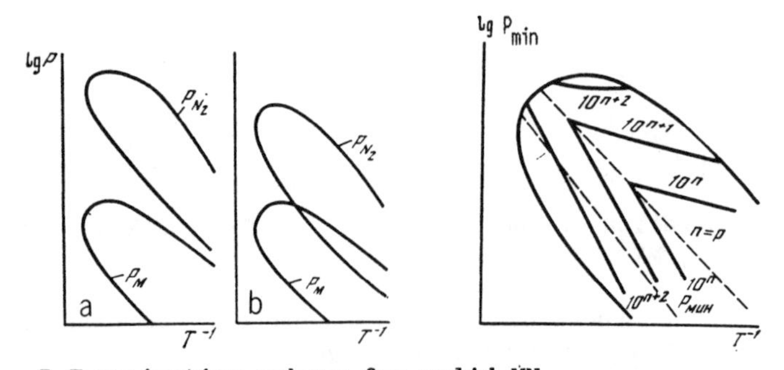

Fig. 13. *P-T*-projection scheme for solid MN
a - the compound has a congruently-subliming composition; b - the compound has no such composition

Fig. 14. *P-T*-projection scheme for solid MN including the composition lines of the samples with a constant carrier concentration (numbers labeling the curves)

Pb atoms). *n*-type conductivity was observed in PbS at low S_2 pressures and *p*-type conductivity was observed at high values of P_{S_2} with a sharp transition at intermediate sulfur pressures.

Analogous dependencies of the carrier concentration in PbTe on the partial pressures of the Te_2 and Pb components were obtained in study [14] where the authors calculated the *P-T-X* microdiagram of PbTe taking the intrinsic carrier concentration of the PbTe samples at 77 K to be a measure of the deviation of the composition from stoichiometry. The vacancy formation energies of Pb and Te were calculated from the *P-T-X* diagram; these were equal to 0.3 and 1.2 eV, respectively.

***P-T*-diagrams.** The *P-T*-projection of the phase diagrams of A^4B^6 compounds showing the temperature dependence of the vapor pressure in equilibrium with the solid compound is shown schematically in Fig. 12 in $\lg P$-$1/T$ coordinates. The curve gives the three-phase equilibrium line of MN, i.e., the solid phase, liquid, and vapor coexist for the points lying on this line. The solid and vapor phases coexist for the field of points, (i.e., the composition stability range of MN) lying within the three-phase equilibrium line, while the liquid and vapor coexist for points lying outside the line. It is clear that the pressures P_{N_2} may vary by several orders of magnitude for different compositions and different temperatures within the composition stability range. The straight line inside the field gives the partial pressure of MN_{gas} in equilibrium with solid MN. As discussed above this pressure is largely constant for the composition stability range of MN and lies between the minimum and maximum P_{N_2} whose values are given by the two dashed lines; the upper dashed line gives P_{N_2} in equilibrium with a pure liquid N. The lower dashed line gives the hypothetical partial pressure of N_2 that would exist in equilibrium with solid MN and M in liquid or solid form depending on temperature in accordance with the equation

$$K_T = P_M(P_{N_2})^{1/2}. \qquad (15)$$

Depending on the ratio between the partial pressures of the M and N components different cases can be realized where the composition stability range will contain the composition of congruently-subliming MN (Fig. 13a), and when no such composition exists (Fig. 13b).

Fig. 14 gives a schematic representation of the relation $\lg P_{N_2}$-$1/T$ for a compound (PbTe-type) whose composition stability range includes the stoichiometric composition. The solidi give the relation between T and P_{N_2} for specific constant carrier concentrations measured at room temperature in *n*- and *p*-samples. The partial pressures for the stoichiometric composition where $p = n$ are given by the dashed line.

Several methods have been used to experimentally determine the three-phase equilibrium lines, and the most important method - the optical absorption technique - will be described in greater detail below. We note that the optical absorption technique first proposed and developed by Brebrick and Strauss [8] made it possible to determine the pressures of Te_2 vapors in equilibrium with Te-saturated CdTe, HgTe, SnTe, GeTe, PbTe, compounds. The temperature dependence of the Te_2 pressure was investigated for the broader composition stability ranges of GeTe and SnTe for different compositions within the composition stability range [15].

In conclusion we note that the data provided here on the micro-diagrams of A^4B^6 compounds can be used not only to select the method of growing crystals with given parameters but also has made it possible to draw certain conclusions regarding the nature of defects and even to attempt estimating the formation energy of these defects responsible for deviation from stoichiometry and to estimate the magnitude of this deviation. Below we will show that in order to correctly estimate the deviation of the composition from stoichiometry we will require an experimental verification of certain assumptions made by the authors of studies on phase microdiagrams of A^4B^6 compounds; we will also need to incorporate other, independent investigation techniques.

1.4 Crystalline Structure of A^4B^6 Compounds

Compounds consisting of group IV and VI elements belong to the A^4B^6 semiconductor compounds. The nine binary compounds of the (Ge, Sn, Pb) - (S, Se, Te) system will, depending on the composition, temperature, and pressure, crystallize in the NaCl, GeSe, and α-GeTe structural types where the GeSe and α-GeTe type structures can be considered a deformed NaCl-type structure.

In normal conditions PbTe, PbSe, PbS, and SnTe will crystallize in the NaCl-type structure (space group $Fm3m$, structural type $B1$); when $T > 630$ K the high-temperature phase β-GeTe will also crystallize.

Only the low-temperature ($T < 630$ K) modification α-GeTe (space group C_{3v}^5 - $R3m$, structural type A7) will crystallize in the rhombohedral structure. Its structure can be represented as the result of the rhombohedral deformation (along the <111> axis) of the NaCl-type lattice with Ge atoms at the face-centered cell sites and with Te atoms in the (x, x, x) positions when $x = 0, 466$.

The GeSe-type structure belongs to the rhombic system and belongs to the space group D_{2h}^{16} - P_{nma} (structural type B29). Such an orthorhombic lattice can be considered a deformed NaCl-type where each atom has a coordinate representation in the form of a heavily distorted octahedron. This structural type includes, aside from GeSe, the compounds GeS, Sn, SnSe, as well as the third modification of GeTe (the γ-phase) that is stable in normal conditions with an elevated concentration of Te (above 50.6 at.%).

Primary data on the crystalline structure of A^4B^6 type compounds are given in Table 2.

Until very recently lead and tin tellurides and their solid solutions were classified as primarily ionically bonded compounds based on the nature of their chemical bond [19, 20]. Indeed the NaCl-type structure is characteristic of ionic compounds; the distances between neighboring atoms in PbTe, SnTe, and $Pb_{1-x}Sn_xTe$ correspond more closely to the ionic radii of Pb^{2+}, Sn^{2+}, and Te^{2-} than the covalent radii of the same atoms; like ionic crystals the high-frequency and static dielectric constants of these materials differ significantly and, finally, the PbTe and $Pb_{1-x}Sn_xTe$ compounds, like other compounds with an ionic bond, are not rendered amorphous under ion or electron bombardment [21]. However investigations of the charge carrier scattering mechanisms in PbTe, PbSe, and PbS [18] have revealed that carrier scattering in these compounds is due primarily to long-wave acoustic vibrations characteristic of the covalent bond. A comparative analysis of all experimental data makes it possible to assign these semiconductors to compounds with a mixed ion-covalent bond;

Table 2

Crystalline structure of A^4B^6 compounds

Compound	Space group	T, K conditions p, kbar	Lattice parameters*, Å	Reference
PbTe	$Fm3m$	$a = 6.46$	$T > 0$ $p \leqslant 40 \div 50$	[16, 17]
PbSe	$Fm3m$	$a = 6.12$	$p < 43$	[16, 17]
PbS	$Fm3m$	$a = 5.94$	$p < 25$	[16, 17]
SnTe	$Fm3m$	$a = 6.31$	$T > 75 \div 77$ $p < 10 \div 30$	[16, 17]
SnSe	$Pnma$	$a = 4.46$ $b = 4.19$ $c = 11,57$	$T < 813$	[16]
SnS	$Pnma$	$a = 4.34$ $b = 3.99$ $c = 11.20$	$T < 875$	[16]
GeS	$Pnma$	$a = 4.40$ $b = 3.65$ $c = 10.44$	$T < 863$	[16]
GeSe	$Pnma$	$a = 4.38$ $b = 3.82$ $c = 10.79$	$T < 893$	[16—19]
α - GeTe	$R3m$	$a = 5.99$ $\alpha = 88.4°$	T 630 ÷ 700 $p < 35$ $<$ 50.4 ат. % Te	[16—19]
β - GeTe	$Fm3m$	$a = 6.01$	$T > 630 \div 700$	[16—19]
γ - GeTe	$Pnma$	$a = 4.36$ $b = 4.15$ $c = 11,76$	$T < 630 \div 700$ $>$ 50.4 ат. % Te	[16—19]

*The lattice parameters are averaged over the literature data and correspond to normal conditions

unfortunately there is no precise information on the relative percentages of the bonds in lead and tin chalcogenides.

According to extensive published data (see, for example, [22]) the PbTe cubic crystal lattice parameter is independent of the crystallization temperature and the fabrication technique and in normal conditions is 6.460 ±0.0002 Å. However precision X-ray diffraction measurements carried out by the authors of the present study [23] have revealed that the lattice parameter of bulk PbTe saturated monocrystals is a function of the crystallization temperature and when the temperature rises from 550 to 700°C this parameter changes by 0.002 Å (see Section 2.3). Moreover, by using the new laser epitaxy technique the authors of the present study were able to obtain on BaF_2 substrates monocrystalline PbTe films with a crystal lattice parameter a = 6.4683 Å [24] which significantly exceeds previously-published maximum values of a = 6.4600. This increase was attributed to a reduction in the PbTe lattice defect concentration due to a reduction in vacancy concentration resulting from pulsed laser irradiation [24]. It is interesting to note that the PbTe autoepitaxial monocrystalline films grown by means of the laser epitaxy technique had a near standard parameter a = 6.4608 Å which may indicate that the layers inherit the structural characteristics of the substrate.

Unlike the bilateral composition stability range of PbTe the composition stability range of SnTe is located entirely on the side of excess tellurium due to the high vacancy formation enthalpy of tellurium compared to the corresponding value for tin (see Table 1). Within the unilateral composition stability range the cubic crystal lattice parameter of SnTe according to [13, 25] drops with growth of Te concentration and can be calculated by the formula

$$a(\mathrm{SnTe}) = 6.3278 - 3.54(x - 1/2). \qquad (16)$$

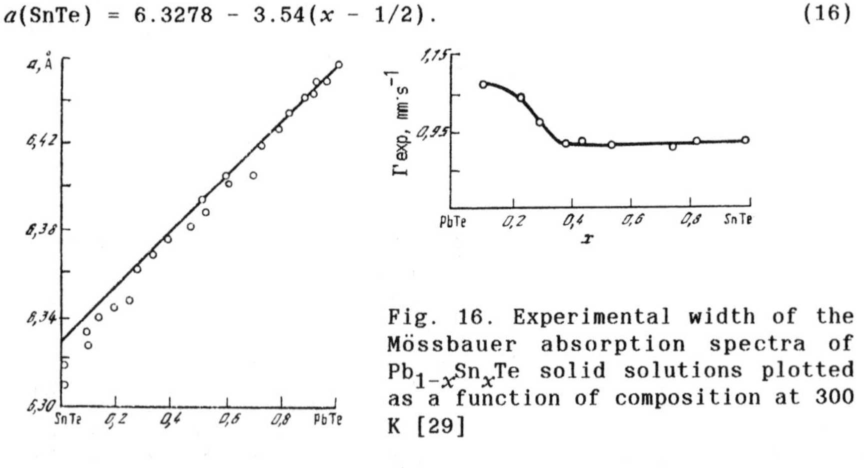

Fig. 16. Experimental width of the Mössbauer absorption spectra of $Pb_{1-x}Sn_xTe$ solid solutions plotted as a function of composition at 300 K [29]

Fig. 15. The crystal lattice parameter of $Pb_{1-x}Sn_xTe$ solid solutions plotted as a function of the composition x [16]
The straight line represents compositions with minimum deviation from stoichiometry

where x is the atomic percentage of tellurium.

In normal conditions the PbTe and SnTe isomorphic compounds (i.e., isostructural compounds having an identical chemical bond) have a continuous series of substitutional solid solutions with an NaCl-type crystal lattice. The crystal lattice period of the solid solutions of the PbTe-SnTe system is highly dependent on the ratio of the tin and lead concentrations as well as the tellurium concentration [22, 23, 26, 27] (Fig. 15). The analysis [26] of literature data on the change in the lattice period a for $(Pb_{1-x}Sn_xTe)_{1-y}Te_y$ as a function of the composition y together with experimental investigations of this relation on monocrystal films with small deviation from stoichiometry made it possible to write the following relation:

$$a(Pb_{1-x}Sn_xTe) = 0.133(1-x) - 1.7\cdot10^{23}\ p + 6.327, \tag{17}$$

where x is the alloy composition; p is the whole concentration [cm^{-3}]; 6.327 Å is the lattice parameter of tin-saturated SnTe.

An investigation of short-range order in p-$Pb_{1-x}Sn_xTe$ with a composition x = 0.10, 0.20, 0.24, and 0.32 by the diffusive X-ray scattering technique [28] revealed that the nature of the relative position of the Sn and Pb atoms changes fundamentally with growth of x: an alloy with x = 0.10 tends to form Pb-Sn atomic pairs at the same time that a tendency towards formation of Pb-Pb and Sn-Sn pairs is characteristic of the remaining compositions.

In order to establish the local symmetry of the ^{119}Sn atomic configuration in the $Pb_{1-x}Sn_xTe$ crystal lattice in normal conditions the Mössbauer spectra of ^{119}Sn in alloys with $0.1 \leqslant x \leqslant 1$ were investigated [29-31]. As we know this technique is based on the fact that when the cubic symmetry of the Mössbauer nucleus is distorted an electrical field gradient appears on this nucleus and this produces quadropole line splitting in the Mössbauer spectrum. The Mössbauer line splitting [31] or broadening [29, 30] detected at room temperature in $Pb_{1-x}Sn_xTe$ solid solutions with $x < 0.3$ (Fig. 16) revealed a reduction in the local crystal field symmetry in the vicinity of the Mössbauer nuclei of ^{119}Sn which the authors of study [31] believed could be attributed either to the displacement of the ^{119}Sn atoms from the lattice sites or to a reduction in the symmetry of the entire crystal. However the latter conclusion is not supported by X-ray analysis data available in the literature [16]. The change in the Mössbauer spectra for ^{119}Sn when $x < 0.3$ could be considered experimental proof of the fact that with an Sn concentration of less than 30% in $Pb_{1-x}Sn_xTe$ the Sn atomic configuration becomes a distorted octahedron. Such distortion could reduce the structural stability of the $Pb_{1-x}Sn_xTe$ solid solutions with a low concentration of Sn.

1.5. Phase Transitions in the PbTe-SnTe System

The problem of determining the role of various crystal lattice imperfections in phase transitions has recently become one of the most important problems in the physics of phase transitions. This is the result of the appearance of a number of theoretical studies (see, for example, [32]) that have attempted to modify the thermodynamical treatment phase transitions developed previously for ideal (defect-free) crystals for the case of real crystals, i.e., systems with crystal lattice defects. It is clear that a theoretical investigation of the influence of defects on the properties of real crystals, particularly their phase transitions, is quite important for a correct interpretation of experimental data. Indeed, without properly understanding the role of defects in the anomalies of physical properties it is impossible to compare theory which applies to an ideal crystal to experimental data obtained from actual crystals. Moreover, as indicated by the authors of study [32] there is reason to believe that in many cases experimentally-observed anomalies in physical properties are completely attributable to the existence of defects.

In order to test this statement we thought it best to summarize in this volume the extensive experimental data devoted to anomalies in the number of physical properties including phase transition data in the entire system of $Pb_{1-x}Sn_xTe$ solid solutions where a change in x from zero to unity causes a significant (three orders of magnitude) expansion in the composition stability range of the solution, and then to analyze the derived picture of the physical properties from the corresponding viewpoint.

The existence of a low-temperature phase transition in SnTe was first discussed in studies [33, 34]. The authors of these studies discovered a minimum on the temperature dependence of the coefficient of thermal expansion at 75-77 K and interpreted this minimum to be a structural transition from the high-temperature cubic modification to a low-temperature modification, presumably a rhombohedral modification [34].

X-ray diffraction investigations of the temperature behavior of the $(642)K_\alpha$ and $(820)K_\beta$ diffraction lines obtained by Ni radiation [35] confirmed the existence of a phase transition at 70 K from the cubic to the rhombohedral modification (rhombohedral angle $\alpha = 89.88^\circ$ at 5 K) in SnTe with a hole concentration $1.2 \cdot 10^{20}$ cm^{-3}.

Further X-ray [36] and neutron diffraction [37, 38] analyses of SnTe revealed that the rhombohedral deformation of the initial cubic structure in the second order phase transition occurs due to displacement of the Sn and Te atoms along the <111> axis, forming a structure with symmetry $C_{3v}^5(R3m)$; displacement of the sublattices is 0.09 Å at $T = 25$ K [38].

The crystal structure of SnTe was later investigated at low temperatures in study [39]; an entire series of experiments were carried out in this regime: galvanometric measurements, differential thermal analyses, and at T = 16 K the Laue and Debye diffraction patterns were recorded for SnTe monocrystals and powders with a Te concentration of 50.7, 50.4, and 48.8 at.% with a charge carrier concentration of $p \approx 10^{21}$ cm^{-3}. The diffraction line intensity redistribution discovered with a drop in temperature together with the broadening and particularly the splitting of the (h00) lines into three components made it possible for the authors of study [39] to assign orthorhombic symmetry to the low-temperature modification of SnTe and to calculate the parameters of the elementary cell of its crystal lattice: a = 6.274 Å; b = 6.288 Å, and c = 6.309 Å (±0.005 Å). The study did not determine the transition order. Vallasiades' convincing X-ray data therefore did not make it possible to assign rhombohedral symmetry to the low-temperature phase of SnTe.

The theoretical investigation of the displacement phase transition in narrowband semiconductors carried out in study [40] using the Kristofel-Konsin model [41] demonstrated the possibility of the existence of two critical temperatures in a specific range of free carrier concentrations where one critical temperature corresponds to the transition to the disproportional phase.

The assumption of the existence of more than one phase transition in SnTe [42] based on an investigation of the Raman spectra in SnTe with $p = 1.0 \cdot 10^{20}$ cm^{-3} was experimentally confirmed in study [43] which investigated the Mössbauer spectra of this material. The authors of these studies employed the temperature dependence of the Debye-Waller factor to establish two phase transformations for SnTe: one structural transition at 160 K and a second, presumably ferroelectric, phase transition in the 15-30 K range where a jump in the Debye-Waller factor of 12% was observed.

An analysis of available literature data on the low-temperature features of SnTe indicates the existence of at least two phase transitions in this material, although the sequence of the change in symmetry of the crystal lattice when passing through the critical currents cannot be considered to be established beyond doubt.

The PbTe crystal lattice conserves cubic symmetry to liquid helium temperatures [44, 45]. The authors of studies [46, 47] in analyzing the structural stability of the A^4B^6 compounds from the ionic viewpoint attributed the stability of the PbTe cubic lattice to the significant ionicity of the $Pg^{2+}Te^{2-}$ chemical bond. However this analysis is clearly without basis since introducing a small quantity (~1 at.%) of certain impurities which would not significantly change the ionicity of the compound nonetheless initiates in PbTe a structural instability (~0.1 at.% Te, $T_{p.t.} \approx 70$ K [48]), superconductivity (≤1.5% Te, $T_c \leq 1.5$ K [49]), and a ferroelectric phase transition (≤1 at.% Ge or Ga, T_c = 0-60 K [50, 45]). The sensitivity of PbTe observed in experiment to the introduction of impurities may be attributed, in the opinion of the authors of studies [45, 47, 49], to interimpurity electrostatic interaction and the overlap of the deformation fields around the impurity atoms.

In $Pb_{1-x}Sn_xTe$ solid solutions ($0 \leq x \leq 1$) anomalies were discovered at temperatures below 300 K in the temperature dependencies of a number of physical parameters: the electrical resistance and the Hall coefficient [51, 46], the thermal capacity [52], the velocity of acoustic propagation [52], magnetic susceptibility [53, 54], thermo-EMF [53, 55], etc. Generalized data from the literature on these features are found in Fig. 17. The authors of the cited studies interpreted these effects as the manifestation of phase transformations, and in certain cases as structural phase transformations although direct X-ray and neutron-diffraction proof of the latter have been quite limited [38, 61, 64].

Table 3

Features of the physical properties and phase transitions in $Pb_{1-x}Sn_xTe$

Composition x	Recording technique	Critical temperatures K	Quantity and nature of features	Reference
x>0.36	Theory, literature analysis	T>0	1 ferrotransition	[46,62]
x≤0.17	Theory	T<0	1 ferrotransition	[63]
0.44	Neutron diffraction and inelastic	54	1 ferrotransition to the	[38]
0.54	scattering	48	rhombohedral phase	[61]
1.0		98		[64]
0.29-1.0	Electrical measurements	32-98	-	[65]
0.44-1.0	Electrical measurements	52-98	1	[66]
0.35-0.70		12-51		[67]
0.15	Helicon waves, electrical measurements	24	1 second order transition	[68]
0.18	Magnetic susceptibility, thermo-EMF	10-26	1 transition	[33,34]
0.18-0.20		20-26		
		and >78	2 transitions	[54]
0.2-0.8	Electrical measurements, thermal	0-200	2-3 transitions	[51,52]
	capacity, acoustic velocity	-		[69]
0.21-0.35	Neutron diffraction		none detected	[37].

The authors of studies [51, 52, 69] also carried out a number of investigations of the physical properties of $Pb_{1-x}Sn_xTe$ using a wide range of physical techniques. They measured the change in acoustic velocity, thermal capacity and reflection in the IR as well as electrical conduction in $Pb_{1-x}Sn_xTe$ monocrystal samples with a composition $0.2 \leq x \leq 0.75$ in the 4.2-300 K temperature range. Based on experimental data the authors of studies [51, 52] concluded that at least two structural phase transitions exist across the entire range of test materials: the low-temperature transition (10-100 K) was interpreted as a ferroelectric transition and it was assumed with regard to the nature of the high-temperature structural transition that it is either a transition to a disproportional phase or is a local reconfiguration of the lattice in the vicinity of lattice impurities or defects.

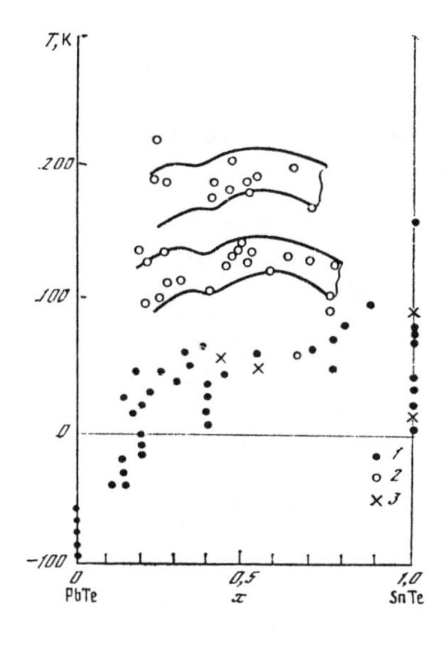

Fig. 17. *T-X*-diagram of the physical properties of $Pb_{1-x}Sn_xTe$ obtained from an analysis of literature data
1 - results from studies [33-37, 39-43, 56-70]; 2 - results from acoustic velocity and thermal capacity measurements from study [52]; 3 - results from diffraction analyses in studies [38, 61, 64]

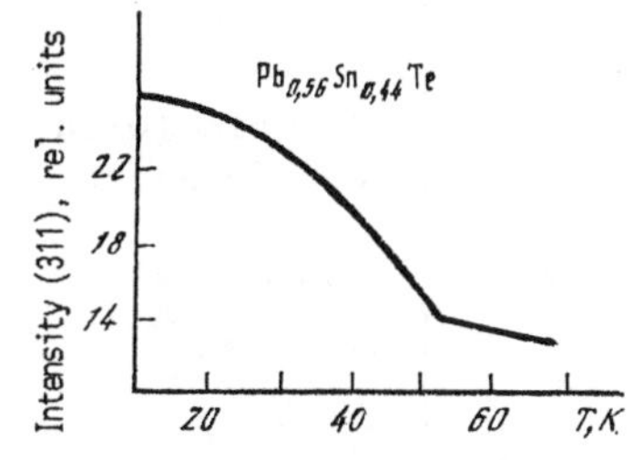

Fig. 18. The temperature dependence of the integral X-ray interference (311) intensity obtained by neutron diffraction analysis [46]

Among the few studies in which the authors have succeeded in investigating the structural characteristics of $Pb_{1-x}Sn_xTe$ at low temperatures it is important to cite the work by Kobayashi et al. that employed inelastic slow neutron scattering [64] and neutron diffraction [36] techniques (Table 3). The temperature dependence of the integral X-ray interference intensity (311) of polycrystalline $Pb_{0.56}Sn_{0.44}Te$ obtained in neutron scattering analyses [38] is given in Fig. 18. The authors of study [38] interpreted the anomaly in the X-ray interference intensity (311) that is weak for a face-centered-cubic lattice at 54 K as proof of a second order phase transition from the cubic high-temperature modification to the low-temperature rhombohedral modification. We should note that this relation makes it possible to establish only a change in symmetry but is not a sufficient condition for establishing its type.

The last group includes studies [70-22] carried out at the Physics Institute of the Academy of Sciences in which low-temperature X-ray diffractometry techniques were used on $Pb_{0.78}Sn_{0.22}Te$ monocrystals to first detect three structural phase transitions (presumably second order transitions) near 60, 130, and 220 K, and the following sequence of temperature variation in the symmetry of their crystal lattice was established:

220K		130K		60 K	
Cubic phase	→	Orthorhombic proportional phase	→	Orthorhombic disproportional phase	→ Monoclinal phase

Table 4 contains the results of the calculations carried out in studies [71, 72] on the parameters of the crystal lattices of all polymorphic modifications of $Pb_{0.78}Sn_{0.22}Te$.

Table 4

Elementary cell parameters of the low-temperature modifications of $Pb_{0.78}Sn_{0.22}Te$

T, K	Symmetry group	Cell parameters* Å*j*, angular deg.
295	Cubic	a = 6.431
197	Rhombic	a = 4.537
		b = 4.540
		c = 6.416
78	Rhombic (with a disproportional superstructure)	a = 4.523
		b = 4.532
		c = 6.407
8	Monoclinal	a = 4.528
		b = 12.798
		c = 4.522
		β = 89.93°

* The linear and angular parameters of the elementary cells were calculated accurate to ±0.002 Å and±0.02°

Two or three anomalies exist in the temperature dependencies of a number of physical properties in the $Pb_{1-x}Sn_xTe$ solid solutions at $0.15 \leqslant x \leqslant 1.0$ in the 0-300 K temperature range; in the majority of cases these are interpreted as phase transitions where the lower temperature anomaly is attributed by many researchers to a ferroelectric phase transition (see, for example, [50]).

A comparison of the low-temperature anomalies in the physical properties of the $Pb_{1-x}Sn_xTe$ solid solutions to the expansion of the composition stability ranges of these solutions with growth of the SnTe concentration (see Figs. 9 and 17) clearly demonstrates how as the homogeneity increases the temperature of the lower ferroelectric phase transition grows together with the temperature of the anomalies in many physical properties accompanying this transition. We believe that this correlation indicates an increase in the lability of the crystal lattice as its defect concentration increases.

2. EXPERIMENTAL ANALYSES OF THE COMPOSITION STABILITY RANGES OF CADMIUM AND LEAD TELLURIDES

This section includes information on methods of fabricating saturated crystals, techniques for determining the type and concentration of intrinsic defects and experimental results obtained for CdTe and PbTe.

2.1. Methods of Fabricating Saturated Crystals

As discussed above in order to determine the boundaries of the MN compound it is necessary to obtain saturated MN crystals, i.e., crystals maximally saturated by N or M components at different temperatures, yet below the melting point T_m of the MN compound.

There are several methods of fabricating saturated crystals. One involves crystal fabrication by direct melt cooling in quasi-equilibrium conditions. This method is not widely used and is employed in most cases to obtain a composition corresponding to the maximum melting point.

In the second technique the samples saturated by the M or N component are fabricated by annealing MN crystals of any initial composition in the presence of a two-phase (MN + M) or (MN + N) mixture where the composition of the derived crystal is determined by the annealing temperature. In this technique the time required to achieve equilibrium depends on the diffusion rate in the solid phase which drops exponentially with the annealing temperature. This means that the annealing time jumps sharply if the annealing temperature is reduced. At the end of the annealing process the sample is quickly quenched in order to conserve the composition characteristic of the selected temperature. This method is called the "frozen equilibrium" method.

The third method of fabricating the saturated samples is based on the use of internal precipitation of the second phase. In this method the crystals enriched with M or N are annealed at the temperature where their solubility is less than the total excess M or N component in the sample. When equilibrium is achieved the quantity of M or N component dissolved in the lattice will correspond to the solidus at the given annealing temperature, while the remaining excess component will appear as microprecipitates. The quenching rate of the sample after annealing must be sufficient to avoid further precipitation. The primary advantage of the internal precipitation method is that the rate of achieving equilibrium at a given annealing temperature is many times greater than the rate achieved in the "frozen equilibrium" method. A drawback of this technique is the fact that it is not suitable for high solidus temperatures due to retrograde solubility of the defects and the impossibility of achieving sufficiently high quenching rates that would make it possible to avoid advanced precipitation of the excess M or N components. The "efficient" quenching temperature of PbTe is T_{eff} = 600°C. This means that for all PbTe samples annealed at higher temperatures an estimate of the composition stability range based on the charge carrier concentration will yield a charged vacancy level corresponding to 600°C. Moreover the values for the defect concentration of samples heat-treated by diffusion can differ from the lattice defect concentration resulting from material crystallization in near-equilibrium conditions.

The most reliable method of obtaining saturated crystals over a broad solidus temperature range is solution growing. Growing from solution is a universal method of obtaining crystal compounds with a high melting point and a high vapor pressure, since growing from solution occurs at significantly lower operating temperatures and pressures which makes it possible to effectively control the composition of the growing crystals. With this method a significantly smaller concentration of electrically- and optically-active, uncontrolled impurities is introduced to the growing crystal compared to the case of growing in a higher temperature range. Our PbTe and CdTe saturated samples were grown from solutions whose vapor composition remained virtually constant.

In order to determine the range (solidus) of any compound it is necessary to fabricate a series of saturated crystals using any of the techniques discussed above at different temperatures and then to determine the composition of each crystal, i.e., to determine the degree of deviation of its composition from stoichiometry.

2.2. Techniques for Experimental Analysis of the Composition Stability Range

Modern analysis techniques can be used to determine the concentration of impurities in a 10^{22} to 10^{10} cm^{-3} range and to establish the type of chemical bond. The primary such methods include tunneling spectroscopy, mass-spectroscopic analysis, X-ray photoelectron and auger-spectroscopy and X-ray microanalysis which are widely used to investigate semiconductor surfaces and combine high sensitivity with good local resolution. However it is much more difficult to determine the concentration of excess or sub-stoichiometric atoms of the primary element of some matter. Direct chemical analysis techniques make it possible to determine the super-stoichiometric quantity of a component with only ±0.01 at.% accuracy.

Until recently no direct experimental method had been developed for determining the degree of deviation from stoichiometry, i.e., determining the total concentration of intrinsic defects in a solid for the majority of A^2B^6 and A^4B^6 semiconductor compounds of variable composition. In our view this methodological lapse is primarily due to the fact that the overwhelming majority of researchers followed Brebrick, De Nobel, and Wagner in formulating their studies (see, for example, [11]); these investigators had introduced the practice of identifying the experimentally-determined free carrier concentration of a semiconductor of variable composition with the total concentration of intrinsic defects as early as the 1950s. This "method" of determining the deviation of a semiconductor composition from stoichiometry naturally does not account for the complex nature of interaction between the defects that can result in their complex formation and possible transitions to an electrically inactive state. The simplest example of complexing of charged intrinsic defects and their transition to an uncharged state is the formation of the neutral cadmium divacancy $(V_{Cd})_2$ in CdTe whose state is stabilized by the partially covalent bonds of the tellurium atoms. In addition to the neutral double associations similar to $(V_{Cd})_2$ more complex neutral formations such as $(V_{Cd})_n$ can also exist.

It is easy to see that a phase microdiagram drafted by the Brebrick method will not represent reality since it corresponds to a metasolidus drafted for the case where the composition is identified with the carrier concentration. Certain authors have stated that such an information discrepancy is quite acceptable in the application of semiconductors of variable composition since in the majority of cases, as noted in study [73], the question of the exact quantity of excess constituent atoms is not the important question; rather, the important question is what concentrations of electrically active centers and charge carriers can be achieved when growing monocrystals of pure semiconductor compounds and how can the desired concentrations be achieved. However recent research has revealed the narrow aspect of such a criterion for evaluating and predicting the physical parameters of a semiconductor compound of variable composition. It has been established experimentally that a number of physical properties of such material including the physical density, the transition temperature to a ferroelectric state, the quantum efficiency, and the electrical activity of the doping impurity are determined by the total concentration of intrinsic defects [74-77].

In light of these experimental results it is easy to understand why the problem of the experimental determination of the actual size of the composition stability ranges has become so important for the physics of real crystals in general and semiconductor physics in particular.

Since no single method used in isolation can yield the correct representation of the total concentration of intrinsic defects we believe that in order to establish the dimensions of the composition stability ranges of CdTe and PbTe it is most advisable to use a set of independent crystallochemical and physiochemical techniques that as a group makes it possible to estimate the concentration of both electrically-active and electrically-inactive intrinsic defects.

We can divide into two groups the techniques we employed to investigate the nature and total concentration of intrinsic defects responsible for deviations from stoichiometry. The first group includes methods of investigating the physiochemical and crystallochemical properties of the materials that depend on the total concentration of intrinsic defects:

1. precision measurements of relative changes in the physical density of the crystals;

2. measurement of the optical density of the vapors of the compound and the components in equilibrium with the crystals;

3. precision measurements of the crystal lattice parameters;

4. determination of changes in the average electron concentration in the sublattices of the metal and the chalcogenide in the crystal lattice.

The second group includes methods of investigating the electrically-active defects:

1. determining the conductivity type by thermo-EMF;

2. determining the free charge carrier concentration and their mobility by the Hall method.

We also estimated the quantum efficiency of luminescence of the material in order to obtain information on the microprecipitates of the excess component which function as luminescence quenching centers [76].

The methods in the second group are rather standard and therefore it is most advisable to examine in detail only the methods in the first group.

Precision measurements of the physical crystal density. The differential hydrostatic weighing technique [78] was employed for precision measurements of the difference of the physical density of crystals of stoichiometric composition (the reference) and crystals with maximum possible deviation from stoichiometry; in this method, unlike previous techniques [79, 80], only a single fluid bath is used to measure the density difference, thereby eliminating the need to carefully control the temperature of the working fluid or the influence of capillary action. As a result the method is independent of the weighing accuracy of the reference and sample masses as well as the density of the working fluid. The sample mass could also vary from weighing to weighing, which was significantly advantageous when it was necessary to eliminate impurities or oxide films from the sample surface.

The density was measured on an assembly (Fig. 19) consisting of microanalysis weights, a working fluid bath and a temperature control system. The SMD-1000 microanalysis weights (maximum load of 1 g) made possible measurements with an accuracy of $5 \cdot 10^{-5}$ g. The temperature control system reduced temperature variations to $2 \cdot 10^{-2}$°C. Dibutyl phthalate was used as the auxiliary fluid. The error in determining the change in density of the stoichiometric and nonstoichiometric samples whose weight was close to the maximum load of the weighing device was $5 \cdot 10^{-5}$ g·cm^{-3}.

The change in density $\Delta\rho/\rho$ in the test samples was calculated by the formula

$$\Delta\rho/\rho = 1 - W(\omega - \omega_1)/\omega(W - W_1), \qquad (18)$$

where ω is the weight of the stoichiometric sample (reference) in air, ω_1 is the weight of the stoichiometric sample (reference) in the fluid, W is the weight of the sample with deviation from stoichiometry in air, W_1 is the weight of the sample with deviation from stoichiometry in the fluid. The measurement accuracy in these conditions was 30-40%.

Measurement of the optical density of vapors. This method measures the variation in optical density characterizing light absorption at a given wavelength as a function of the concentration of matter in the system. The method is widely used in spectrophotometry to

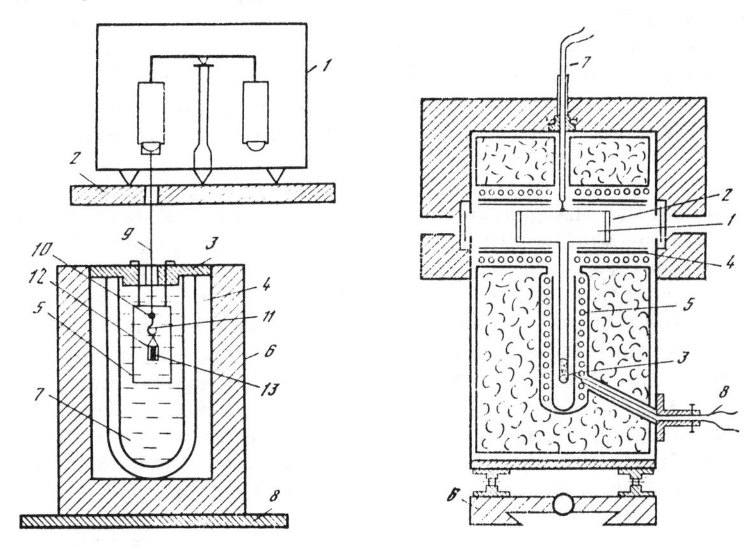

Fig. 19. Measurement scheme for pyncnometric determination of crystal density
1 - measurement weights; 2 - weight support; 3 - cover; 4 - Dewar vessel; 5 - copper cylinder with perforated bottom; 6 - thermostat; 7- working fluid; 8 - thermostat support; 9 - platinum wire Ø 0.02 mm; 10 - platinum sphere; 11 - platinum wire hook Ø 0.05 mm; 12 - sample holder; 13 - sample

Fig. 20. Assembly for measuring the optical density of vapors
1 - crystal cell; 2 - window; 3 - lateral extension piece; 4, 5- heaters; 6 - support; 7, 8 - thermocouples

analyze the composition of systems and in the case of gas mixtures to determine the partial pressures of the separate components in the mixture. This method was first proposed and applied to determining partial pressures in equilibrium with solid semiconductor compounds by Brebrick and Strauss [8] and was later developed by S.A. Medvedev et al. [81].

An assembly consisting of the following primary parts is used to measure optical density: a stabilized light source; an optical cell containing the sample; an oven for heating the cell; a spectrometer or a device containing filters and an optical system for coupling the analyzed luminous flux to the spectrometer. A low pressure mercury lamp or an incandescent lamp with lamp current stabilization of 0.1% is used as the light source. The optical cell (Fig. 20) consists of optical vitreous silica cell 1 22 mm in diameter and between 20 and 100 mm in length with sealed vitreous silica windows 2 and a lateral extension piece 3 7 - 10 mm in diameter and 10-15 cm in length into which a test sample between 0.01 and 1 g is placed. The optical cell is heated in the compound oven with separate heaters 4 and 5 mounted on adjustable support 6. Platinum-platinum radium or chromel-aluminum thermocouples are used to monitor the temperature in the oven: one thermocouple is located in the center of the upper section of cell 7 while the second thermocouple is located in the lower section of lateral extension piece 8. The heaters must maintain the temperature so that the lateral extension piece containing the substance remains cooler than the upper section of the cell in order to avoid condensation on the optical windows.

Uncontrolled changes in the temperature of the lateral extension piece containing the substance of, say, 5°C will produce an error of 1-2% of the measured optical density and therefore it is necessary to maintain a strictly constant temperature in the optical section of the cell and on the lateral extension piece during the entire measurement process. Electronic stabilizers are used for this purpose to maintain a temperature of ±(0.5-1)% across the entire operating temperature range up to 1000°C. A variety of methods are used to reduce the influence of background radiation from the heating oven on the readings of the recording element, including introducing platinum screens into the oven and reducing the effective oven aperture to a few millimeters.

In order to obtain reliable results it is necessary to carry out measurements of the optical density of vapors over the monocrystals since the intergrain boundaries in the polycrystalline materials may contain excess components.

The test samples are placed on the lateral extension piece of the optical cell which must first be degassed by long-term heat treatments in a vacuum. The cell containing the sample is then heated in the vacuum to eliminate possible contaminants from the crystal surfaces. After slow cooling the cell is sealed in a vacuum and placed in the oven.

In order to measure optical density the optical cell is heated to a given temperature and then is maintained at this temperature to establish equilibrium so that the temperature of the upper section of the cell is higher than the temperature of the extension piece containing the sample; the optical density at the working wavelength is then measured. In order to improve the sensitivity of the method employing a spectrometer the wavelength at which maximum absorption of the test vapor component occurs is selected as the working wavelength. For example, the maximum optical density of Te_2 vapors was recorded at 4165 Å and at 3250 Å for PbTe. Calibrated data obtained from measurements of the optical density of component vapors in equilibrium with the pure component as a function of the temperature of the extension piece and the optical cell in identical experimental conditions were used to convert the optical density to the partial pressure of the component. Existing $p = f(T)$ relations for pure components obtained independently were employed to calculate the optical density.

Brebrick and Strauss' method made it possible for these authors not only to investigate the composition and measure the pressure of A^4B^6 compound vapors but also to determine the boundaries of a composition stability range with dimensions greater than 1020 cm^{-3} (GeTe and SnTe). This method was subsequently expanded to a significant degree by the authors of study [81] and in combination with other techniques allow the authors of the present study to investigate more narrow composition stability ranges using CdTe and PbTe samples.

X-ray diffraction analysis techniques for investigating structural characteristics. This group of techniques includes: standard methods of determining the space group of crystals; measurements of the crystal lattice parameters and the integral intensity of X-ray diffraction losses; methods of evaluating crystal imperfection involving determination of the substructural elements of the crystal lattice and a method for determining the average electron concentration at the crystal sublattice sites.

1. The a parameters of CdTe and PbTe cubic crystal lattices were determined either by determining Debye diffraction patterns employing chambers 143 mm in diameter or by goniodiffractometry using standard general purpose diffractometers. The accuracy in measuring the a parameters using an individual line in the range of large diffraction angles $\theta \approx 70^\circ$ in the first case was better than ±0.001 Å and in the second case was better than ±0.0005 Å. In both methods the final values of a were determined by extrapolating the $a = f(\theta)$ relation to $\theta = 90^\circ$ which improved the accuracy of determinations of a to $\pm 2 \cdot 10^{-4}$ Å.

2. The integral intensities of X-ray diffraction losses in both methods were measured with a relative accuracy of better than 2%.

3. The structural integrity of the monocrystals was determined by the angular half-width of the diffraction lines obtained by gonio-

diffractometry when simultaneously rotating the sample and the counter ($\theta/2\theta$-curves) and when rotating the sample only with a fixed counter (ω-curves). The physical broadening β of the $\theta/2\theta$- and ω-curves were used as criteria for structural perfection of the crystals.

We know that the quantity β in the general case is related to the dimensions L of the coherent scattering regions and the inhomogeneous lattice microdeformation $\varepsilon = \pm\Delta d/d$, where d is the interplanar spacing, and the angular disorientation α (the latter may result from bending of the planes in the vicinity of dislocations and the rotation of separate substructural elements with respect to one another). Due to the different dependence on the diffraction angle θ, it is possible to differentiate separate components of physical broadening β by recording either certain reflections at one radiation wavelength or recording a single reflection at several wavelengths.

It is possible to identify from the measured integral half-widths B of these and other curves at several reflection orders such as (200), (400), and (600) the corresponding values of the physical broadenings β_θ and β_ω by approximating the experimental profile with a Gaussian or Cauchy function. Instrument broadening is estimated by means of near-perfect monocrystals (such as vitreous silica crystals).

We know [82] that when approximating the diffraction profile by a Gaussian function the physical broadening of the $\theta/2\theta$-curves is determined by the relation:

$$\beta_\theta^2 = \lambda^2/L_n^2 \cos^2\theta + 16\,\mathrm{tg}^2\,\theta\varepsilon_n^2. \tag{19}$$

Equation (19) contains the dimensions of the coherent scattering region and the lattice deformation component in the direction of the diffraction vector **g** coinciding with the normal **n** to the reflecting planes (L_n and ε_n, respectively). We also know [83] that the physical broadening β_ω of the ω-curves when recording a reflected signal in an angular range significantly less than the spreading of the reciprocal lattice site in the direction of the diffraction vector **g** (in our experiments this condition was achieved by narrowing the width of the slit in front of the counter to 50 μm) depends on the size of the coherent scattering region and the deformation component of the lattice on the reflecting plane (L_t and ε_t) as well as the angular disorientation α of the separate elements of the substructure with respect to the diffraction vector **g**:

$$\beta_\omega^2 = K_\alpha + K_L\lambda^2/\sin^2\theta + K_\varepsilon\,\mathrm{tg}^2\,\theta, \tag{20}$$

where

$$K_\alpha = 4\ln 2/a^2 = \alpha^2, \tag{21}$$

$$K_L = 4\ln 2/\pi L_t^2, \tag{22}$$

$$K_\varepsilon = 8\ln 2\varepsilon_t^2, \tag{23}$$

a is a quantity independent of θ and λ; θ is the diffraction angle.

The quantities L_n, L_t, ε_n, ε_t and α were determined by relations (19)-(23) from β_θ and β_ω.

Determining the average electron concentration in the crystal sublattices. We demonstrated previously based on sample superconducting compounds of variable composition (Nb_3Sn NbSe) [84, 85] that an X-ray diffraction analysis of the structural characteristics of the material can yield valuable information in investigating the crystallochemical properties of compounds of variable composition. This method makes it possible to answer the question of whether or not the deviation mechanism of the compound is a vacancy mechanism and if it is it is then possible to use the results of these experiments to, first, reliably determine the prevalent type of vacancies formed and, second, establish the dynamics of the change in their concentration under the influence of a variety of variable factors (for example, by varying the crystallization temperature or the annealing temperature and duration of the grown crystals).

The essence of the method used in studies [84, 85] is a separate analysis of the variation in the site occupation of each of the crystal sublattices by estimating the change in the structural amplitudes $F(h, k, l)$ of X-ray diffraction interference and then the variation in the atomic scattering factors f_i, since the latter are proportional to the average volumetrical electron concentration at the sites of the corresponding sublattices and can yield information on changes in the vacancy subsystem of the crystal.

An experimental solution of this problem requires correct measurement of the integral intensities of a number of X-ray diffraction interference beams which in the final analysis is determined by the space group of the test crystal. The accuracy of the experimental procedure is dependent on maintaining a constant "total luminance" of X-ray radiation impinging the sample. It is clear that this requirement is most easily satisfied by carrying out the X-ray diffraction analyses on polycrystals by the Debye method since in this case it is possible to record all X-ray diffraction interference beams simultaneously.

Underlying the "polycrystal" variant of this method is the integral intensity equation $E(h, k, l)$ of X-ray diffraction interference beams with indices (h, k, l) familiar in the case of polycrystals:

$$E_{hkl} = E_0 \frac{e^4\lambda^3}{m^2c^4V^2} \frac{1+\cos^2\theta}{8\sin\theta} pF^2_{hkl}, \tag{24}$$

where E_0 is the intensity of the incident X-ray beam; e, m is electron charge and mass, respectively; λ is the wavelength of the diffracted X-ray radiation; c is the speed of light; V is the elementary cell volume; θ is the diffraction angle; p is the frequency factor; F_{hkl} is the structural amplitude of the X-ray diffraction interference beams with indices (hkl) demonstrating to what degree the intensity scattered by the test structure is greater than the intensity scattered by a single electron with identical values of λ and θ.

The structural amplitude $F(h, k, l)$ for the X-ray diffraction interference (h, k, l) is expressed by the formula

$$F_{hkl}=\sum_{i=1}^{n} f_i \exp\left[i2\pi (hx_i + ky_i + lz_i)\right], \tag{25}$$

where n is the number of atoms in the elementary cell; f_i is the atomic factor of the i-atom which indicates to what degree the scattering intensity by the i-atom in a given direction at a given wavelength of the incident radiation is greater than the scattering intensity by a single atom in identical conditions (the exponent in equation (25) accounts for the geometry of n atoms in the elementary cell); x_i, y_i, z_i are the coordinates of the i-atom in axial units.

If we know the space symmetry group of the test crystal (and therefore the form of F_{hkl}), we also know the coordinates (x_i, y_i, z_i) of all n atoms in the elementary cell, and after substituting the values (x_i, y_i, z_i) into the formula for F_{b1kl} we obtain the expression for F_{hkl} through f_i. Substituting this expression of F_{hkl} into equation (24) for the integral intensity E_{hkl} demonstrates that with fixed values of E_0, λ, and θ the integral intensity of the given diffraction line is dependent only on the atomic scattering factors.

Combining several E_{hkl} equations each of which is reduced to a function of the atomic scattering factors f_i can in a number of cases make it possible to obtain separate expressions for each f_i through the experimentally-determined integral intensities of several X-ray diffraction interference beams.

Finally since we know that the scattering factor f_i of any i-atom is equal to

$$f_i = \Phi Z_i, \tag{26}$$

where Z_i is the total number of electrons of the i-atom, while Φ is a tabulated function dependent on the wavelength λ and diffraction angle θ, in the final analysis any change in the atomic factor f_i averaged over the reflecting volume of the crystal, corresponds to the change in the electron concentration (averaged over the same volume) at the sites of the i-sublattice, i.e., the change in vacancy concentration V_i.

By measuring E_{hkl} of a specific set of X-ray diffraction interference beams it is possible to establish with constant coefficient accuracy the nature of the evolution of the vacancy subsystem of a compound under the influence of any factors. The accuracy of the technique is determined by the measurement error of E_{hkl} and is 3-5% in standard conditions.

We altered this method for PbTe which has an NaCl-type cubic crystal lattice by substituting the coordinates (x_i, y_i, z_i) of the Pb and Te atoms in the PbTe elementary cell into the tabulated expression for the structural amplitude of the space group *Fm3m* which includes the PbTe crystal lattice.

After some simple transformations the expressions for $F(h, k, l)$ appeared as:

$$F_{hkl} = 192\,(f_{Pb} + f_{Te}) \text{ for lines with even } h, k, l; \tag{27}$$
$$F_{hkl} = 192\,(f_{Pb} - f_{Te}) \text{ for lines with odd } h, k, l; \tag{28}$$

here f_{Pb} and f_{Te} are the atomic scattering factors of lead and tellurium, respectively.

Substituting the derived expressions for F_{hkl} into equation (24) for the integral intensity E_{hkl} yields in the case of lead telluride the following relations:

- for lines with even h, k, l

$$E_{hkl} = E_0 \frac{e^4\lambda^3}{m^2c^4V^2} \frac{1+\cos^2\theta}{8\sin\theta} p\,[192\,(f_{Pb} + f_{Te})]^2; \tag{29}$$

- for lines with odd h, k, l

$$F_{hkl} = E_0 \frac{e^4\lambda^3}{m^2c^4V^2} \frac{1+\cos^2\theta}{8\sin\theta} p\,[192\,(f_{Pb} - f_{Te})]^2; \tag{30}$$

here the conventions E_0, e, m, c, λ, V, θ, and p are the same as in (24).

It follows from these expressions that with fixed values of E_0, λ, V, and θ in order to determine f_{Pb} and f_{Te} separately it is sufficient to carry out a measurement of the integral intensities E_{hkl} of two neighboring lines with even and odd values of (h, k, l) and then use the formulae obtained by algebraic transformations of expressions (29) and (30) to determine, with constant multiplier accuracy, the separate values of the quantities f_{Pb} and f_{Te} proportional to the average electron concentration in the lead and tellurium sublattices, respectively. The fundamentally important results derived using this method from an investigation of the composition stability range of PbTe are described in Section 2.4 of this chapter.

We did not use this method, which was quite successful for investigating the composition stability ranges of Nb_3Sn [84], $NbSe_2$ [85], and PbTe [23] for investigating the composition stability range of CdTe. The reason is that after the appropriate transformations of the structural amplitudes we could not derive separate expressions for the atomic scattering factors f_{Cd} and f_{Te} through the integral intensities E_{hkl} of any of the X-ray diffraction interference beams resolved in this structure.

2.3. Investigation of the Composition Stability Range of CdTe

In 1959 De Nobel became the first to estimate the CdTe range by measuring the pressure of the gas phase in equilibrium with cadmium telluride crystals [86]. These calculations were carried out assuming that the predominant dissociation products of CdTe in the gas phase are Cd atoms and Te_2 molecules, and that the dominant intrinsic defects in the CdTe crystals are cadmium vacancies V_{Cd} at a low cadmium vapor equilibrium pressure in the gas phase and interstitial cadmium Cd_i at a high cadmium vapor pressure. De Nobel also assumed that the cadmium vacancies in this material determine the hole conductivity, and that the interstitial cadmium atoms determine the electron conductivity.

Fig. 21 gives the *T–X*-diagram of CdTe from 0 K to the melting point T_m. The range at low temperatures ($T < 0.5\ T_m$) was not established experimentally but rather was determined by extrapolation and calculation from equations of the type [73]

$$[Cd_i]^2 = K \cdot p_{Cd}. \tag{31}$$

As indicated by the diagram given in Fig. 21 the composition corresponding to maximum T_m of the CdTe solidus contains excess Te and lies halfway between the far boundaries of the range: between 49.99985 at.% Te at 1170 K and 50.00085 at.% Te at 1350 K.

Therefore consistent with De Nobel the prevalent form of intrinsic defects in cadmium telluride are Fresnel defects $[V_{Cd}][Cd_i] = K_F$, while the maximum deviation from stoichiometry in these crystals does not exceed $\pm 8.5 \cdot 10^{-4}$ at.% Te (i.e., $\sim 10^{17}$ cm^{-3}).

The subsequent estimate of the composition stability range of CdTe generated by Brebrick in 1964 within the framework of the same Frenkel defect yielded a value of $\sim 10^{-3}$ at.% based on measurement results of the Hall carrier concentration; here the magnitude of the deviation from stoichiometry was equal to the concentration of electrically-active intrinsic defects only.

Naturally monitoring deviation from stoichiometry by investigating only the electrical properties of the crystals made it impossible for the authors of study [87] to determine the true deviation from stoichiometry and, consequently, the width of the composition stability range. Therefore we thought it best to employ independent techniques to fix both electrically-active and electrically-inactive types of defects in the material in order to establish the degree of deviation from stoichiometry in grown crystals and to determine the width of the composition stability range.

A technology for fabricating CdTe crystals with maximum deviation from stoichiometry was developed under the direction of S.A. Medvedev at the Physics Institute of the Academy of Sciences for these applications; precision measurements of small density deviations were employed in conjunction with measurements of crystal lattice parameters and the optical density of Te_2 vapor over cadmium telluride in conjunction with traditional electrical measurements to investigate their properties.

We know [73, 88, 89] that the structurally-sensitive properties of CdTe depend on its fabrication conditions. Due to the physiochemical features of this compound the magnitude and type of conductivity of undoped crystals may vary from 10^{-1} ohms·cm for *n*-type conductivity to 1 ohm·cm for *p*-type conductivity. Therefore it is very important to establish the degree of maximum possible deviations from stoichiometry since this would make it possible to evaluate the maximum levels of certain physiochemical properties of CdTe. Crystals having saturated compositions are normally grown from solutions [89] and therefore study [90] investigated the possibility of growing undoped CdTe crystals from dilute solutions of Cd(Te) components with maximum deviation from stoichiometry and also investigated their physiochemical properties.

The initial materials included metallic - 99.999% Cd, rectification Te (followed by zone purification) and CdTe obtained by synthesis and directional crystallization at a controlled Cd vapor pressure. The CdTe + Cd and CdTe + Te quantities calculated by the diagram with a small residual CdTe (5-7%) were placed into the vitreous silica vessel that was sealed at a vacuum of $\sim 5 \cdot 10^{-5}$ mm/mercury. The vessel was then placed in a refractory brick unit which was in turn placed in a vertical oven (Fig. 22). An additional heater was placed over the upper optical wall of the vessel to generate the necessary temperature and concentration gradient. The temperature gradients maintaining the necessary supersaturation between the upper and lower ends of the vessel were selected experimentally.

The replenishment crystal floated continuously on the liquid metal surface since the CdTe density is less than that of liquid Cd and Te. When saturation is achieved an auxiliary heater is activated and this can be used to regulate the temperature gradient in the solution. As soon as the concentration gradient is reached in the solution, crystal formation begins on the lower vessel wall.

Subsequent growth from the nucleus occurred in the solution volume. In order to reduce turbidity currents in the fluid a gauze barrier was included in the vessel to provide more stable growing conditions. The height of the growth cell varied from 0.5 to 1.0 cm. After the end of the growing process in order to maintain the working temperature the oven was rotated to separate the crystals from the source solution. The remaining solution droplets were eliminated by treating the crystals in a vacuum with fixed evacuation for 24 hours at 200-250°C. The experiment lasted a total of four days.

Monocrystal platelets with smooth specular reflection surfaces 5×7×0.3 mm were grown from cadmium solutions containing 95-98 at.% Cd at 800-850°C; specular-reflection smooth platelets 5×3×0.2 mm were

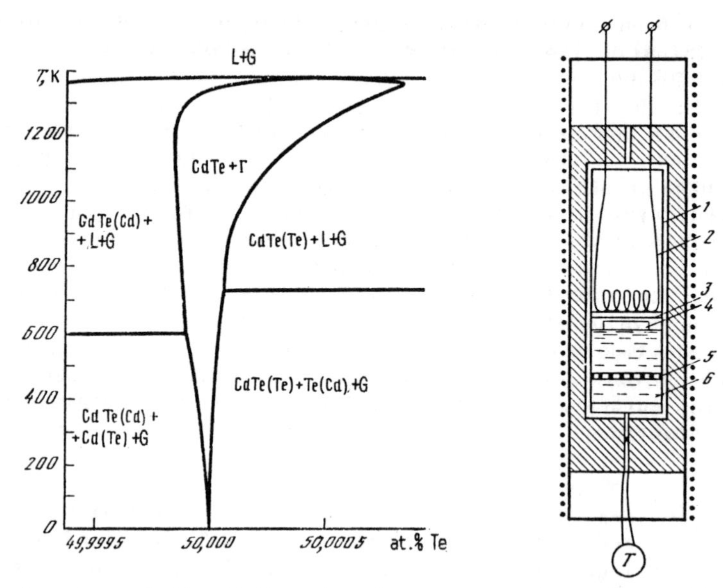

Fig. 21. Portion of the *T–X*-projection representing the entire range of solid CdTe (the gas phase line is not shown)

Fig. 22. The vitreous silica apparatus for growing crystals from solutions
1 - vitreous silica vessel; 2 - auxiliary heater; 3 - optical glass; 4 - replenishment crystal; 5 - gauze barrier; 6 - growth cell; T - thermocouple

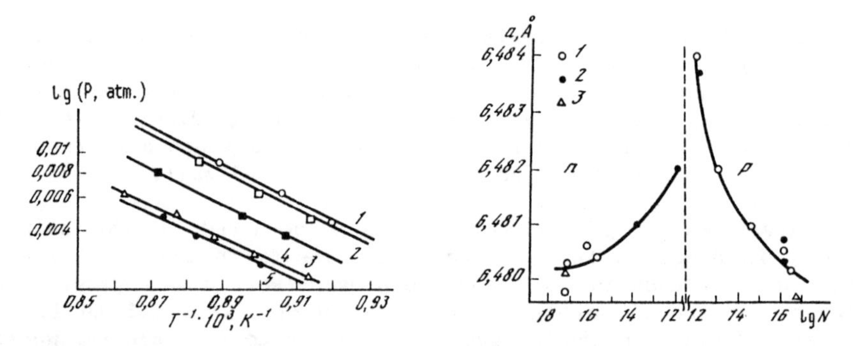

Fig. 23. The temperature dependence of the partial pressures of Te_2 vapor over CdTe
1, 2 - the crystals are grown from tellurium solutions; 3 - the crystals are grown from the gas phase; 4, 5 - the crystals are grown from cadmium solutions

Fig. 24. The lattice parameter plotted as a function of carrier concentration in CdTe crystals grown from the melt (1), gas phase (2), and solutions (3)

also grown from tellurium solutions containing 82-86 at.% Te at 750-780°C. Twins characteristic of A^2B^6 compounds were observed on the surfaces of certain crystals. The solution height in the vessel was ~1 cm. In the first case the temperature gradient was 12-16 deg/cm and in the second case it was 20-24 deg/cm.

X-ray structural analyses have revealed that all grown CdTe monocrystals had a cubic structure (space group T_d^2) and the most developed platelet facets had a {111} orientation.

The lattice parameters determined by the Debye scattering diagram technique (using a chamber 143 mm in diameter, Cu-radiation) by extrapolating the parameter values to a diffraction angle θ = 90° are given below (also given for comparison purposes is the value of the lattice parameter of a near-stoichiometric crystal obtained by the float zone method):

Growing conditions	a, Å
grown from a Cd-rich solution	6.480 ±0.001
grown from a Te-rich solution	6.482 ±0.001
float zone method ("stoichiometric" composition)	6.482 ±0.001

In order to estimate the deviation from stoichiometry in saturated crystals the optical density of Te_2 vapor over CdTe was determined [81]. Experimental curves of the temperature dependence of the partial pressures of Te_2 over the derived CdTe crystals are given in Fig. 23 (curves 1, 2, 4, 5). A straight line is provided for comparison (curve 3); this was obtained for a crystal with minimum deviation from stoichiometry grown from the gas phase. The observed difference in the Te_2 pressures over the CdTe characterizes the deviation from stoichiometry in crystals of various conductivity. The approximate calculations of deviations from stoichiometry [91] revealed that the difference in the concentrations between the crystals of "stoichiometric" composition and crystals with maximum deviation from stoichiometry is ~10^{20} cm^{-1} on both the Cd-rich and the Te-rich sides.

Since the derived platelets had a relatively small cross-sectional area it was not possible in the majority of cases to obtain standard samples suitable for Hall measurements. Therefore measurements were carried out using the Van der Paue method employed for platelets of arbitrary shape. In certain cases only the resistivity and the conductivity type were measured due to small platelet size.

It was established based on these measurements that crystals grown from cadmium solutions, as would be expected, had n-type conductivity an those grown from tellurium solutions have p-type conductivity. The primary measurement results are given in Table 5.

The derived results reveal the existence of a composition stability range in CdTe with dimensions of the order of 1 at.% at 800°C; the total concentration of intrinsic defects in CdTe was several orders of magnitude greater than the concentration of electrically-active defects.

Table 5

CdTe crystal parameters as a function of growth temperature

T, °C	Type of conductivity	ρ_{293}, ohm·cm	μ_{292}, cm²/(V·s)	n_{293}, cm⁻³
850	n	0.08	780	$1 \cdot 10^{17}$
800	n	0.02	625	$5 \cdot 10^{17}$
780	p	4.0	78	$2 \cdot 10^{16}$
750	p	10	55	$1 \cdot 10^{18}$

We continued our investigation of the composition stability range of CdTe and the nature of intrinsic defects in CdTe in study [74] where we investigated CdTe crystals grown not only from dilute solutions [90] but also from the melt [88] and the vapor phase [92]. Growing from the melt made it possible to obtain large bulk monocrystals rather easily while the vacuum sublimation method improved the chemical purity and structural perfection of the grown crystals, while the solution growing technique yielded crystals with maximum deviation from stoichiometry which made it possible to unambiguously solve the problem of the width of the composition stability range.

The electrical properties of the grown crystals were investigated in studies [93, 94].

An investigation of the difference in the Te_2 vapor pressures over crystals grown from melts with maximum possible composition deviations and crystals with a near-stoichiometric composition revealed that the maximum defect concentrations in these crystals were $\sim 10^{20}$ cm^{-3}.

The lattice parameters were determined in order to refine these results for monocrystals with given deviations from stoichiometry and known electrical properties. Fig. 24 gives the cubic lattice parameter of CdTe crystals obtained from the melt, component solutions and the vapor phase plotted as a function of carrier concentration. These data reveal that regardless of the crystallization method there is a correlation between the parameter a and the carrier concentration. According to the derived data, undoped cadmium telluride crystals with deviation from stoichiometry are realized in a more compressed lattice compared to the stoichiometric composition, and the degree of compression correlates with the concentration of electrically-active centers. Since the lattice parameters have identical values on the boundaries of the p- and n-regions we can expect that this is the maximum compression in the given structure. In this regard the minimum value of the parameter a was also observed in n- and p-type crystals obtained from dilute solutions of the components having, as noted in study [90], maximum deviations from stoichiometry, the implemented range of compositions can be considered the range of existence of cadmium telluride.

The λ-like nature of the variation in the crystal lattice parameter established within its range of existence made it possible to assume that in cadmium telluride the prevalent defect responsible for deviations from stoichiometry on either side is of an identical nature.

In order to establish the nature of this defect we carried out precision measurements of changes in material density in the transition from a stoichiometric composition to maximum deviation from stoichiometry and the derived experimental quantities $\Delta\rho^{exp}$ were then correlated with the corresponding values of $\Delta\rho^{X\text{-}ray}$ calculated from the X-ray data.

Pyncnometric measurements revealed that

$$\Delta\rho^{exp} = \rho_n - \rho_{st} = 0.0026 \pm 0.0008\ g\cdot cm^{-3}, \tag{32}$$

$$\Delta\rho^{exp} = \rho_p - \rho_{st} = -0.0025 \pm 0.0008\ g\cdot cm^{-3}, \tag{33}$$

where ρ_{s5}, ρ_n and ρ_p are the density of the materials of stoichiometric composition and with maximum deviation from stoichiometry in the n- and p-regions, respectively (here and henceforth a near-stoichiometric material in the p-range with a carrier concentration of 10^{12} cm^{-3} and a = 6.484 Å will be used as the stoichiometric material). According to the X-ray data the density of the defect-free material in both cases would increase by $\Delta\rho^{com}$ = +0.008 g/cm³.

If we represent the quantity $\Delta\rho^{exp}$ as the sum

$$\Delta\rho^{exp} = \Delta\rho^{def} + \Delta\rho^{com}, \tag{34}$$

where $\Delta\rho^{def}$ is the change in density due to the appearance of defects (vacancies or institial atoms), for a material of n-type conductivity $\Delta\rho_n^{def}$ = -0.0054 g/cm³, while for a material of p-type conductivity $\Delta\rho_p^{def}$ = -0.0105 g/cm³. We have taken as constant (independent of the composition) the concentration of such growth defects as microcracks, microvoids, etc. which depend on the crystal fabrication technology, since all pyncnometric measurements were carried out on crystals obtained by the same melt growing technique.

The negative values of $\Delta\rho_n^{def}$ and $\Delta\rho_p^{def}$ reveal that the prevalent defect with deviation from stoichiometry in the n- and p-regions takes the form of vacancies. A recalculation for vacancy concentrations for n- and p-materials with maximum deviation from stoichiometry yields a value of $2\cdot10^{19}$ and $5.7\cdot10^{19}$ cm^{-3}, respectively, which is approximately two orders of magnitude greater than the concentration of electrically-active centers in the same crystals. The derived defect concentrations are within an order of magnitude of the values calculated from measurements of the optical density of Te_2 vapors [92].

The results from previous investigations of the physical properties of CdTe made it possible for the authors of studies [78, 86] to propose a crystallochemical model of this material in which the primary defect in its entire range is the Fresnel defect in the cadmium sublattice $[V_{Cd}^0]\ [Cd_i^0] = K_F$. According to this model the prevalent defect in the *n*-region is Cd_i while the prevalent defect in the *p*-region is V_{Cd}; these same defects are responsible for the corresponding electrical properties of the crystals. As a result the range of the cadmium telluride was previously identified with the range of variation in the electrically active carriers and was $\sim 10^{-3}$ at.% [87].

Our data are in qualitative agreement with this model in the *p*-region but clearly contradict the data from studies [80, 88] for the *n*-region. However since the existence of Cd_i responsible for the formation of the donor centers can clearly be taken as proven by Kroger [95] in the *n*-type material (his calculations of the depth of the donor level of Cd_i coincided with the experimental results), we concluded that in addition to the prevalent type of V_{Te} defects it is reasonable to assume the formation of a quasi-interstitial cadmium atom that can implement its own donor properties on the boundaries of complex formation from the V_{Te} vacancies.

The concepts discussed above support the assumption that in addition to the Frenkel defect in cadmium telluride there will also be the Shottky defect $[V_{Te}][V_{Cd}] = K_S$; this defect will clearly be prevalent in compositions with deviations from stoichiometry (on both sides). According to these concepts the range of cadmium telluride will be of the order of 10^{-1} at.%.

The significant difference between the total defect concentration established in the present study and the defect concentration determined previously from electrical measurements reveals that in both *p*- and *n*-materials only a small portion of the defects that occur from deviations from stoichiometry will be ionized; in *n*-type material the nonionized defects are tellurium vacancies whose ionization energy is 0.43 [96] or 0.6 eV [97] and clearly their appearance has no significant influence on the electrical properties of the material. The discrepancy between analogous values in *p*-materials can be attributed to the formation of associated vacancy complexes $(V_{Cd})_n$.

According to our data the width of the composition stability range on both the Cd-rich and the Te-rich sides is $\sim 10^{-1}$ at.%; there are 100 electrically-inactive defects for every charged defect. The prevalent defect in CdTe on the Cd-rich side takes the form of electrically inactive tellurium vacancies while on the Te-rich side the prevalent defects are cadmium vacancies which makes it possible to assume a Shottky defect in CdTe in addition to the Frenkel defect.

Our values for the composition stability range of CdTe were later expanded by studies [98, 99] where this quantity based on measurements of the dew point of the components (Cd and Te) as well as their vaporization points was $7 \cdot 10^{20}$ cm^{-3}.

We therefore believe we have proved the lack of validity of estimating the composition stability range of a compound of variable composition based on the free charge carrier concentration, since the concentration of intrinsic defects in an electrically inactive state can exceed that of the charged defects by several orders of magnitude. As discussed above in our view the reason for this is the formation of neutral complexes of intrinsic defects. One recent study [100] has attempted to attribute the discrepancy between the concentrations of electrically-active and inactive intrinsic defects in cadmium telluride by proposing the existence of an acceptor interstitial tellurium Te_i and donor antistructural defect Te_{Cd} with a dominant role of the acceptor cadmium vacancies V_{Cd} (material of *p*-type conductivity with excess Te was used) whose maximum concentrations according to the authors were: $[V_{Cd}] = 2 \cdot 10^{20}$ cm^{-3}, $[Te_i] = 1.3 \cdot 10^{20}$ cm^{-3}, and $[Te_{Cd}]$ * $1.7 \cdot 10^{20}$ cm^{-3}. According to the assumption of these authors the donor antistructural defects Te_{Cd} can compensate the acceptor cadmium vacancies.

Leaving aside the issue of to what degree it is reasonable to introduce experimentally-unconfirmed types of intrinsic defects in concentrations comparable to the concentrations of the prevalent cadmium vacancies as was done in study [100] we emphasize that today the popular issues involve the problems of the form of neutral intrinsic defects and the interrelationship between the electrically active and inactive forms of such defects.

2.4 Investigation of the Composition Stability Range of PbTe

As noted above Brebrick who was the first to investigate the composition stability limits of PbTe [11] employed data on the carrier concentration to determine the saturation compositions of this compound, assuming that the difference in the lead and tellurium concentrations in the solid compound was equivalent to the difference in the electron and hole carrier concentrations. Brebrick also assumed that the predominant point defects are vacancies in the two PbTe sublattices, while each V_{Pb} vacancy (or V_{Te} vacancy) is related to a single acceptor (or donor) level. Hall measurements of the carrier concentration at 300 K and 77 K were carried out by Brebrick on samples whose saturation concentration was obtained by annealing at specific temperatures in the presence of excess Pb and Te, followed by quenching to room temperature. Since the results from our study on cadmium telluride [74] revealed that only a comprehensive approach to investigating point defects using a number of independent methods will make it possible to carry out comprehensive investigations of the defect structure of semiconductor compounds of variable composition we believed that it was best to estimate the composition stability range of PbTe and establish the prevalent type of defects using several independent techniques by investigating samples whose equilibrium saturation composition would be dictated by the fabrication technique used to produce the material [23]. In order to achieve this goal we carried out the following experiments: 1) we grew saturated PbTe monocrystals from saturated solutions in Pb and Te at different crystallization temperatures; 2) we investigated the crystallization temperature dependence of the crystal lattice parameter of the saturated monocrystals; 3) we determined the change in the atomic scattering factors of Pb and Te in the PbTe crystal lattice as a function of crystallization temperature; 4) we investigated the Te_2 vapor density over the saturated PbTe monocrystals; 5) we carried out Hall measurements of the PbTe samples at 298 K and 77 K.

The PbTe crystal growth conditions including the blend composition and the crystallization temperature were selected in accordance with the *T-X* diagram of the Pb-Te system (see Fig. 9). The raw materials included SVCh S-0000 brand lead and "extra" brand tellurium purified by the zone float technique in hydrogen. The calculated quantities of Pb+Te for achieving saturated solutions at T_{cryst} of 500, 550, 600, 650, 700, 750, 800°C were placed in silica vessels which were evacuated and heated in an oven equipped with a rotation mechanism. Crystallization was achieved by maintaining a constant temperature gradient $\Delta T = T_1 - T_{cryst} = 50°C$ on the solution column (Fig. 25). The vessel was left in the oven for 24 hours to saturate the solution and for synthesis at temperature T_1; an auxiliary heater was then activated to create a gradient in the vessel and growing began with T_{cryst} maintained to ±1°C for 100 hours. At the end of the process without reducing the temperatures the oven was rotated to remove the solution from the grown crystals. After cooling the remaining solution was removed from the crystals by oxygen treatment. Vacuum vaporization of the residual Te was also employed at 200°C when the PbTe was grown from Te-solutions. The derived samples were bulk crystals of cubic form with characteristic dimensions of the natural {100} faces of several square millimeters; the monocrystalline nature of the samples was monitored by means of Laue diffraction patterns.

This method made it possible to fabricate equilibrium saturated samples corresponding to the solidus points at a given crystallization temperature both with excess lead PbTe(Pb) and excess tellurium PbTe(Te). All samples grown from the solution in Pb had *n*-type conductivity while those grown from the solution in Te had *p*-type conductivity.

The cubic crystal lattice parameters *a* are determined for the derived PbTe(Te) and PbTe(Pb) saturated samples [23]. X-ray diffraction measurements of *a* were carried out on the natural {100} faces and the {100} cleavage plane chips of the monocrystals as well as on polycrystals fabricated by grinding bulk monocrystals followed by annealing for 20 hours at 250°C to eliminate the crystal lattice stresses resulting from this process. The values of *a* were obtained by extrapolating the function $a = f(\theta)$ to the diffraction angle $\theta = 90^{\circ}$ and the error here was less than $2 \cdot 10^{-4}$ Å.

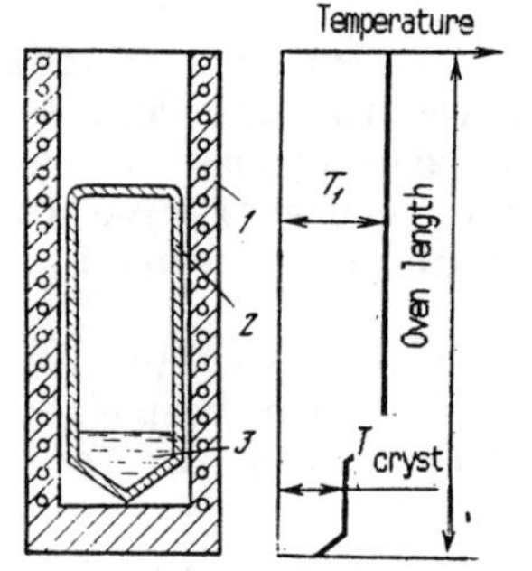

Fig. 25. The working vessel and its position with respect to the temperature profile of the oven for fabrication of lead telluride from component solutions
1 - oven; 2 - vessel; 3 - solution

Fig. 26. The crystal lattice parameter a plotted as a function of the crystallization temperature for PbTe(Te) (a) and PbTe(Pb) (b)

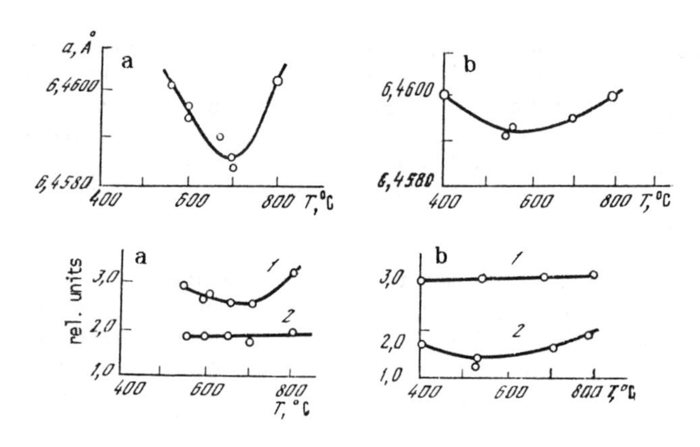

Fig. 27. The atomic scattering factors f_{Pb} (1) and f_{Te} (2) plotted as a function of the crystallization temperature for PbTe(Te) (a) and PbTe(Pb) (b)

Fig. 28. The temperature dependencies of the optical density of vapors over PbTe(Te) and PbTe(Pb) crystals obtained at various crystallization temperatures
1 - PbTe(Te), 700°C; 2 - PbTe(Te), 620°C; 3 - PbTe(Te), 595°C; 4 - PbTe(Te), 800°C; 5 - PbTe(Pb), 800°C; 6 - PbTe(Pb), 700°C

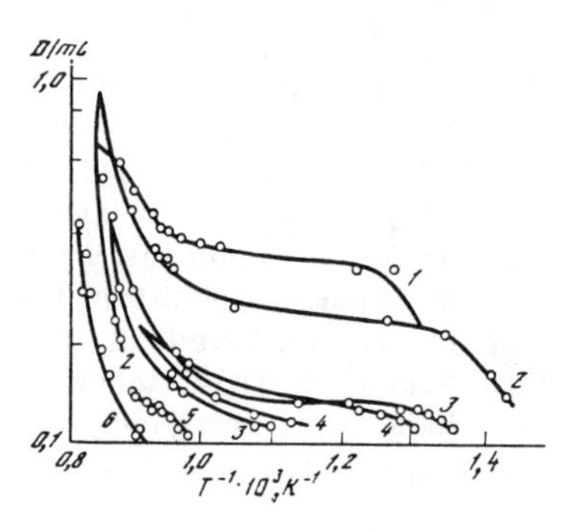

The dependencies of the parameter a on the crystallization temperature derived for PbTe(Te) and PbTe(Pb) (Fig. 26) revealed that on both sides of stoichiometry as the crystallization temperature increases the parameter a of the saturation samples corresponding to solidi on both the excess Pb and the excess Te sides initially drops and then grows; the minimum value of a for PbTe(Te) corresponds to a crystallization temperature of 700°C and for PbTe(Pb) the minimum value of a is 550°C.

The nature of the variation in the parameter a from T_{cryst} detected for the PbTe(Te) and PbTe(Pb) samples accounting for our inverse dependence of the crystal lattice parameter of CdTe on the vacancy concentration (see Section 2.3) made it possible to draw the following conclusions:

1. Lead telluride is a bilateral phase of variable composition with a maximum composition stability range at 550°C for excess Pb and at 700°C for excess Te;

2. The prevalent type of defect responsible for deviation from stoichiometry in both directions has the same vacancy nature.

In order to independently identify the prevalent defects and establish their dynamics as a function of the crystallization temperature for each of the PbTe(Te) and PbTe(Pb) saturated samples we experimentally estimated the variations in the integral intensities E_{hkl} of several pairs of neighboring (on the diffraction angle scale) X-ray diffraction interference beams, one of which has even indices and the other has odd indices (hkl).

The temperature dependencies of the change in the atomic scattering factors f_{Pb} and f_{Te} (Fig. 27) proportional to the change in the average electron concentration at the sites of the corresponding sublattices were drafted based on the derived results using the method described in Section 2.2. Calculation error was less than 3-5%.

The dependencies of f_{Pb} and f_{Te} on T_{cryst} for PbTe(Te) and PbTe(Pb) samples shown in Fig. 27 reveal that the deviations from stoichiometry that occur in lead telluride result from the formation of lead vacancies in the PbTe(Te) samples and telluride vacancies in the PbTe(Pb) samples. Therefore on the basis of these results we can conclude that the prevalent intrinsic defects in PbTe are Shottky defects $[V_{Pb}][V_{Te}] = K_S$ and according to Fig. 27 the maximum concentration of lead vacancies is achieved in PbTe(Te) samples at 700°C, while the maximum concentration of tellurium vacancies in the PbTe(Pb) samples is achieved at 500°C. We should note that the temperatures determined in this manner for the largest section of the T-X diagram for PbTe are in good agreement with the values obtained from the $a(T_{cryst})$ relations (see Fig. 26).

We also investigated the nature of the intrinsic defects in PbTe responsible for deviations from stoichiometry by measuring the optical density of the tellurium vapors over the saturated PbTe crystals. Brebrick was the first to apply this method to determining the partial pressures of Te_2 in equilibrium with PbTe crystals and the two-phase blends PbTe + Te, PbTe + Pb and then determined the three-phase equilibrium line showing that the partial pressures of Te_2 along the solidus on the tellurium side.

Study [23] was the first to employ the Te_2 optical absorption method to estimate the degree of deviation of the PbTe composition from stoichiometry by determining the pressure of the excess (superstoichiometric) tellurium over PbTe monocrystals of various composition. The optical absorption of nonsaturated vapors of the superstoichiometric Te_2 was measured for this purpose at a wavelength λ = 410 nm by cooling the PbTe condensate from 930°C to 350°C at a rate (~200 deg/hr) to avoid reverse diffusion of the excess Te_2 to solid PbTe that already crystallized at 924°C. The measurements were repeated until the optical absorption values of the superstoichiometric tellurium over crystals of identical composition were completely recovered. The derived curves of the temperature dependence of the optical density of excess Te_2 for six PbTe samples with different crystallization temperatures are shown in Fig. 28.

The entire series of curves (see Fig. 28) has a complex temperature dependence of the optical density. An analysis of these relations reveals that the sections in the 930-870°C temperature range can be considered the set of optical densities of PbTe and Te_2 vapors. According to study [8] when T < 870°C the optical density of the saturated PbTe vapors at λ = 410 nm becomes quite low and therefore the second section of the curve is a plateau in the 870-450°C range (the solidification point of tellurium) and this can be interpreted only as the change in the optical density of the unsaturated Te_2 vapors over the solid PbTe. The third section of the curve corresponds to the Te_2 vapor pressure over the solid Te. This suggests that the height of the quasi-plateau of the second section of the curve represents the degree of deviation of the test PbTe(Te) crystals from stoichiometry. Assuming that with noncongruent vaporization all nonstoichiometric tellurium atoms go from the lead telluride to the vapor phase, we carried out an approximate calculation of the composition stability range for three PbTe(Te) samples grown at 800, 700, and 620°C (see Fig. 28, curves 4, 1, 2). We carried out the calculation using the experimental values of the optical density of Te_2 vapors obtained for T = 679°C where the contribution of the PbTe vapors was already negligible (approximately half way on the quasi-plateau in Fig. 28). The derived data are given in Table 6, where ΔD_λ is the optical density of the superstoichiometric Te_2 normalized to one gram of sample and 10 cm of optical cell length; p_{Te_2} is the Te_2 vapor pressure; r is the scaling factor.

An estimate of the composition stability range of PbTe on the Te side by the Te_2 optical absorption method accounting for both the electrically active and the neutral intrinsic point defects V_{Pb} yielded a value of ~10^{20} cm^{-3} at the same time that the Hall concentration of the carriers in the same samples was ~10^{18} cm^{-3}. It follows from Table 6 that the maximum size of the composition stability range on the excess tellurium side corresponds to T = 700°C which is in good agreement with results from X-ray structural analyses (see Figs. 26 and 27).

Table 6

Deviations from stoichiometry of the PbTe(Te) samples and the charge carrier concentrations

T_{cryst} °C	ΔD_λ, g^{-1}	$p_{Te_2} = r[\Delta D_\lambda]$, atm.	No. g·mol Te_2 at 679°C	No. of excess Te atoms cm^{-3}	Hole concentration cm^{-3}
600	21	0.06	$2.4 \cdot 10^{-5}$	$2.4 \cdot 10^{20}$	$1 \cdot 10^{18}$
700	35	0.1006	$4.04 \cdot 10^{-5}$	$3.9 \cdot 10^{20}$	$2.2 \cdot 10^{18}$
800	4.4	0.0118	$4.74 \cdot 10^{-6}$	$4.4 \cdot 10^{19}$	$1.3 \cdot 10^{18}$

The two order of magnitude deviation for the PbTe(Te) samples in the total concentration of thermodynamically equilibrium defects - Pb vacancies - and the charge carrier concentration - charged Pb vacancies - can be explained rather easily if we assume as in the case of cadmium telluride above the possibility of the formulation of neutral complexes from intrinsic defects. In PbTe(Te) vacancy clusters can serve as such complexes. The existence of these clusters was later confirmed in experiments on PbTe and $Pb_{1-x}Sn_xTe$ doping (see Section 3.2).

Therefore our comprehensive experimental investigations of cadmium telluride and lead telluride have revealed that both compounds are bilateral phases of variable composition where on both sides of stoichiometry the prevalent defects are Shottky defects. The true composition stability ranges were established for the first time for both compounds: $\sim 10^{19}$ for CdTe and $\sim 10^{20}$ for PbTe and it is demonstrated that the total concentration of thermodynamically equilibrium intrinsic defects - vacancies - determining the width of the composition stability range is two orders of magnitude greater than the concentration of electrically-active vacancies.

3. METHODS OF CONTROLLING DEVIATION FROM STOICHIOMETRY

In the preceding sections CdTe and PbTe were used as examples to demonstrate that the prevalent intrinsic point defects in A^2B^6 and A^4B^6 compounds are thermodynamically-equilibrium vacancies in the metallic or chalcogen sublattices. The concentration of these defects is at least two orders of magnitude greater than the level of uncontrolled impurities in the so-called "pure" undoped crystals; in the narrowband PbTe semiconductors the defects are ionized already at the liquid helium temperature and these completely determine the carrier concentration. It follows that the problem of controlling the properties of such semiconductors is largely a problem of controlling the type and quantity of intrinsic defects/vacancies that are responsible for the deviations from stoichiometry.

The most natural method for changing the composition of A^4B^6 compounds within the composition stability limits that has been widely used in practice is crystal annealing in the presence of a blend containing an excess of one of the components. As will become clear from this discussion annealing in an appropriate atmosphere makes it possible to reduce the charge carrier concentration in the material. However it is not free of significant drawbacks.

Methods of controlling deviation from stoichiometry include purification of the initial components and the compounds themselves by eliminating extraneous impurities. The primary purification methods include standard zone float techniques for the initial materials and sublimation of binary compounds (with the exception of those that break down substantially in the gas phase) and hence are not described in our survey.

The technological annealing parameters are determined by the features of the phase microdiagrams of the compound in solid form and depend on the fabrication method of the material.

3.1. Annealing

Melt growing techniques, primarily the Bridgman and Czochralski techniques described in detail in the surveys by Harman and Preier [101, 102] are used most often to obtain PbTe and PbSe compounds and solid solutions based on these compounds. The melt techniques make it possible to obtain large "pure" undoped crystals up to 5 cm in diameter and up to 10 cm in length. The composition of these $Pb_{1-\delta}Te$ crystals in accordance with the phase diagram contains an excess (δ) of Te appearing as vacancies at the Pb sites, while the material has a high hole carrier concentration of 10^{18}-10^{20} cm^{-3}.

The $Pb_{1-x}Sn_xTe$ and $Pb_{1-x}Sn_xSe$ solid solution crystals obtained by melt techniques also have inhomogeneity in the composition x which can reach 10%. In order to equalize the x composition and reduce the initial hole concentration in the $Pb_{1-x}Sn_xTe$ alloys homogenizing isothermic annealing is used in the vapors of a metal-rich $(Pb_{1-x}Sn_x)_{1+\delta}Te$ blend, where δ is the excess metal. During the annealing process the initial $(Pb_{1-x}Sn_x)_{1-\delta}Te$ crystals change their composition in "δ" by diffusion of the metal from the gas phase until equilibrium is achieved with the vapor phase with metal saturation of the samples:

(35)

As a result of this annealing the vacancy concentration in the metallic sublattice drops and the crystal composition approaches stoichiometry. Normally wafer samples a few millimeters in thickness cleaved from bulk crystals are annealed. The technological annealing parameters, the temperature, time period of the process and the annealing results are highly dependent on the size of the annealed samples, the level of uncontrolled impurities, and the surface condition of the annealed samples. In order to reduce the hole concentration in the p-$Pb_{0.8}Sn_{0.2}Te$ alloy to p_{77} = 10^{17} cm^{-3} normally samples at T = 600°C are annealed for several weeks in the presence of a $Pb_{0.51}Te_{0.49}$ powder.

S.A. Medvedev obtained the best results from annealing of PbTe [103] in which monocrystals grown from a material first purified by repeated sublimation were annealed. Investigations revealed that the samples had a high mobility (p_{77} = $2.1 \cdot 10^{17}$ cm^{-3}, μ_{77} = $5.2 \cdot 10^4$ $cm^2/(v.s)$; n_{77} = $3.8 \cdot 10^{17}$, μ_{77} = $7.3 \cdot 10^4$ $cm^2/(v \cdot s)$) as well as efficient photoluminescence and cathodoluminescence.

In order to evaluate the quality of crystals that have been annealed and their suitability for fabrication of injection lasers, the Physics Institute of the Academy of Sciences investigated the structural and luminescence characteristics of the monocrystals used as active media in lasers [104]. The initial monocrystal material for the substrates was fabricated by crystallization from the gas phase using a blend with x from 0.17 to 0.22. Measurements of the electrical parameters of the initial $Pb_{0.78}Sn_{0.22}Te$ material revealed that the grown crystals have p-type conductivity with a carrier concentration $6 \cdot 10^{19}$ cm^{-3}. Precision measurements of the crystal lattice parameter revealed a local spread of the values from 2.4420 ±0.0006 Å to 6.4210 ±0.0006 Å where the greatest specific weight was 6.4310 ±0.0006 Å. Such a spread could correspond to a local change in the atomic weight of the lead in the $Pb_{1-x}Sn_xTe$ solid solution from 0.87 to 0.71 for the most probable $Pb_{0.78}Sn_{0.22}Te$ composition. Here we assume that the vacancy concentration is constant throughout the crystal volume since crystallization occurred at a constant temperature. In order to fabricate laser heterostructures material with a composition $Pb_{0.83}Sn_{0.17}Te$ was employed; this material was exposed to long-term (30 days) homogenizing annealing in the presence of a blend of the same composition in x yet containing 1% excess metal. As a result of the annealing process the hole concentration was reduced to $2.4 \cdot 10^{17}$ cm^{-3}. The local spread of the crystal lattice parameter was from 6.4394 ±0.0006 Å to 6.4383 ±0.0006 Å which corresponds to a change in the atomic percent of the lead from 0.84 to 0.83. The material began to luminesce and its cathodoluminescence at T = 95 K is given in Fig. 29. Investigations of the diffraction curves and the oscillation curves for two-crystal spectrometer operation revealed that annealing causes a significant increase in the half-widths of the X-ray diffraction interference beams (600) from 0.09° to 0.15° and the rocking curves (200) from 0.2° to 1° (Fig. 30).

These increases in the half-widths indicate that the monocrystal nature of the material undergoes significant changes during annealing: the average size of the monocrystal microblocks drops (from 3500 Å to 1500 Å) and their disorientation increases (polygonization occurs). However this process is accompanied by a significant increase in the microhomogeneity of the crystal blocks themselves which is responsible for the luminescence properties.

Annealing is therefore a complex process and in addition to reducing the carrier concentration and homogenizing the composition it improves the structural characteristics of the monocrystals. This fact limits the application of annealed crystals for fabrication of laser diodes. Moreover annealing can be used only for bulk crystals and cannot be applied to epitaxial structures containing layers of different composition or a *p-n* junction, since the junctions become washed out from the annealing process.

In spite of these drawbacks annealing in a blend of appropriate composition has been widely used to control deviations from stoichiometry in A^4B^6 since due to the small vacancy formation energies rather low annealing temperatures of 500-600°C are required to "cure" the defects in these materials. The most efficient technique for controlling the defects is to regulate the composition of the compound in the actual growing process from the gas phase by doping the crystals or the epitaxial layers with its own intrinsic components.

Unlike the melt and solution growing techniques which produce saturated samples, the gas phase methods make it possible to obtain crystals and layers of various composition within the composition stability range as a function of the ratio of the partial pressures of the components over an A^4B^6 or A^2B^6 crystal (see Section 1). Using an additional vapor source of a selected component (most often a metal) it is possible to manipulate within specific limits the composition of the vapor phase and therefore the composition and properties of the growing crystal. The limits on the applicability on this method of controlling deviations from stoichiometry are determined by the *P-X* phase diagrams and the vaporization features of the specific compounds.

It is established in study [92] that when growing CdTe from the vapor phase it is possible by controlling the Cd vapor pressure to obtain *p*-CdTe crystals with a carrier concentration from $p = 10^{16}$ cm^{-3} to $p = 10^{12}$ cm^{-3}. The pressure P_{Cd} could be manipulated only within a very narrow range of values $P_{Cd\ pure}/P_{Cd\ min} = 0.83-1.2$, i.e., crystallization is significantly degraded with growth of P_{Cd}. The crystal growth rate is directly dependent on the material composition: the greater the deviation from stoichiometry the slower the growth rate. This is due to the excess Cd(Te) component that suppresses vaporization. The suppression effect is quite strongly manifest in the presence of excess Cd atoms. This sublimation method is then suitable for obtaining only *p*-CdTe crystals with variable deviation from stoichiometry.

Study [105] employed open vaporization to derive monocrystal epitaxial *p*-PbTe (111) layers on BeF_2 substrates having a concentration from $p_{77} = 1 \cdot 10^{16}$ cm^{-3} to $p_{77} = 5 \cdot 10^{17}$ cm^{-3} depending on the excess tellurium; *n*-type layers could not be obtained.

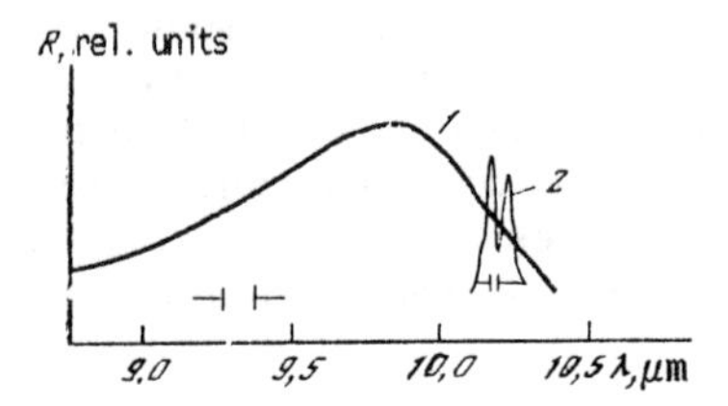

Fig. 29. The cathodoluminescence spectrum at 95 K of an annealed $Pb_{0.83}Sn_{0.17}Te$ sample (curve 1) and stimulated radiation from a diode fabricated from this sample at 77 K (curve 2)

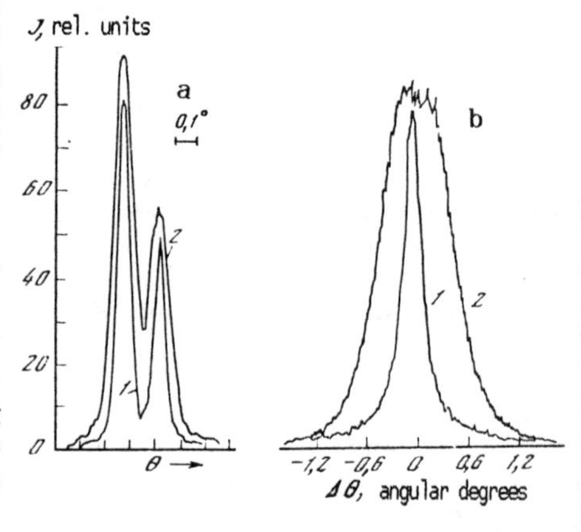

Fig. 30. The diffraction curves of the X-ray interference beams (600) (a) and the rocking curves of the X-ray diffraction beams (200) (b) before and after annealing (curves 1 and 2, respectively) (the maxima of the diffraction curves are offset to assist in the comparison)

The authors of study [106] investigated the dependencies of the composition of vapor-grown PbS, PbSe, and $Pb_{0.8}Sn_{0.2}Te$ crystals on the composition of the initial blend at a variety of vaporization and condensation temperatures. By varying the excess lead in the blend and altering the growth temperature it was possible to obtain n-type PbSe crystals with concentrations from $4.5 \cdot 10^{18}$ cm^{-3} to $8 \cdot 10^{18}$ cm^{-3} together with crystal containing grown p-n junctions. It was not possible to obtain p-type crystals when using an Se-rich blend of the $Pb_{0.48}Se_{0.52}$ composition due to the sharp drop in the growth rate due to insufficient lead in the vapor phase. An analogous reduction in growth rate and a degradation in crystallization are observed when using a $(Pb_{1-x}Sn_x)_{0.48}Te_{0.52}$ blend.

Therefore by controlling the vapor pressure of the intrinsic components for growing crystals from the gas phase it is possible in certain cases to obtain bulk crystals and layers with controlled electrical properties. We note that the crystals obtained in these conditions have a more uniform structure than annealed crystals.

The use of electromagnetic radiation represents a fundamentally new method of controlling deviation from stoichiometry in compounds of variable composition; this radiation can be used to manipulate and improve the crystallization process itself.

3.2. The Influence of Electromagnetic Radiation on the Composition Stability Range of Lead Telluride

The issues of the influence of electromagnetic radiation on the growth processes and properties of semiconductors are the focus of special attention today. We can say that a new field has developed in semiconductor technology that employs the advantages deriving from luminous radiation action on the growth processes. The Physics Institute of the Academy of Sciences has been conducting priority research under the direction of S.A. Medvedev [107, 108] in this field; this research has focused on the autoepitaxial growing of Si by optical heating of the substrate. It was discovered that the Si growth rate (1.3 μm/min) when heating the substrate by a concentrated luminous flux is significantly higher than the Si growth rate for the case of heating by HF current or a resistance furnace. By irradiating the substrate with light it was possible to obtain Si layers up to 100 μm in thickness without stacking faults or growth traces on the layer surfaces; a doping impurity level up to 10^{19} cm^{-3} did not degrade surface quality in this case. As demonstrated by K.V. Kiseleva et al. [109] the epitaxial Si layer growth technique that employs luminous flux heating of the substrate is free of the sources responsible for generating at the interface and in the film volume such structural imperfections as stacking faults and dislocations. In this case the structural imperfections of the epitaxial layers are simply dislocations inherited from the substrate; their density in the layer is approximately one-half the corresponding level in the substrate. The authors attributed these features of the epitaxial growth of Si and the improved structural properties to successful purification of the growth surface by the high-power luminous flux and the development of more favorable crystallization conditions.

Further research at the Physics Institute of the Academy of Sciences resulted in the development of the photostimulated epitaxy technique [109, 110, 111]; a detailed description of this method lies beyond the scope of the present study. We simply point out that the essence of the method involves luminous flux irradiation of the gas-phase epitaxial growth process of a semiconductor onto a substrate that does not contain its components. When this method is used to grow Si [107, 109], CdTe [110], $Pb_{1-x}Sn_xTe$ [110, 111] layers an increase in the growth rate was discovered as well as an improvement in their structure [108, 112] and a reduction in the epitaxy temperature [112]. We explained these effects based on the mechanisms generated by the formation of the free atoms of the components in the crystallizing compound in the growth zone under the influence of light irradiation of a specific wavelength and intensity [113]. The photodissociation of PbTe during its deposition under light irradiation at $\lambda \leqslant 0.45$ μm was confirmed experimentally [114] and it was demonstrated that photodissociation in the gas phase near the growth surface and on the growth surface itself plays a significant role in the kinetics of the PbTe epitaxial process.

A critical aspect of the photostimulated epitaxy technique was the influence of light irradiation on the growth processes of semiconductor compounds of variable composition resulting in the capacity to regulate their composition. The composition of such compounds can be regulated by using light of appropriate wavelength and intensity to excite the electron shells of specific atoms during the crystallization of the compound. This excitation process will then change the valance properties of the component and, therefore, the number of atoms of this type contributed to the lattice. This factor then becomes responsible for the specific capabilities to control the formation of intrinsic defects, since exciting the electron subsystem causes changes in the interaction potentials of the atoms forming the crystal lattice [115].

Intense UV radiation action during the growth process can change the effective interaction potential, since some of the atoms crystalize at a different potential: one corresponding to the given excitation level. Experiments have revealed that under irradiation by a xenon lamp at 0.21-1.2 μm it was possible to expand the range of compositions of the CdTe layers obtained in the gas epitaxy process. In this process CdTe layers with a controlled conductivity type and a carrier concentration from $p_{300} = 1 \cdot 10^{17}$ cm^{-3} to $n_{300} = 4 \cdot 10^{16}$ cm^{-3} [111] were obtained by altering the Cd vapor pressure. A comparison of these quantities to the range of carrier concentrations for CdTe ($p_{300} = 1 \cdot 10^{16}$ cm^{-3} - $n_{300} = 1 \cdot 10^{14}$ cm^{-3}) obtained in identical growth conditions but without radiation (Section 3.1) reveals that under photostimulation the range of the concentrations of charged vacancies expands on both the Cd and the Te sides.

In certain irradiation cases we can justifiably claim altering the composition stability limits of the compound. Growing $Pb_{0.8}$, $Sn_{0.2}Te$ layers in photostimulated sublimation conditions (growth temperature 760°C) using a Pb-rich blend produced material of the desired composition with n-type conductivity. By altering the excess Pb from 1-9% by weight in the (PbSnTe + Pb) source it is possible to control the carrier concentration in the layers from $p_{77} = 9.1 \cdot 10^{18}$ to $n_{77} = 1.5 \cdot 10^{18}$ cm^{-3} on a reproducible basis [116]. We note that in accordance with the equilibrium T-X phase microdiagram, the $Pb_{0.8}Sn_{0.2}Te$ material with n-type conductivity could not be obtained at any growth temperature, i.e., the n-type layers obtained had a composition stability range broader than in the case where luminous irradiation was not used.

X-ray structural analyses revealed a high structural quality of the grown layers. No broadening of the diffraction or rocking curves were observed; these normally occur in annealed monocrystals (Table 7). It is important to note that the broadening of the composition stability range under the influence of luminous irradiation and the realization of the $Pb_{0.8}Sn_{0.2}Te$ compositions with $n \approx 10^{18}$ cm^{-3} did not increase the precipitates based on the excess Pb, since these layers had effective luminescence after growing, which made it possible to employ such layers as active laser media in the ~10 μm spectral range.

An auxiliary source of Pb vapor was used to alter the composition of the gas phase in study [117] for growing PbTe and PbSe layers by the photostimulated epitaxy technique; here the carrier concentration in the monocrystal layers can be regulated over a broad range: for PbTe layers from $p_{77} = 1.3 \cdot 10^{18}$ cm^{-3} to $n_{77} = 10^{18}$ cm^{-3}, while for PbTe from $p = 6.8 \cdot 10^{18}$ cm^{-3} to $n = 7.2 \cdot 10^{18}$ cm^{-3}. Growing layers of both materials in identical growth conditions yet without luminous irradiation did not make it possible to reliably obtain PbSe and PbTe layers with such significant deviations from stoichiometry; the layers themselves were polycrystalline layers.

It is clear from these examples that irradiating the growth zone of A^2B^6(CdTe) and A^4B^6(PbSe, PbTe, $Pb_{0.8}Sn_{0.2}Te$) compounds in the gas epitaxy process causes a broadening of the range of implemented compositions with different deviations from stoichiometry which has made it possible to obtain perfect monocrystal layers of n- and p-type conductivity of these compounds and to use these layers as active media for lasers.

Table 7

Results from diffractometric analysis of $Pb_{0.8}Sn_{0.2}Te$

Material	Half-width of diffraction curve $\theta/2\theta$	Half-width of rocking curve	Luminescent properties
$Pb_{0.8}Sn_{0.2}Te$ epitaxial layer on PbTe	0.07	0.07	Yes
$Pb_{0.8}Sn_{0.2}Te$ epitaxial layer on BaF_2	0.085	0.025	Yes
Nonannealed $Pb_{0.8}Sn_{0.2}Te$ crystal	0.09	0.2	No
Annealed $Pb_{0.8}Sn_{0.2}Te$ crystal	0.15	1.0	Yes

We propose that the irradiation causes a change in the equilibrium between the vacancy charged point defects and the neutral vacancy complexes which set the composition stability limits. In p-type material with significant deviation from stoichiometry where the prevalent defects are vacancies the composition in the vicinity of the composition stability limits is determined by the relation

$$V_{Me} \underset{h\nu}{\overset{\text{excess Te, } \Delta E_{act.}}{\rightleftarrows}} (V_{Me})_n^0, \quad T = \text{const.} \tag{36}$$

Employing light with $h\nu \geqslant \Delta E_{act}$ for irradiation eliminates the activation limitations, shifts the equilibrium towards a lower number of vacancy neutral clusters, i.e., it results in a potential increase in

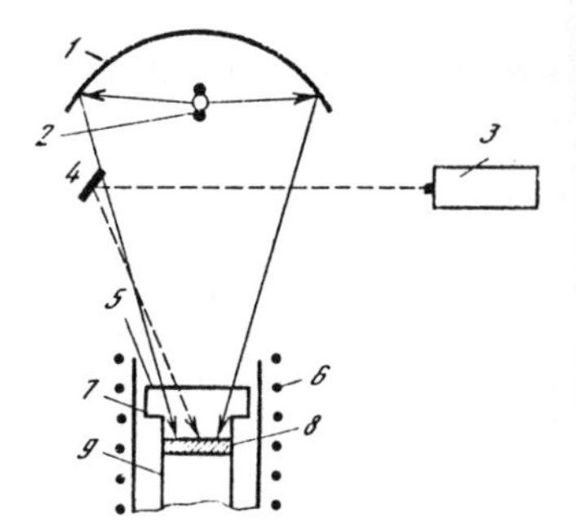

Fig. 31. Experimental configuration for investigating the influence of pulsed laser radiation on the epitaxy of lead and tin chalcogenides
1 - elliptical reflector; 2 - xenon lamp; 3 - laser; 4 - mirror; 5 - optical silica window; 6 - resistance heater; 7 - blend; 8 - substrate; 9 - vitreous silica apparatus

the number of charged point defects. In the case of photostimulated epitaxy of PbTe or $Pb_{1-x}Sn_xTe$ the additional source of one of the compound components in the growth zone (such as Pb or a doping impurity) expands the range of compositions of the compound due to interaction with these additional charge vacancies.

Therefore electromagnetic irradiation can be used to alter the composition stability range of $Pb_{0.8}Sn_{0.2}Te$ and to reproduce material with properties that are not attainable by other techniques.

Xenon lamps producing a polychromatic luminous flux between 0.21 μm and 1.2 μm were employed in these experiments as the light source; filters were used to isolate the desired spectral range. Lasers can be used to provide additional capabilities.

Recently study [24] was the first to carry out a comprehensive experimental investigation of the influence of pulsed laser radiation on the epitaxial growth processes and the structural characteristics of layers of PbTe and $Pb_{0.8}Sn_{0.2}Se$ semiconductor compounds. Many studies have been devoted to laser applications to annealing implanted Si, GaAs layers [118]. The prospects for this application are obvious. However there have been only a few attempts to employ laser irradiation to directly influence growth processes and no experiments have yet been devoted to compounds of variable composition, which represent the test materials in this case.

The epitaxial growth process was employed in conditions of sublimation transport in the quasi-closed volume of the vitreous silica reactor. BaF_2 (111) and PbTe (100) were used as the substrates. The experimental assembly is shown in Fig. 31. The primary feature of this assembly is that two radiation sources are employed: a CW polychromatic source and a pulsed source. A DKSR-3000 lamp radiating in the 0.2-1.5 μm range was used as the CW source. The luminous flux density was less than 100 w/cm^2. The study also employed neodymium and nitrogen lasers. Their primary specifications are as follows: neodymium laser (LTIPCh-5) wavelengths: λ_1 = 1.06 μm, λ_2 = 0.53 μm, λ_3 = 0.35μm; pulse duration: $\tau 2\text{и} \sim 10^{-8}$ sec; maximum pulse power: $W_1 \sim 10^6$ w, $W_2 \sim 10^5$ w, $W_3 \sim 10^4$ w; repetition rate f = 12.5 Hz; spot diameter d = 3 mm; nitrogen laser (LGI-21): λ = 0.337 μm, $\tau_\text{и} \sim 10^{-8}$ s, $W \sim 10^3$ w f = 25 Hz, d = 3 mm. The experimental conditions made it possible to carry out epitaxy both in the absence of luminous irradiation and with laser and (or) lamp irradiation. The temperatures of the substrate and the blend measured directly during the experiment when the same material was being deposited were maintained at the same level independent of the irradiation conditions. We obtained monocrystal films with the substrate orientation up to 200 μm in thickness and with an area of ~1 cm^2.

The structural characteristics of the films - the lattice parameter a and the substructure elements - were determined in the two-crystal spectrometer mode. The measurement accuracy for measuring the lattice parameter for large diffraction angles ($\theta > 70°$) was $\pm 1 \cdot 10^{-3}$ Å for $Pb_{0.8}Sn_{0.2}Se$ and $\pm 2 \cdot 10^{-4}$ Å for PbTe. The values of a given in the study were obtained by extrapolating the $a(\theta)$ relation to $\theta = 90°$; in this case the error was $\pm 2 \cdot 10^{-5}$ Å. The number of dislocations in the films was determined by the chemical etching pits.

The primary experimental results are given in Table 8 and can be summed up as follows:

Table 8

Structural parameters of layers fabricated under laser irradiation

	Luminous irradiation		Growing parameters		Structural data			
Layer, substrate	Lamp, λ=0.25-1.5 mcm	Laser, λ=0.35 mcm	T of substrate °C	Growth rate mcm/min	Lattice parameter	Substructure parameters		Dislocation density cm^{-2}
	No	No	500	1	-			
PbTe(111)	Yes	No	800	2.8	6.4657			
on	Yes	Yes	780	3.3	6.4692	-	-	-
BaF_2(111)	Yes	Yes	800	3.9	6.4675			
	Yes	Yes	830	4.5	-			
	No	No	800	1.2	6.4602	$4.19 \cdot 10^3$	$1.17 \cdot 10^{-4}$	$5 \cdot 10^6$
PbTe(100)	Yes	No	790	3.8	6.4604	$4.35 \cdot 10^3$	$1.2 \cdot 10^{-4}$	$2 \cdot 10^5$
on	Yes	No	800	3.8	6.4606	$3.86 \cdot 10^3$	$1.21 \cdot 10^{-4}$	$4 \cdot 10^5$
PbTe(100)	Yes	No	790	3.9	6.4606	$4.22 \cdot 10^3$	$1.00 \cdot 10^{-4}$	
	Yes	Yes	800	3.8	-	-	-	$4 \cdot 10^4$
	Yes	Yes	780	3.8	6.4608	$4.88 \cdot 10^3$	$1.28 \cdot 10^{-4}$	$7 \cdot 10^4$
	No	No	500	1	6.095			
$Pb_{0.8}Sn_{0.2}Se$	Yes	No	680	1.3	6.103	-	-	-
(111) on	Yes	Yes*	680	2.2	6.106			
BaF_2(111)	Yes	Yes*	680	2.2	6.105			

* λ = 0.33 μm

1) An increase in the growth rate is observed in the vicinity of the laser spot. The rate of growth of $Pb_{0.8}Sn_{0.2}Se$ on (111) BaF_2 at a growing temperature $T = 680°C$ increased from 1.3 μm/min without the laser to 2.2 μm/min under laser irradiation at λ = 0.0.337 μm; for PbTe growing on (111) BaF_2 ($T = 800°C$) this rate jumped from 2.8 μm/min to 4.5 μm/min under laser irradiation at λ = 0.35 μm. Irradiation at 1.06 μm and 0.53 μm did not result in significant changes in the growth rate and henceforth all results will apply to experiments with λ = 0.337 and 0.35 μm.

2) An electron microscope investigation of the surface of the grown films with ×1000 magnification has revealed that their surfaces are mirror-smooth without any noticeable growth traces or relief details in the vicinity of the laser spot.

3) Joint laser and xenon lamp irradiation resulted in a significant increase in the crystal lattice parameter of both compounds without any change in the lattice symmetry type (NaCl-type). For an autoepitaxial PbTe layer (direction of growth: <100>) a = 6.4608 Å; for PbTe and $Pb_{0.8}Sn_{0.2}Se$ layers on BaF_2 substrates (growth direction: <111>) a = 6.4683 Å and a = 6.1055 Å, respective.y. These quantities exceed the maximum values of the parameter 6.4601 Å for PbTe and 6.095 Å for $Pb_{0.8}Sn_{0.2}Se$ we obtained for bulk monocrystals and epitaxial layers grown in the absence of electromagnetic irradiation. We note that the latter are in good agreement with the literature data [22, 119]. An evaluation of the substructural elements of the autoepitaxial PbTe layers - the composition stability range L_n and the elastic deformation range of the lattice $\varepsilon_n = \Delta d/d$ (d is the interplanar spacing normal to the film surface) has revealed that growth of a is not accompanied by degradation of these characteristics (at the laser spot $L_n = 4.8 \cdot 10^3$ Å and $\varepsilon_n = 1.28 \cdot 10^{-4}$; outside the spot: L_n = $4.2 \cdot 10^3$ Å, $\varepsilon_n = 1.21 \cdot 10^{-4}$).

4) The number of dislocations in the vicinity of the laser spot diminishes. With autoepitaxy of PbTe on a PbTe(100) substrate with a dislocation density of $5 \cdot 10^6$ cm^{-2} their mean density in a lamp-illuminated film was $2 \cdot 10^5$ cm^{-2} and at the laser spot the density was $4 \cdot 10^4$ cm^{-2}.

In examining these results a significant fact is that the epitaxy process is influenced by UV laser irradiation (λ = 0.337, 0.35 μm) with a very short pulse duration ($\tau_И \sim 10^{-8}$ s) and a comparatively low power density (in our experiments less than 10^5 w/cm²). We note that in our conditions a single monolayer will form in $\tau \sim 10^{-2}$ s $\gg \tau_И$. Under nanosecond laser annealing the pulse power densities are $\sim 10^7$–10^8 w/cm² are two to three orders of magnitude greater than the levels we employed.

The effects that occur during pulsed annealing are related primarily to melting resulting from absorption of radiation. Estimates reveal that even a maximum power density of $\sim 10^5$ w/cm² is 20 to 30 times smaller than that at which the film surface would begin to melt. Moreover the change in temperature occurring at the instant of laser pulse action is less than 50°C. The temperature rapidly (in ~10^{-6} s) relaxes to the initial value, so its average value remains constant during the epitaxy process. Nonthermal laser action is clearly the predominant mechanism in our experiments.

One possible mechanism influencing the epitaxial growth is photodissociation of the PbTe, Te_2, and PbSe gas phase molecules (E_{dis} = 2.7-3 eV) [112]. This sheds light on the specific nature of laser action for laser wavelengths whose energy quantum is greater than the dissociation energy of these molecules. Two series of autoepitaxial monocrystal PbTe (100) layers were grown from a *p*- and *n*-PbTe blend that is either saturated with Te or Pb to investigate the influence of laser action on the composition stability range of lead telluride.

The first series was obtained from the combined action of a 3 kw xenon lamp and a pulse laser with λ = 0.35 μm; the epitaxial layer growing temperatures varied from 520 to 800°C; the lamp power for the second series was increased to 10 kw while the growing temperature of the layers varied from 460 to 620°C. The laser radiation (λ = 0.35 μm) characteristics were held constant in both cases.

The cubic crystal lattice parameters were determined accurate to $\pm 2 \cdot 10^{-5}$ Å for all grown PbTe mirror-smooth layers up to 200 μm in thickness (Table 9).

Fig. 32 shows the substrate temperature dependence of the crystal lattice parameter *a* of monocrystal autoepitaxial PbTe *n*- and *p*-type layers obtained under xenon lamp action and in combined laser and lamp irradiation. For comparison purposes Fig. 32 gives graphs of *a* plotted as a function of *T* for saturated PbTe monocrystals grown from solutions in Pb and Te in an identical temperature range without electromagnetic irradiation [23].

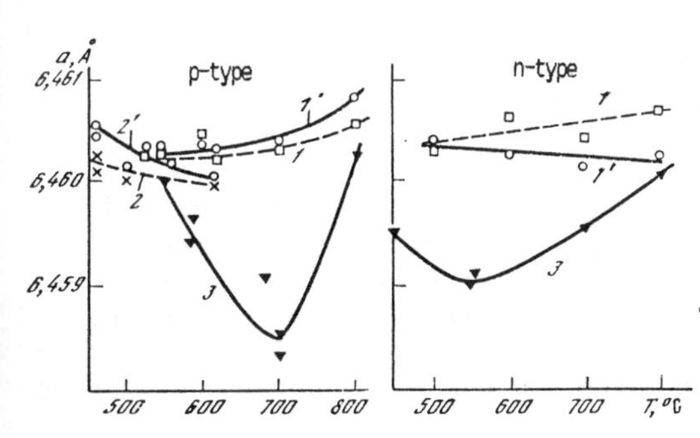

Fig. 32. The lattice parameter *a* of *p*- and *n*-PbTe layers plotted against the growing temperature
1 and 2 - 3 and 10 kw lamp irradiation of the growth zone, respectively;
1′ and 2′ - combined laser and lamp 3 and 10 kw irradiation of the growth zone, respectively;
3 - growth in zero radiation

It is clear from Fig. 32 that with light irradiation or combined (or simultaneous) light and laser irradiation all *a*(*T*) curves are shifted towards higher values of *a*. This means that near-stoichiometric compositions have been achieved across the entire temperature range. It is important to note that no other existing technological methods in the literature have yet produced PbTe with such high lattice parameter values. A comparison of the curves demonstrates that the nature of the dependence of the lattice parameter of the layers on the crystallization temperature changed significantly in the case of electromagnetic irradiation: the composition stability range is significantly expanded and retrograde solubility of the components in solid PbTe occurs across the entire temperature range.

Table 9

Crystal lattice parameter of PbTe plotted as a function of growing temperature for the case of electromagnetic irradiation

Growing temperature °C	Lattice parameter of layer PbTe, A				Lattice parameter of bulk monocrystals, Å
	Lamp (3 kw)	Laser (10 kw)	Laser (λ=0.35 mcm) and lamp (3 kw)	Laser λ=0.35 mcm) and lamp (10 kw)	
460	—	6,46024	—	6.46061	—
520	6.46028	—	6.46030	—	—
540	—	6.46026	—	6.46028	—
550	—	—	—	—	6.46020
560	6.46018	—	6.46018	—	—
585	—	—	—	—	6.45940
590	—	—	—	—	6.45960
600	6.46039	—	6.46036	—	—
620	—	6.46024	—	6.46026	—
660	—	—	—	—	6.45900
	6.46035	—	6.46036	—	—
700	—	—	—	—	6.45820
	—	—	---	—	6.45830
780	6.46060	—	6.46080	—	—
800	—	—	—	—	6.46035

The angular half-widths $(\Delta\theta)_{1/2}$ of the X-ray diffraction interference beams (600) and (800) were measured to evaluate the degree of perfection of the crystal lattice of the derived PbTe layers (Table 10). A comparison of the values of $(\Delta\theta)_{1/2}$ and a revealed that as the composition of the layers approaches stoichiometry there is a significant improvement in their crystal lattice.

The derived results reveal that pulsed laser action at λ = 0.3 5μm is not equivalent to UV irradiation from a xenon lamp and does not result in the free atom formation mechanism. The pulsed laser causes local energy release at the defects, thereby forming high plasticity states (for example, due to possible cavitational microvoid collapse). This causes the medium to become increasingly dense and reduces the total number of vacancies, thereby producing an increase in the lattice parameter.

An analysis of these experimental results reveals that the action of electromagnetic radiation on the crystallization process of compounds of variable composition influences their composition stability range. The possibilities for controlling the composition within this range and also altering the composition stability limits

Table 10
Results from X-ray diffraction analyses of PbTe layers obtained under electromagnetic irradiation

Layer lattice parameter, *a*, Å	Half width of the diffraction lines (600), angular degrees	
	Lamp (3 kw)	Lamp (10 kw)
6.46024	-	0.0321
6.46026	-	0.0314*
6.46026	-	0.0308
6.46028	-	0.0303*
6.46060	0.0430	-
6.46061	-	0.0302*
6.0410	0.0410	-

*Layers fabricated under light and laser irradiation

depend on the specific irradiation parameters: the spectral composition of the light, the luminous flux intensity and the radiation conditions (CW or pulsed) radiation. Implementation of these technological capabilities creates new methods of controlling the properties of semiconductors of variable composition. This refers to the development of a new photostimulated doping technique whose application has made it possible to obtain epitaxial layers and structures having efficient luminescence properties immediately after the growing process. Select experimental results from an investigation of the dependence of the structural and luminescence properties of A^4B^6 layers on the conditions of photostimulated doping by intrinsic components or extraneous impurities are given in the next section.

4. CERTAIN FEATURES OF CONTROLLING THE ELECTRICAL AND LUMINESCENCE PROPERTIES OF LEAD-TIN TELLURIDES

4.1. Aspects of Doping Semiconductors with a Broad Composition Stability Range

One of the most important and popular problems in the materials processing of semiconductors is the investigation of doping processes in narrowband semiconductors. The practical application of such research includes the fabrication of laser diodes based on these materials for operation in the IR. These processes include: growing structurally complete doped substrates with a given conductivity and carrier concentration; fabricating a stable *p-n*-junction in semiconductor structures and increasing the quantum efficiency in the active region of the laser. Issues of equal importance include investigating the doping processes of A^4B^6 components from the viewpoint of understanding the behavior of the impurities and their interaction with the intrinsic defects in compounds of variable composition.

The significant deviations from stoichiometry in lead and tin chalcogenides and the related high vacancy concentration make it possible to expect certain specific features of the impurity dissolution mechanism and, consequently, certain features of their influence on the semiconductor and optical properties of A^4B^6.

Lead and tin tellurides and solid solutions based on these compounds, as demonstrated above, are bilateral phases of variable composition realized from the formation of Shottky-type intrinsic defects, i.e., structural faults of the type $(Pb, Sn)_{1-\delta}Te_{1-\gamma}$, where δ and γ characterize the deviations from stoichiometry and largely determine the type of conductivity and the carrier concentration and mobility as well as photosensitivity and the luminescence properties of the material. It follows that in developing doping techniques for such compounds it is necessary to maintain a specific relation between the concentration of impurity introduced and the dimensions of the composition stability range at a significantly lower level of uncontrolled impurities. In our doping experiments (see below) the purity level of the initial materials with respect to uncontrolled impurities was less than 10^{16} cm^{-3}. This is a new situation compared to elementary semiconductors and A^3B^5 compounds in which it was possible to ignore the intrinsic defects when designing doping techniques due to their narrow composition stability range (see Table 1) and only the electronic properties were accounted for in the selection of the impurity.

According to the literature the doping action of a single impurity in type A^4B^6 compounds may be significantly different depending on whether or not this compound is a bilateral or unilateral phase of variable composition. For example, In is a donor in PbTe and it is an acceptor in SnTe, i.e., with an increase in the In concentration the free hole concentration grows; their mobility drops compared to undoped SnTe [133]. In turn the dissolution of the impurity in the *p*- and *n*-regions of the bilateral phase follows different mechanisms. Consequently in order to develop the doping techniques it is necessary to investigate the nature of the intrinsic defects as well as the dissolution mechanisms of the impurities and their interaction with the intrinsic defects within the framework of the given material fabrication technique.

These concepts are not often included in most studies devoted to doping in A^4B^6 and therefore the literature often contains the term "anomalous" impurity behavior as well as contradictory experimental data and inconsistent theoretical models.

In this section we will present results from experiments carried out at the Physics Institute of the Academy of Sciences devoted to the doping of lead and tin tellurides that account for the interaction of the doping atoms with the vacancy subsystem of the compound, although

Table 11

The influence of impurities on the conductivity type of A^4B^6 compounds

Compound	Compound fabrication technique	Doping impurities		Influence of impurities on conductivity type			Ref.
		Doping method: diffusion	In-growth doping	Donor	Acceptor	Amphoteric	
PbS	Grown from gas phase	Ag, Cu, Ni, Bi Sb	-	Donors	-	-	[120]
PbSe	Grown from gas phase	Sb	-	Donor	-	-	[120]
$Pb_{1-x}Sn_xSe$ (x=0-1.01)	Grown from gas phase	Cd, Zn, Hg, In, Ga, Bi, Sb	-	Donors	-	-	[125]
PbTe	Sintering followed by annealing	-	Bi, Nb, Sb, BrAg	Donors	Acceptor	-	[121]
"		-	Na,Tl + + NaTl	-	Acceptors Acceptor	-	[122] [128]
"	Melt-grown	-	Cr, TiMn	Donors	-	Amphoteric	[123]
"	Melt-grown		Ga, In	Donors		Amphoteric	[124, 131]
"	Melt-grown + annealing in Pb or Te vapors	-	Ag, As	-	-	Amphoteric	[134]
$Pb_{1-x}Ge_xTe$ (x= 0.05-0.3)	Sintering	-	In	Donor	-	-	[122]
$Pb_{1-x}Sn_xTe$ (x=0-0.3)	Grown from gas phase	Cd	-	-	-	Amphoteric	[129]
$Pb_{1-x}Sn_xTe$ (x=0-0.22)	Grown from gas phase	-	In, Ag	Donors	-	-	[130]
$Pb_{1-x}Sn_xTe$ (x=0.2)	Melt-grown	Cd	-	Donor	-	-	[127]
$Pb_{1-x}Sn_xTe$ (x=0.2)	Grown from gas phase (epitaxial layers)	-	Zn	Donor	-	-	[126]
$Pb_{1-x}Sn_xTe$ (x=0.2)	Thermal sputtering of polycrystalline films	-	In	-	-	Amphoteric	[132]
SnTe	Sintering	Cu, Ga, In, Sb, Pb, Bi	-	Donors	-	-	[135]
"	"	In	-	-	Acceptor	-	[133]
"	"		In, Sb, Cu, Bi, Ge	Donor	-	-	[125]

other, possibility no less important aspects of doping will not be considered here. Before discussing our results we will provide a brief literature survey of results from doping of A^4B^6 compounds.

According to the references shown in Table 11 the group I, II, III, IV, and V doping impurities are introduced into A^4B^6 compounds by a variety of techniques during the crystallization process as well as by diffusion or ion implantation into the finished crystals. The table clearly indicates that the influence of the impurities on the conductivity of the material is largely dependent on the method of fabricating the samples, although for the majority of compounds, regardless of the valency of the impurity, the doped material acquires electron conductivity. One additional feature of doping A^4B^6 compounds is that certain impurities behave amphoterically by settling at two different sites of the crystal lattice of the host crystal [120]. For example, impurity Cu and Ni atoms can occupy both the Pb atomic sites and the interstitial sites in the PbS, thereby influencing the physical properties of the PbS in different ways. When the Cu is occupying the Pb atomic sites it functions as an acceptor and in the interstitial sites it functions as a donor; in the same material Ni located at the interstitial sites is a donor and at the Pb atomic sites it is electrically neutral. Obviously the distribution of the impurity atoms among the different crystallographic lattice sites will be determined by the type and concentration of the prevalent intrinsic defects, i.e., the impurities will manifest different properties in *p*- and *n*-undoped samples.

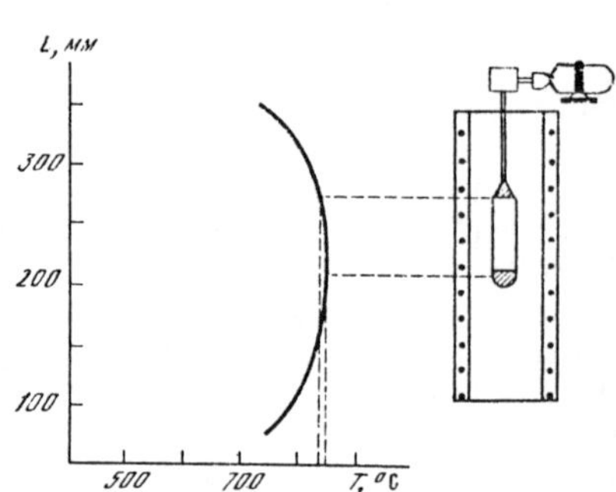

Fig. 33. The working vessel and its position with respect to the temperature profile of the oven for growing doped PbTe monocrystals

On balance we can claim (see Table 8) that doping the lead-tin chalcogenides with foreign impurities causes a significant change in the physical properties of the host materials.

4.2. Experimental Results from Doping Lead-Tin Tellurides

We investigated the behavior of group I (Ag, Au), III (In, Tl), and V (Bi) impurities in $Pb_{1-x}Sn_xTe$ (x = 0-0.24) in our studies [76, 77, 104, 116, 136, 137] in order to control the electrophysical properties of this semiconductor which is a bilateral phase with a composition stability range of 10^{19}-10^{21} dependent on x and the crystallization temperature. In the present study we will briefly describe results from the bismuth- and thallium-doping process and will provide detailed results for indium and silver, since the behavior of these two impurities is quite interesting from both the theoretical and practical viewpoints. We discovered a number of unusual effects for the first time from doping a solid solution of lead and tin tellurides with indium: stabilization of the Fermi level with respect to the doping action of the indium at a record low free electron concentration (10^{14}-10^{15} cm^{-3}) [130]; long-term photoconductivity [176], and a nontrivial change in the low-temperature phase transitions recorded in undoped PbSnTe [72]. In the silver doping of PbTe and $Pb_{0.8}$, $Sn_{0.2}Te$ we detected for the first time an indirect (through the vacancy subsystem) mechanism of doping impurity action on the electrical properties of a narrowband semiconductor [77, 136].

We employed polycrystalline $Pb_{1-x}Sn_xTe$ containing doping additives as the blend for growing doped monocrystals and epitaxial layers of lead-tin telluride solid solutions by the gas-phase technique; this material had a specific Pb and Sn ratio and was first obtained by direct synthesis in vitreous silica evacuated vessels by employing vibrational mixing of the melt to improve synthesis conditions. X-ray diffraction analysis was used to establish that the synthesized material was a single-phase material throughout the volume and of homogeneous composition. The monocrystals were grown in sealed vitreous silica vessels in ovens with a temperature gradient to allow for sublimation transport in the vessel [137].

The position of the vessel in the furnace and the temperature distribution are both shown in Fig. 33. The vessel was transported during the growing process at a rate of 0.15 mm/hr which is near the crystal growth rate so that the temperature gradient between the growth surface of the crystal and the blend remained constant and the growth process itself occurred at an identical temperature.

The derived crystals were 10 mm in diameter, 15 mm in length and consisted of one or several monocrystal blocks with one or a few {100} faces exposed to the growth surface.

Local X-ray spectral analysis on the MAR-2 and "Camebax" was employed to detect the impurity in the monocrystal samples (impurity detection sensitivity was ~0.1-0.5 at.% with a measurement accuracy of 20-30%). Impurity concentrations below 0.1 at.% were determined by quantitative spectral analysis; the sensitivity of this method to the impurities was ~10^{-4} at.% with an accuracy of 10-20% of the measured quantity. The electrical properties of the crystals were investigated by employing the Hall technique at 300 and 77 K.

Doping of PbTe with Bi and Tl. An investigation of the electrical characteristics of lead telluride monocrystals doped during the bismuth growth process revealed (Table 12) that Bi is a donor; a good correlation is observed between the quantity of Bi introduced and the electron concentration in the material up to 2 at.% Bi. In crystals with a higher Bi concentration X-ray phase analysis revealed traces of a second Bi-rich phase. The observed correlation between the solubility limit of Bi (~$2 \cdot 10^{20}$ cm^{-3}), and our width of the composition stability range ($1.4 \cdot 10^{20}$ cm^{-3}) of PbTe and the maximum achievable carrier concentration (n_{300} = $2.1 \cdot 10^{20}$) supports the contention that in the low concentration range (up to 2 at.%) in *p*-type lead telluride Bi occupies the vacancy sites, thereby generating electron conductivity.

Table 12

The properties of Bi- and Tl-doped PbTe monocrystals

Impurity	Impurity concentration in the crystal, at.%	Conductivity type at 77 K	Carrier concentration at 77 K, cm^{-3}	Carrier mobility at 77 K $cm^2/(V \cdot s)$	Lattice parameter, Å
Bi	0	*p*	$5 \cdot 10^{18}$	$1,0 \cdot 10^3$	6.4606
	0.53	*n*	$2.1 \cdot 10^{19}$	$2.8 \cdot 10^3$	6.4608
	0.67	*n*	$6.7 \cdot 10^{19}$	—	6.4600
	0.87	*n*	$7.2 \cdot 10^{19}$	$3.6 \cdot 10^2$	6.4580
	1.8	*n*	$1.6 \cdot 10^{20}$	4	—
	2.6	*n*	$2.1 \cdot 10^{20}$	1,5	6.4580
Tl	$7 \cdot 10^{-2}$	*p*	$1.1 \cdot 10^{19}$	$5.8 \cdot 10^3$	6.4599
	$1.5 \cdot 10^{-1}$	*p*	$2.5 \cdot 10^{19}$	$1.67 \cdot 10^3$	»
	$3 \cdot 10^{-1}$	*p*	$5 \cdot 10^{20}$	4	»
	$7 \cdot 10^{-1}$	*p*	—	—	6.471

Table 12 also indicates that the thallium impurity in PbTe monocrystals behaves as a simple acceptor impurity. The PbTe crystals had stable *p*-type conductivity across the entire range of Tl concentrations from 0.07 to 0.7 at.%.

Bi- and Tl- doped PbTe monocrystals were used [137] as substrates in *n*-PbTe(Bi) - *p*-$Pb_{0.8}Sn_{0.2}Te$ - *p*-PbTe(Tl)-type epitaxial structures.

The microanalyses of the component and impurity distribution in the grown epitaxial structures of the first type revealed that in the successive application of *p*-$Pb_{1-x}Sn_xTe$ and *n*-PbTe(Bi) epitaxial layers to a *p*-PbTe(Tl) substrate the tin diffuses from the $Pb_{1-x}Sn_xTe$ active layer to the adjacent doped *n*-PbTe(Bi) and *p*-PbTe(Tl) layers (Fig. 34). As indicated by Fig. 34 the Tl in the *p*-PbTe substrates reduces Sn diffusion from the *p*-$Pb_{0.8}Sn_{0.2}Te$ active layer to the substrate, thereby eliminating the shift of the radiation wavelength to the shortwave region, which occurs when employing substrates fabricated from an undoped material.

The Sn diffusion causes the structure to become a variable band structure and the sharp jump in the lattice parameters of the $Pb_{0.8}Sn_{0.2}Te$ and PbTe equal to 0.4% vanishes and is now distributed over several tens of microns, which effectively eliminates the formation of mismatch dislocations in the vicinity of the technological layer boundary. This could explain the absence of degradation of test

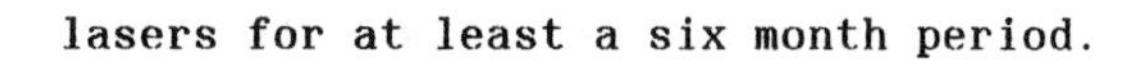
lasers for at least a six month period.

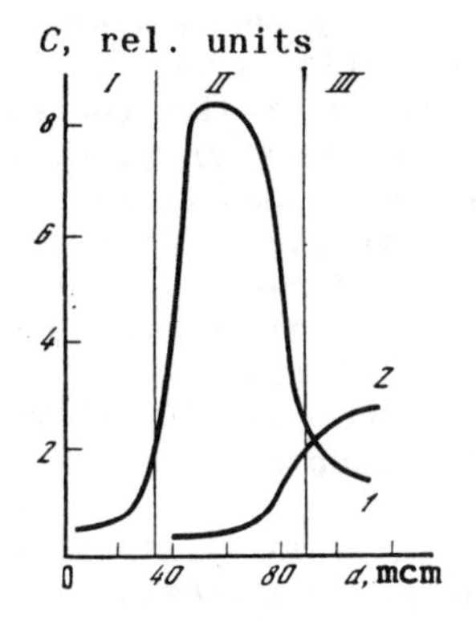

Fig. 34. Tin (1) and bismuth (2) distribution curves for PbTe(Bi (III) - $Pb_{0.8}Sn_{0.2}$ (II)-PbTe(Te) (I)

During oriented overgrowth the bismuth also diffuses from the upper n-layer to the active $Pb_{1-x}Sn_xTe$ layer. However due to the proximity of the ionic radii of lead (1.26 Å) and bismuth (1.20 Å) this has no significant influence on variations in the lattice parameters in the test structures. By measuring the EMF arising in the vicinity of the p- n-junction from the excitation of the nonequilibrium carriers by the electronic probe of the microanalyzer the position of the p- n-junction in the structure was determined and it was established that the p- n-junction does not shift from the active region to the larger broadband layers of the structure (both upper and lower). Overall we can conclude that Tl and Bi impurities produce PbTe material with a degenerate carrier concentration and stable p- and n- (respectively) conductivity and that the charge carrier concentration in these materials follows the concentration of the impurity introduced up to its solubility limit which is within an order of magnitude of the total vacancy concentration in the undoped material.

In-doping. We know that the indium doping of $Pb_{1-x}Sn_xTe$ solid solutions is intended to reduce the free carrier concentration which normally is $N_p \simeq 10^{18}$-10^{19} cm^{-3} (77 K) with a total charged and uncharged intrinsic defect concentration of 10^{20}-10^{21} cm^{-3} (see Section 2.4). We carried out comprehensive investigations of the electrical and crystallochemical characteristics of $Pb_{0.8}Sn_{0.22}Te$ bulk monocrystals in order to determine the features of the doping process of such a material and establish the nature of the interaction of the impurity atoms and the intrinsic defects (vacancies) of the crystal lattice (matrix) [108]. The experimental material in this section will be presented in accordance with study [139].

The test materials were grown from the gas phase in sealed vitreous silica vessels in an 800-830°C temperature range. The samples were monocrystal slabs 15 mm in diameter and 10-20 mm in length. The In concentration in the monocrystals introduced in the form of InTe during their growth process was determined by X-ray spectral analysis on a "Camebax" microanalyzer accurate to 3-5% of the measured quantity. It was established that beginning at 2 at.% the indium is nonuniformly distributed through the slab.

A DRON-2.0 diffractometer employing CuK_α-radiation was used to measure the lattice parameters and to conduct the X-ray phase analysis. Pyrolytic graphite was used as the monochromator. The lattice parameters were calculated to better than ±0.0005 Å.

In order to establish the solubility range of indium in $Pb_{0.78}Sn_{0.22}Te$ the study carried out a precision X-ray phase analysis of these solid solutions with an impurity concentration from ~2 to 4 at.%. No second phase was detected in samples with an indium concentration $N_{In} \leqslant 2$ at.% including pure indium precipitates or precipitates of its compounds InTe and In_2Te_3 whose appearance was considered to be very likely according to the phase diagram of the In-Te system [139]. New diffraction lines were observed in the X-ray diffraction patterns of the $Pb_{0.78}Sn_{0.22}Te$ samples with $N_{In} \approx 4$ at.%; according to [140] these were identified as the (220), (311), and (331) X-ray interference beams of the In_2Te_3 crystal structure; the quantity of the second phase of In_2Te_3 did not exceed 0.01% in the samples.

It was therefore established that the solubility limit of indium in $Pb_{0.78}Sn_{0.22}Te$ for our selected doping technique and growing temperature of 800°C is near 3-4 at.% In and exceeding this concentration results in the formation of a second phase in the form of In_2Te_3 [138].

However according to the literature [141, 142] the solubility limit of indium in the p-$Pb_{1-x}Sn_xTe$ crystal lattice is largely dependent on the growing conditions of the materials and the method of introducing the indium doping impurity into the materials. Therefore doping in liquid-phase epitaxy conditions at lower growth temperatures ($T_{Kp} \approx$ 500-600°C) reduces the solubility limit of indium to 1 at.% [41]. When introducing In by diffusion to p-$Pb_{1-x}Sn_xTe$ when the impurity will diffuse in the finished crystal along the preformed matrix, the solubility limit of In does not exceed 0.5 at.% [142].

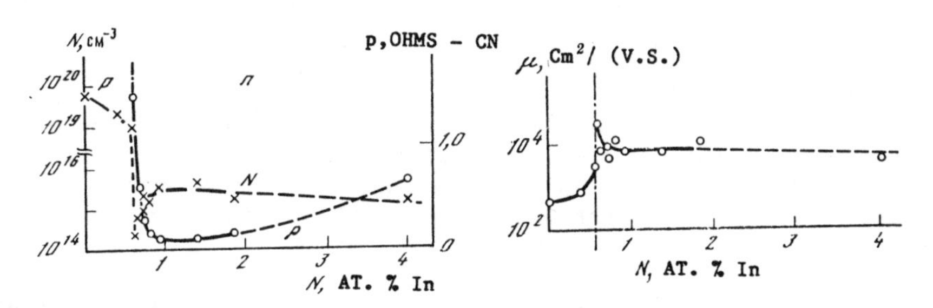

Fig. 35. The free carrier concentration and the resistivity of indium-doped $Pb_{0.78}Sn_{0.22}Te$ solid solution plotted as a function of indium concentration at 77 K

Fig. 36. The carrier mobility of the indium-doped $Pb_{0.78}Sn_{0.22}Te$ solid solution plotted as a function of indium concentration at 77 K

These data accounting for the T-X phase microdiagrams of $Pb_{1-x}Sn_xTe$ (see Fig. 9) reveal that the solubility limit of In in this material increases with growth of deviation of the crystal composition from stoichiometry and, consequently, is determined by the "volume" of the vacancy subsystem in the crystal matrix.

The resistivity, Hall coefficient, and crystal lattice parameter were measured for the $Pb_{0.78}Sn_{0.22}Te$ monocrystals as a function of In concentration in the samples.

The undoped monocrystals always had *p*-type conductivity with an average carrier concentration of $6 \cdot 10^{19}$ cm^{-3} at 77 K. As we see from Fig. 35 introducing the indium doping impurity from 0 to 0.6 at.% causes a drop in the hole concentration from $6 \cdot 10^{19}$ to $1 \cdot 10^{19}$ cm^{-3}. At 0.6 at.% In a conductivity inversion from *p* to *n* occurs in the monocrystals and the electron concentration reaches $2.2 \cdot 10^{14}$ cm^{-2} and the resistivity then rises from $3 \cdot 10^{-4}$ to 1.5 ohms·cm. In the indium concentration range of 0.6-0.8 at.% the electron concentration grows to $3 \cdot 10^{15}$ cm^{-2} and the resistivity drops to 0.1 ohm·cm. A further growth in the In concentration to 2 at.% has virtually no influence on the electron concentration nor the resistivity of the material.

The calculated Hall electron mobility $\mu = R\sigma$ (Fig. 36) grows from $5 \cdot 10^2$ to $3 \cdot 10^3$ $cm^2/(v \cdot s)$ with growth of the indium concentration from 0 to 0.6 at.%. After the conductivity inversion and an indium concentration of approximately 0.6 at.% in the monocrystals the Hall electron mobility is $3.2 \cdot 10^4$ and then diminishes to $1 \cdot 10^4$ $cm^2/(v \cdot s)$ with an increase in the In concentration to 0.8 at.%.

A further growth in the concentration of the doping impurity will not cause a significant change in the Hall electron mobility.

These electrical measurements therefore reveal that two critical indium concentration points exist within the In solubility range (~3.5 at.%) in $Pb_{0.78}Sn_{0.22}Te$ monocrystals: $N_{In} = 0.6$ at.% which is where the conductivity inversion of $Pb_{0.78}Sn_{0.22}Te$ from *p* to *n* occurs and $N_{In} \approx 0.8$ at.% where the free electron concentration becomes stable at $\sim 10^{15}$ cm^{-3}.

The limiting electron concentration at 10^{18} cm^{-3} (300 K) which cannot be increased by increasing the concentration of the doping indium impurity (up to 6 at.%) was first discovered in study [143] by doping PbTe. In attempting to explain this phenomenon V.I. Kaydanov proposed a quasi-local indium impurity level model that stabilizes the energy position of the Fermi level and makes it insensitive to the introduction of additional donor or acceptor impurities when the concentration of these impurities does not exceed the concentration of the added indium. However the hypothesis of a quasi-local indium level (or impurity zone) could not explain, for example, that the occupation and emptying of this zone is not manifest in its position [143] although the repulsion between electrons would cause an energy shift in the case of occupation.

In order to explain the Fermi level stabilization effect in PbTe the authors of studies [144-146] proposed a different model based on the dissociation of a neutral (with respect to the metallic sublattice of the matrix) impurity in the In^{2+} state to the more advantageous charged In^{2+} and In^{3+} states. The primary cause of instability of the In^{2+} ion in the authors' opinion is that the In^{1+} and In^{3+} ions are charged with respect to the Pb^{2+} sublattice and their energy drops due to polarization of the surrounding area having a high dielectric constant ($\varepsilon_0 \approx 400$ at 300 K [147, 148]). As proof of the absence of In^{2+} states in PbTe the authors of studies [144-146] provided measurement results of the magnetic susceptibility performed on PbTe samples with an indium concentration of 10^{20} cm^{-3} [144] that revealed that the number of In^{2+} states with unpaired spins does not exceed $5 \cdot 10^{17}$ cm^{-3} and the sample remains diamagnetic through helium temperatures. However the model of the coexistence of the In^{1+} and In^{3+} charge states cannot, in our view, be considered valid since it is based on an unproved assumption that the indium is an impurity with a negative Hubbard energy [144] (in this case the ionization potential of the impurity atom does not grow but rather diminishes with the consistent departure of electrons).

The failure of efforts undertaken in studies [143-146] to explain the Fermi level stabilization effect in PbTe(In) is the result, as will become clear below, of their ignoring the interaction of In atoms with the intrinsic defects in the matrix lattice from their consideration of the crystallochemical mechanisms.

In addition to these electrical measurements we also investigated the dependence of the cubic crystal lattice (NaCl-type) parameter a on the In concentration on the same $Pb_{0.78}Sn_{0.22}Te$ samples at 300 K. The measurements were carried out on {100} cleavage faces for the monocrystals since this was the only case that made it possible to avoid the influence of surface treatments on the experimental results. The derived dependencies (Fig. 37) revealed that with growth of N_{In} from 0 to 0.7 at.% a weak growth of the lattice parameter a is observed (concentration region I), after which the lattice parameter in the 0.7-0.8 at.% (transition region) drops by ~0.004 Å. With a further increase in the indium concentration to 2 at.% an insignificant reduction in the value of a occurs (region II).

A comparative analysis of the experimental dependencies of the electrical characteristics and crystal lattice parameter (see Figs. 35-37) reveals that, first, the entire concentration range of indium solubility in $Pb_{0.78}Sn_{0.22}Te$ can be divided into two ranges and the impurity will have a different influence on the physical properties of the solid solution in each of these ranges: from 0 to 0.7 at.% (concentration range I) and from 0.8 to ~3.5 at.% (concentration range II) and, second, in the transition range of 0.7-0.8 at.% In there will evidently be a qualitative change in the crystallochemical dissolution mechanism of the indium atoms in the crystal lattice of the matrix: the solvent.

It is easy to understand that in order to provide donor properties in the first concentration region (see Fig. 35) the indium must be introduced into the metallic sublattice of the material in the In^{3+} charged state. In this case its two valence electrons will participate in the formation of the chemical bonds in the $Pb_{0.78}Sn_{0.22}Te(In)$ crystal lattice, while the third electron which is not involved in the bond with the tellurium can be used to compensate the V^{2+} charged vacancies of the cation sublattice, each of which, according to studies [149, 150], delivers two holes. These two "free" electrons of the two indium atoms can compensate a single doubly-charged metal vacancy.

According to the experimental data (see Fig. 35) introducing $6 \cdot 10^{19}$ indium atoms into 1 cm^3 (the *p* to *n* conductivity transition point) neutralizes $6 \cdot 10^{19}$ cm^{-3} holes (the initial concentration of free holes in the undoped sample) and the free carrier concentration approaches the intrinsic level ($\sim 10^{14}$ cm^{-3}). This makes it possible to conclude that in the first concentration range the prevalent method of introducing indium to the metallic sublattice of the test material is the occupation of the sublattice sites by indium ions on a level equal with the host elements of the matrix: lead and tin while maintaining a constant metallic vacancy concentration by employing the crystallization temperature and the composition x. Indeed, if the indium atoms were to occupy the vacancy sites as we proposed previously [138] every two indium atoms could neutralize three metallic vacancies which is not in agreement with experiment [138]. It is easy to understand that introducing the impurity atoms into the lattice by the lead and tin mechanism is a direct consequence of the introduction of indium as InTe during the material crystallization process.

Our method of metallic vacancy charge neutralization allows the formation of neutral complexes localized in the lattice, each of which contains a single metallic vacancy and two indium atoms located at neighboring sites in the same sublattice. It is important to know that the model of an indium ion linked to the vacancy which then functions as a "noncentral" ion [151, 152] allows a rather simple explanation of the appearance of spontaneous polarization in $Pb_{0.78}Sn_{0.22}$ Te(In) at an indium concentration of ~0.5 at.% [153].

Within the framework of this mechanism of In interaction with the vacancy subsystem of the material, the weak growth of the crystal lattice parameter a discovered in the first concentration region (N_{In} < 0.7 at.%) with growth of N_{In} (see Fig. 37) can be attributed to an increase in the effective vacancy radius as a result of electron trapping. The observed change in Hall mobility in the same concentration range of N_{In} (see Fig. 35b) is rather easily explained as a change in the concentration of charged scattering centers. Above we pointed out that the collection of experimental data has made it possible for us to propose a change in the crystallochemical mechanism of indium impurity dissolution in the crystal lattice of the $Pb_{0.78}Sn_{0.22}Te$ solid solution in the 0.7-0.8 at.% concentration range of this impurity. We remember that it is precisely in this concentration range of the impurity at 300 K that a sharp jump in the crystal lattice parameter a was observed (see Fig. 37) and in the general case this can be taken as evidence of an indium-concentrated phase transition in the $Pb_{0.78}Sn_{0.22}Te$-In system. Further X-ray diffraction analyses of this system of solid solutions have revealed that the jump in the lattice parameter in the $N_{In} \approx 0.7$-0.8 at.% range is accompanied by a narrowing of the diffraction curves ($\Delta\theta/2\theta$) and the oscillation curves ($\Delta\omega$) by a factor of 1.5 and 3, respectively (Fig. 38). This revealed a reconfiguration of the entire crystal lattice occurring near the critical indium concentration $N_{cr} \approx 0.7$-0.8 at.%. The absence (within experimental error) of indications of a change in crystal symmetry across the entire solubility range of indium (up to ~3 at.%) at 300 K (diffraction line splitting or asymmetry; redistribution of the diffraction line intensities or the appearance of new X-ray interference) reveals that such reconfiguration must be interpreted as an isostructural first order phase transition of the type NaCl → NaCl.

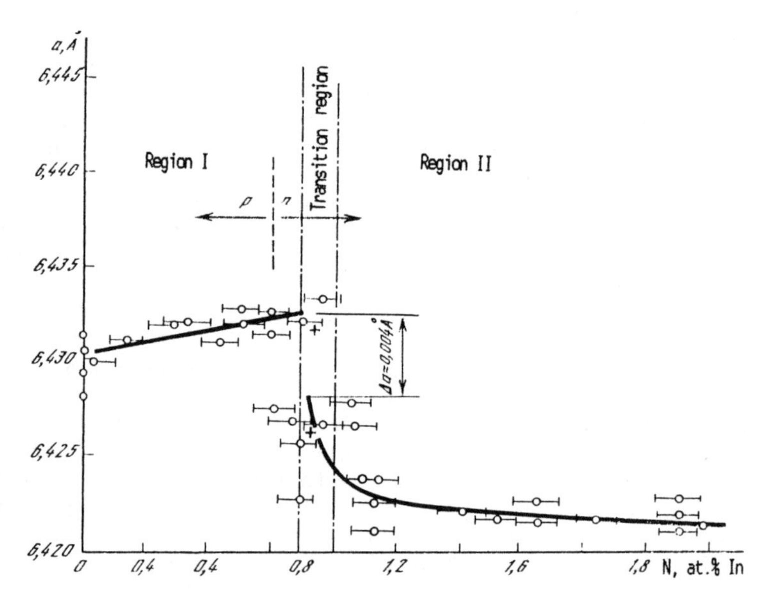

Fig. 37. The crystal lattice parameter of $Pb_{0.78}Sn_{0.22}Te(In)$ solid solutions plotted as a function of indium concentration
The arrows indicate results from an analysis of a two-phase sample

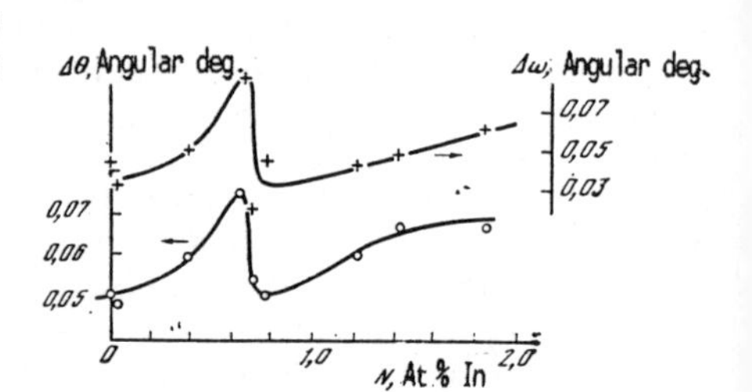

Fig. 38. The angular half-width $\Delta\theta$ of the diffraction curve ($\theta/2\theta$) and the rocking curve $\Delta\omega$ of the (600) reflection from $Pb_{0.78}Sn_{0.22}$-Te(In) plotted as a function of indium concentration

Fig. 39. Transformation of the doublet diffraction line (600) of $Pb_{0.78}Sn_{0.22}Te(In)$ monocrystals plotted as a function of indium concentration (Cu K_{α}-radiation)

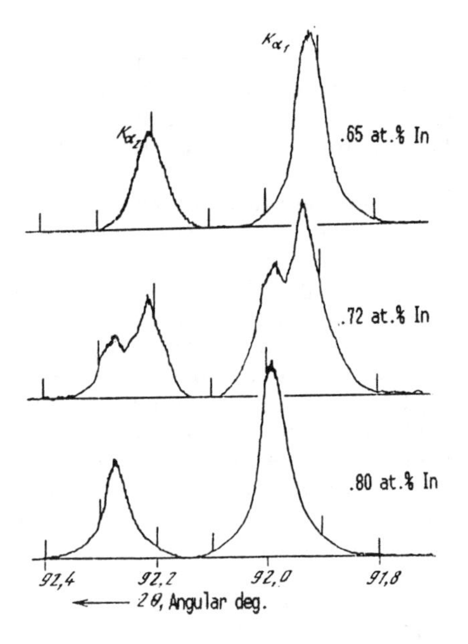

The first order transition unambiguously follows from diffractometry measurements of monocrystals with an indium concentration of 0.72 at.% (Fig. 39) which indicate that in this sample there are coexisting regions with crystal lattice parameter values that differ by an order of magnitude corresponding to the jump in a (N_{In}) at the isostructural phase transition point (see Fig. 37).

The literature contains experiments that examine phase transitions registered by jumps in volume, electrical conductivity, etc. without changes in the crystal lattice symmetry (see, for example, [154-156] and the references cited therein). Such transitions, called isostructural transitions, are electron transitions although they are accompanied by significant structure effects. For example, study [156] has established that an irregular drop in the crystal lattice parameter (by 3.3%) without alteration of symmetry (NaCl → NaCl) occurs in a system of SmS-GdS solid solutions near a 15-16 mol.% concentration of GdS; the valence state of the Gd ion does not change in this case. According to the authors of study [156] this first order transition reflects a change in the charge state of the samarium ion from Sm^{2+} to Sm^{3+} with a smaller radius and this change is analogous to transformations observed in many rare earth element chalcogenides [154].

The following argument supports an indium-concentration phase transition in $Pb_{0.78}Sn_{0.22}Te(In)$ at 300 K in the $0.7 < N_{In} < 0.8$ at.% range. If we assume that the jump in the lattice parameter Δa reflects only its isotropic spontaneous deformation, it is possible, using available pressure dependencies of the lattice parameter $a(P)$ [157] and of the bandgap $E_g(P)$ [158] for $Pb_{1-x}Sn_xTe$ to estimate the value of ΔE_g that will accompany this jump. Clearly identifying the spontaneous composition process of the lattice with its relaxation to a uniform pressure contains a certain idealization of the crystallophysical processes, although this does make it possible to correctly estimate the sign and the order of magnitude of ΔE_g. Our calculation has revealed that the experimental value of $\Delta a = -0.004$ Å will correspond to a narrowing of the bandgap by ~10 meV. According to the literature data the bandgap of $Pb_{0.78}Sn_{0.22}Te(In)$ increases by 10 meV in crystals with $N_{In} \approx 2.7$ at.% [159]. Since the literature does not contain results from a comprehensive measurement of the bandgap of

$Pb_{0.78}Sn_{0.22}Te(In)$ solid solutions over a broad range of In concentrations it is possible to unambiguously determine the order of the transitions based on an experimentally-detected increase in E_g. However our X-ray structural analysis data indicate that the isostructural transition is a first order transition.

An increase in the indium concentration in the second concentration range (N_{In} > 0.8 at.%) causes a smooth reduction in the lattice parameter of the solid solutions with simultaneous growth of $\Delta\theta/2\theta$ and $\Delta\omega$ (see Figs. 37 and 38) and, as revealed by electrical measurements, has a weak influence on the free charge carrier concentration and the carrier mobility (see Figs. 35, 36). This indicates a qualitative change in the nature of indium atom interaction with the atoms in the host crystal lattice compared to the interaction in the first concentration range which, in principle, can be detected by electron spectroscopy techniques. We know that such methods can be used to obtain such information on the nature of the chemical bond of certain atoms due to the dependence of the energies of the internal electron levels on its chemical configuration [160].

A number of studies have demonstrated experimentally that chemical shifts in the lines of such elements as Mg, Al, Cu, and In can be more successfully recorded by the Auger spectra rather than the photoionization spectra. For this reason we carried out an investigation of the change in the nature of the chemical bond of indium in $Pb_{0.78}Sn_{0.22}Te(In)$ solid solutions as a function of its concentration by means of the electron Auger spectroscopy technique.

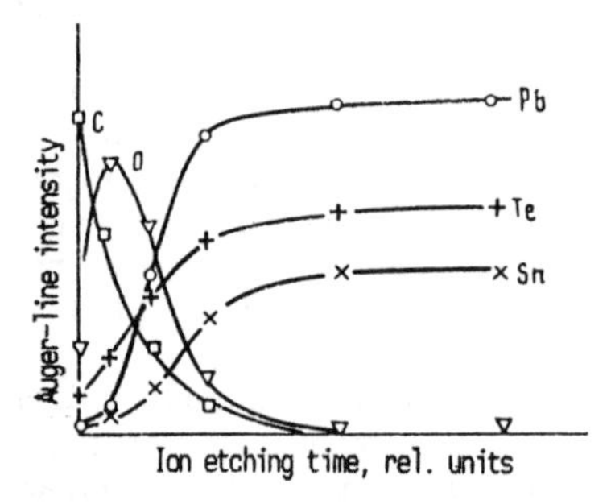

Fig. 40. Depthwise element distribution in $Pb_{0.78}Sn_{0.22}Te(In)$ monocrystals

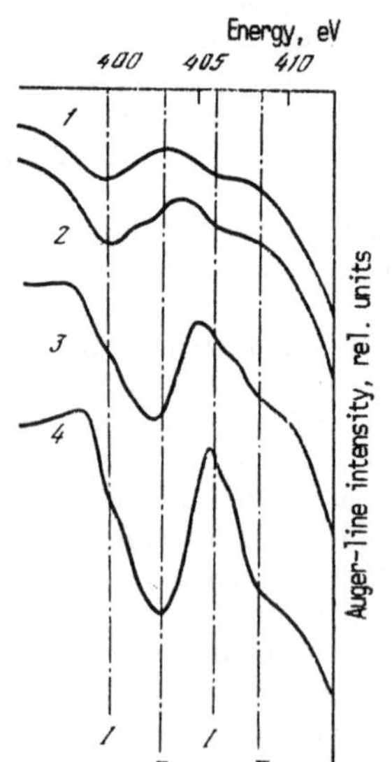

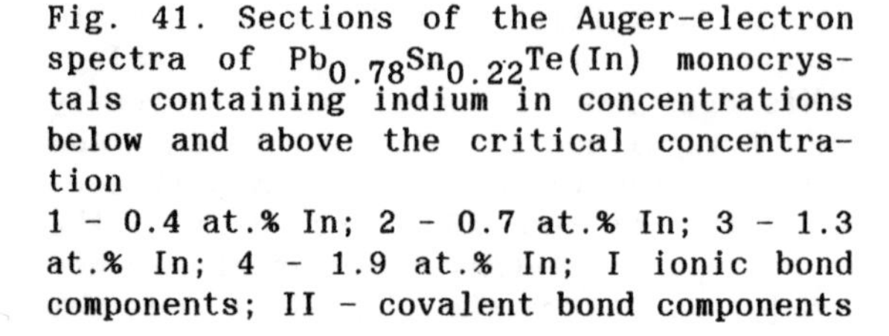
Fig. 41. Sections of the Auger-electron spectra of $Pb_{0.78}Sn_{0.22}Te(In)$ monocrystals containing indium in concentrations below and above the critical concentration
1 - 0.4 at.% In; 2 - 0.7 at.% In; 3 - 1.3 at.% In; 4 - 1.9 at.% In; I ionic bond components; II - covalent bond components

The measurements were carried out on the {100} cleavage faces of $Pb_{0.78}Sn_{0.22}Te(In)$ monocrystals containing In in concentrations of 0.4, 0.7, 1.3, and 1.9 at.% corresponding to the I concentration range, the isostructural phase transition region (the transition range) and the II concentration range. Since the depth of the Auger-electrons does not exceed 10-20 Å [160] the test surface was first purified by a 1 keV argon ion beam at a beam current of 10^{-6} Å in order to eliminate an absorption layer, a possible oxide film and probable loss of elements from the sample bulk.

A recording of the change in the Auger-peak amplitude of certain elements during ion etching revealed the existence of excess quantities of O and C atoms and a shortage of the host elements: Pb, Sn, and Te on the cleavage face surfaces of $Pb_{0.78}Sn_{0.22}Te(In)$ (Fig. 40). For this reason a precision measurement of the energy positions of the indium lines is attempted only after achieving stable intensities of the Auger peaks of Pb, Sn, and Te (approximate estimates yield a depth of approximately 100 Å).

We know that the Auger-line of indium is formed from the $M_4N_{4.5}N_{4.5}$ and $M_5N_{4.5}N_{4.5}$ transitions and therefore in the experimental spectra this element is identified by the doublet line whose components are separated by an energy gap of 6-7 eV and located in the 397-411 eV energy range depending on the nature of the chemical bond [163, 164].

An analysis of the spectra derived in the present study (Fig. 41) reveals that in a monocrystal with an indium concentration of 0.4 at.% the positions of the Auger peaks of this element correspond to energies of $E_1 \approx 400$ and $E_2 \approx 406$ eV (here and henceforth the accuracy of the experimental values given in the text for the Auger-peaks is ±1 eV). In addition to the energies already noted Auger line components of indium at higher energies $E_1' \approx 403$ and $E_2' \approx 408.5$ eV also appear in the spectra of samples with $N_{In} = 0.7$ at.%. These lines become the fundamental lines in the spectra of $Pb_{0.78}Sn_{0.22}Te(In)$ with $N_{In} \geqslant 1.3$ at.%.

The detected shift in the energy positions of the Auger-peaks of indium are interpreted based on results from a comparison of experimental spectra to known tabular data [165]. Table 13 gives the $M_4N_{4.5}N_{4.5}$ transition energies for indium in compounds with different types of chemical bonds. A joint analysis of experimental and tabular data reveals that the Auger-electron lines at $E_1 = 400$ eV and $E_2 = 406$ eV correspond to an indium state in compounds with primarily ionic bonds (In_2O_3, In_2O), while when $E_2' = 403$ eV and $E_2' = 408.5$ eV are standard values for indium compounds with covalent bonds (InTe, In_2Se_3, In_2Te_3).

According to the derived results in the $Pb_{0.78}Sn_{0.22}Te(In)$ solid solution monocrystals with $N_{In} < 0.7$ at.% indium forms bonds of primarily an ionic nature with its neighboring components in the lattice at the same time that in samples with $N_{In} \geqslant 1.3$ at.% a covalent bond type is characteristic of the majority of indium atoms.

Table 13

Energy positions of indium lines in the X-ray photoelectron and Auger-electron spectra of its compounds with group VI elements

Compound	Charge state[1]	Line energies of In, eV		Ref
		X-ray photo-electron spectra	Auger spectra $M_4N_{4.5}N_{4.5}$	
In_2O	+1	444.1	407.0	W 3[3]
In_2O	+1	444.0	—	146
In_2O_3	+3	444.7	406.9	LAK[3]
In_2O_3	+3	444.8	—	146
InS	+2	444.3	408.5	W1[3]
In_2S_3	+3	444.5	407.5	W 3[3]
In_2Se	+1	444.9	—	146
InSe	+2	444.8	408.2	W1[3]
In_2Se_3	+3	444.6	408.5	W1[3]
InTe	+2	444.1	409.4	W1[3]
In_2Te_3	+3	444.8	409.1	W1[3]
PbTe(In)	+3	—	397[2]; 404[2]	164
	+3	—	400[2]; 407[2]	
	+3	444.4	—	146
	+1	443.7	—	
$Pb_{0,78}Sn_{0,22}Te(In)$	+3	—	400[2]; 406[2]	29
	+3	—	403[2]; 408.5[2]	

[1] The charge state represents the number of valence electrons involved in the formation of the chemical bonds (regardless of their nature).
[2] The line energies of In corresponding to the $M_5N_{4.5}N_{4.5}$ Auger transition.
[3] This abbreviated designation of the references is borrowed from study [165].

In accordance with our results [166] the indium impurity is introduced into the $Pb_{0.78}Sn_{0.22}Te$ metallic sublattice in the first concentration range. In this case the indium atoms are surrounded by tellurium atoms in an octahedral configuration while their In-Te bonds are primarily ionic in nature according to the Auger-electron spectroscopy data. Therefore we may expect that in concentrations below 0.7 at.% the indium is located at the metallic sublattice sites of the $Pb_{0.78}Sn_{0.22}Te(In)$ solid solution in the In^{3+} charge state.

At an indium concentration of 0.7-0.8 at.% the interaction between impurity centers reaches a critical value and the nature of the chemical bonds between the indium and the surrounding tellurium sublattice changes from predominantly ionic to predominantly covalent. The change in the nature of the chemical bonding of indium in the matrix lattice is responsible for the isostructural phase transition in the 0.7-0.8 at.% range of In. It is important to again emphasize that the indium was introduced into the samples investigated in the present study in the form of InTe during the crystallization of the $Pb_{1-x}Sn_xTe$ solid solution, which is fundamentally different from the thermodiffusion method of introducing indium [142] where it is redistributed along the existing matrix lattice; in this case our isostructural transition will not occur.

We will now consider indium covalent bond formation in the second concentration region. We know that covalent bonds are formed in InTe and In_2Te_3 compounds in the In-Te system [139]. The primary structural element in these compounds is the tellurium tetrahedron whose center contains an indium atom bound by sp^3-hybrid bonds to its nearest neighbor [139]; these are characteristic of covalent crystals. Therefore in the isostructural phase transition region the first coordination shell of indium is modified and the coordination number drops from 6 to 4. As a result the majority of the indium is located in tetrahedral complexes with covalent bonds.

As demonstrated in study [164] the atomization energy (per single atom) of InTe is greater than the corresponding value for In_2Te_3 which indicates a higher probability of formation of crystal structural elements of the latter compound. The preference given to the formation of In_2Te_3 structural fragments in the $Pb_{0.78}Sn_{0.22}Te(In)$ lattice at an indium concentration greater than 0.8 at.% also derives from the fact that In_2Te_3 has a sphalerite structure with a lattice parameter a = 6.15 Å [140] close to the lattice parameter of $Pb_{0.78}Sn_{0.22}Te(In)$ (unlike InTe whose crystal lattice belongs to the tetragonal crystal symmetry system and has parameters a = 8.44 Å and c = 7.20 Å [139]). The appearance of a second phase in the form of In_2Te_3 discovered in the $PB_{0.78}Sn_{0.22}Te(In)$ samples with N_{In} = 4 at.% is genetically determined by the formation of its structural fragments at lower indium concentrations. An increase in the number of these fragments with an increase in the concentration of indium introduced to the compound is accompanied by an increase in the elastic stresses of the $Pb_{0.78}Sn_{0.22}Te(In)$ crystal lattice as revealed by broadening of the X-ray interference beams (see Fig. 38). The gradual accumulation of In_2Te_3 fragments results in the precipitation of this component in the form of a second phase [138].

In our opinion the In_2Te_3 structural fragments in the $Pb_{0.78}Sn_{0.22}Te(In)$ crystal lattice take the form of Guinier-Preston zones [167, 168] that are coherently bound to the solvent lattice. It is this coherence of the bond between the zone and matrix lattices that is responsible for the components of the Auger-peaks of the indium with an energy corresponding to an ionic bond appearing in the Auger spectra of the $Pb_{0.78}Sn_{0.22}Te(In)$ monocrystals with $N_{In} \geq 1.38$ at.% (see Fig. 41). Clearly such bonds are preserved in the indium atoms located at the zone boundaries at the same time that the indium within these zones forms mostly covalent bonds with its nearest neighbor.

The formation of complexes with crystallochemical parameters similar to those of the solvent lattice that are stable over a broad temperature range is characteristic of heavily-doped Ge(Sb) and Si(As), where in addition to the substitution positions the impurity atoms also appear as tetrahedral Sb_4 and As_4 complexes while Ga_2Se_3 structural fragments clearly appear in GaAs(Se) [169, 170].

Our crystallochemical model of the behavior of indium in a $Pb_{0.78}Sn_{0.22}Te(In)$ crystal lattice [166] can explain stabilization of the Fermi level at an indium concentration $N_{In} > N_{cr} \approx 0.7–0.8$ at.% [171]. Indeed, the formation of In_2Te_3 structural fragments in the $Pb_{0.78}Sn_{0.22}Te(In)$ crystal lattice at $N_{In} > N_{Kp}$ (N_{cr} = 0.7=0.8 at.%) is accompanied by the transition of the indium from an electrically-active state at the metallic sublattice sites where there are only two electrons for the bond with tellurium, to an electrically-passive state where all three outer In eelcns participate in the chemical bonds to Te and therefore make no contribution to the conductivity. As indicated by experiment (see Fig. 35) the transition of the indium to an electrically passive state is not accompanied by the recovery of the acceptor properties of the thermodynamically equilibrium vacancies in the $Pb_{0.78}Sn_{0.22}Te$ metallic sublattice. This is due to the fact, as will become clear below, that the metallic sublattice vacancies in this concentration region become electrically neutral. It is clear that the near-intrinsic free electron concentration (10^{14}–10^{15} cm^{-3}) manifest in this range will be determined by the donor properties of the uncompensated tellurium vacancies.

In order to explain the Fermi level stabilization mechanism by introducing additional (in addition to the indium whose concentration exceeds N_{cr}) electrically active impurities we will consider certain structural and electrical properties of In_2Te_3. The most important feature of this compound is that the stoichiometric vacancies in its metallic sublattice are necessary structural elements whose presence is dictated by valance relations and hence they make no contribution to the electrical conduction of the crystal [172]. According to Ormont [173] the crystallochemical formula of In_2Te_3 can be written as $In_{0.667}[]_{0.333}Te$ ([] designates a stoichiometric vacancy in the indium sublattice). In the structure of this compound one-third of the sites ($\sim 5.5 \cdot 10^{21}$ cm^{-3} [172]) in the metallic sublattice are vacant. According to studies [139, 172] the impurity atoms and the superstoichiometric indium atoms occupy the stoichiometric vacancies in the In_2Te_3 crystals where these atoms do not chemically interact with the surrounding crystal lattice and do not exhibit their doping action. Such a mechanism explains the "undopability" of In_2Te_3 when introducing up to 10 at.% of various impurities (for example, Zn, Sn, and In); here the free electron concentration is maintained at 10^{10} cm^{-3} in the crystal [172].

In our opinion the stabilization of the Fermi level in $Pb_{0.78}Sn_{0.22}Te(In)$ when $N_{In} > N_{cr}$ from the introduction of additional impurities [174] can be attributed to the fact that these impurity atoms are distributed among the thermodynamically equilibrium vacan-

atoms are distributed among the thermodynamically equilibrium vacancies of the In_2Te_3 structural complexes without making any contribution to the total electrical conduction. We can easily see that since there is a single vacancy in the formula of this compound for every two indium atoms, the free carrier concentration in $Pb_{0.78}Sn_{0.22}Te(In)$ will remain stable even after introducing other electrically-active impurities in a quantity equal to half that of the dissolved indium.

It is important to note that the Fermi level stabilization mechanism in $Pb_{0.78}Sn_{0.22}Te(In)$ discussed in the present study based on the formation of In_2Te_3 covalent compound fragments in the matrix lattice [171] is fundamentally different from the mechanism of this phenomenon in $Pb_{1-x}Sn_xTe$ proposed by the authors of studies [144, 146]. The latter model is based on an assumption of the simultaneous existence of In ions in two charged states In^{3+} and In^{1+} resulting from the dissociation of the In^{2+} state that is neutral with respect to the metallic sublattice Me^{2+}. The authors of study [146] take as experimental confirmation of this model the existence of two energy peaks at 444.4 eV and 443.7 eV corresponding to the existence of near-equal In^{3+} and In^{1+} concentrations (see Table 13) in the X-ray photoelectron spectra of a PbTe sample containing 1 at.% In.

An analysis of existing literature data on the energy position of indium lines in X-ray photoelectron and Auger spectra of a number of isoelectronic indium compounds with group VI elements (see Table 13) has revealed that our Auger spectrometry results (see Fig. 41) cannot be explained within the scope of the indium ion multiply-charged state mechanism (In^{3+} and In^{1+}). Indeed, as indicated by Table 13, with the auger experiment accuracy achieved in our study (±1 eV) we should have recorded a deformation (broadening) of the In lines in this case of several tenths of an electron volt rather than a shift of 2.5-3 eV to higher energies.

In conclusion we should note that the results from our study [166] are in complete qualitative agreement with the Auger spectrometry data for PbTe containing 1 at.% In [164]. This allows us to conclude that the Fermi level stabilization mechanism proposed in study [171] is the same mechanism for PbTe(In) and $Pb_{0.78}Sn_{0.22}Te(In)$.

As discussed at the beginning of this section long-term photoconductivity was discovered [175] in indium-doped $Pb_{0.78}Sn_{0.22}Te$ solid solutions having a low, on the order of the intrinsic level (10^{14}-10^{15} cm^{-1}, free electron concentration at $T < 20$ K; this photoconductivity decays exponentially [176] or hyperbolically [177, 178] with time. The authors of study [179] have suggested that the negatively charged chalcogen-tellurium vacancies functioning as Yahn-Teller trap centers of the free electrons are responsible for this effect.

In analyzing these studies we can conclude that the vacancy (chalcogen) model of the Yahn-Teller center in $Pb_{0.78}Sn_{0.22}Te(In)$ crystals proposed in study [179] does not contradict our crystallochemical mechanism of In dissolution in $Pb_{1-x}Sn_xTe$ which holds that a low (on the order of the intrinsic level) free electron concentration (the second concentration region for indium) will be achieved only from uncompensated negatively charged tellurium vacancies since neither the In impurity nor the predominant (in concentration) defects: metallic sublattice vacancies which are related in the In_2Te_3 complexes will have any influence on the electrical properties of the material.

It is important to note that the long-term photoconductivity in study [176] was also observed in a $Pb_{0.78}Sn_{0.22}Te$ sample that was not doped by the indium impurity and where the low free electron concentration was achieved by long-term annealing of the p-$Pb_{1-x}Sn_xTe$ sample in lead vapor. Charge neutralization of the metallic vacancies occurred in this case and the electrical conduction was caused by the uncompensated chalcogen vacancies. We can therefore take it as proven that the Yahn-Teller free electron trap centers are realized due to the negatively charged uncompensated chalcogen vacancies (Te).

The research on long-term photoconductivity in $Pb_{0.78}Sn_{0.22}Te$ carried out in study [180] demonstrating a logarithmic relaxation of photoconductivity at 4.2 K and the appearance of rare negative photoconductivity under illumination above 20-25 K led the authors to conclude that there is significant spreading of the energy of the Yahn-Teller free electron trap center. They related this to the spatial inhomogeneity and nonstoichiometric composition of the A^4B^6 crystals which includes the $Pb_{0.78}Sn_{0.22}Te$ solid solution.

In our view the primary reason for the energy spreading of the Yahn-Teller levels need not be discussed if we remember that the long-term photoconductivity is a physical property inherent in the material itself and is not a result of its low quality. The influence on the spreading of the levels of these centers from the deviation from stoichiometry of the $Pb_{0.78}Sn_{0.22}Te$ solid solution which is, as repeatedly emphasized in this study, a fundamental property of materials of variable composition, can be understood by relating this energy spread to the statistical spread of the dimensions of the complexes of thermodynamically equilibrium tellurium vacancies $(V_{Te})_n$. Indeed we can easily see that the energy position of a charged tellurium vacancy will be highly dependent on the size of the vacancy complex $(V_{Te})_n$.

Ag and Au doping. It seemed to us that doping PbTe and $Pb_{1-x}Sn_xTe$ with group 1 impurities whose valencies are significantly different from those of host components, would have a significant influence on the physical properties of these compounds. Determining the role of the intrinsic defects in the dissolution of these impurities was particularly important. Ag and Au were used as the group I impurities. Study [181] was the first to investigate Au-doped PbTe, while the behavior of Ag was investigated previously in PbTe in studies [121, 134, 182, 183]. Sintered polycrystal PbTe samples containing $0.4 \cdot 10^{19}$ to $2.8 \cdot 10^{19}$ cm^{-3} of silver in the blend were used as specimens in study [121]. The remaining studies employed Bridgemanmelt-grown PbTe monocrystals with a silver blend concentration of 10^{19}–10^{20} cm^{-3}. It is important to note that in these studies, except study [182] the Ag concentration in the test samples was not determined; the quantity of doping impurity in these samples was assumed to be equal to its concentration in the initial blend. According to the cited research the Ag impurity in PbTe is either an acceptor impurity [121, 180, 187] or behaves amphoterically depending on the type of intrinsic defects in the material: donor properties are manifest in the lead-saturated *n*-PbTe and acceptor properties are manifest in tellurium-saturated *p*-PbTe [134]. All studies concluded that the number of electrically active carriers in the silver-doped PbTe coincide in practice with the number of Ag atoms introduced.

Table 14

Properties of PbTe(Ag) monocrystals

Ag Concentration, at.%		Type of conductivity		Resistivity, ohm·cm		Charge carrier concentration, cm^{-3}		Charge carrier mobility, $cm^2/(V \cdot s)$	
Blend	Crystal	300 K	77 K	300 K	77 K	300 K	77 K	300 K	77 K
0	0	*p*	*n*	$3.2 \cdot 10^{-3}$	$1.9 \cdot 10^{-3}$	$3.0 \cdot 10^{18}$	$3.0 \cdot 10^{18}$	$6.5 \cdot 10^{2}$	$1 \cdot 10^{3}$
0.1	$1.55 \cdot 10^{-4}$	*n*	*n*	$1.0 \cdot 10^{-2}$	$4.7 \cdot 10^{-2}$	$4.2 \cdot 10^{17}$	$2.7 \cdot 10^{17}$	—	$4.8 \cdot 10^{1}$
0.2	$2.17 \cdot 10^{-4}$	*p*	*p*	$4.5 \cdot 10^{-3}$	$2.5 \cdot 10^{-3}$	$1.9 \cdot 10^{18}$	$2.7 \cdot 10^{18}$	$7.1 \cdot 10^{2}$	$9.2 \cdot 10^{2}$
0.8	$4.65 \cdot 10^{-4}$	*p*	*p*	$1.3 \cdot 10^{-2}$	$2.3 \cdot 10^{-3}$	$1.5 \cdot 10^{18}$	$2.1 \cdot 10^{18}$	$3 \cdot 10^{2}$	$1.2 \cdot 10^{3}$
0.4	$9.31 \cdot 10^{-4}$	*p*	*p*	—	$1 \cdot 10^{-3}$	$2.6 \cdot 10^{10}$	$2.15 \cdot 10^{18}$	—	$2.9 \cdot 10^{3}$
0.8	$1.86 \cdot 10^{-3}$	*p*	—	$4.62 \cdot 10^{-3}$	$1.16 \cdot 10^{-3}$	$1.58 \cdot 10^{18}$	$2.4 \cdot 10^{18}$	$8.8 \cdot 10^{2}$	$2.24 \cdot 10^{3}$
1	$1.86 \cdot 10^{-3}$	*p*	*p*	$4.0 \cdot 10^{-3}$	$1.8 \cdot 10^{-4}$	$2.8 \cdot 10^{18}$	$4.1 \cdot 10^{17}$	$5.5 \cdot 10^{2}$	$8.3 \cdot 10^{3}$

Table 15

Properties of PbTe(Au) monocrystals

Au Concentration in crystals, at.%	Hole mobility $cm^2/(V \cdot s)$		Hole concentration		Luminescence intensity, rel. units
	300 K	77 K	300 K	77 K	
0	$1.0 \cdot 10^{3}$	$1.0 \cdot 10^{2}$	$1.6 \cdot 10^{18}$	$1.3 \cdot 10^{18}$	12
$1.2 \cdot 10^{-3}$	$8.0 \cdot 10^{2}$	$5.7 \cdot 10^{2}$	$1.7 \cdot 10^{18}$	$5.9 \cdot 10^{18}$	25
$1.28 \cdot 10^{-3}$	$8.8 \cdot 10^{2}$	$4.1 \cdot 10^{3}$	$1.9 \cdot 10^{18}$	$2.6 \cdot 10^{18}$	18
$1.7 \cdot 10^{-3}$	$9.4 \cdot 10^{2}$	$2.4 \cdot 10^{2}$	$2.0 \cdot 10^{18}$	$8.9 \cdot 10^{16}$	14
$2.7 \cdot 10^{-3}$	$7.9 \cdot 10^{2}$	$7.7 \cdot 10^{3}$	$1.7 \cdot 10^{18}$	$2.6 \cdot 10^{18}$	18
$3.4 \cdot 10^{-3}$	$7.8 \cdot 10^{2}$	$5.8 \cdot 10^{4}$	$2.1 \cdot 10^{18}$	$5.6 \cdot 10^{17}$	not measured
$1.2 \cdot 10^{-2}$	$4.5 \cdot 10^{2}$	$2.3 \cdot 10^{3}$	$1.2 \cdot 10^{19}$	$1.5 \cdot 10^{19}$	»
$1.9 \cdot 10^{-2}$	$4.9 \cdot 10^{2}$	$1.6 \cdot 10^{3}$	$1.7 \cdot 10^{19}$	$2.1 \cdot 10^{19}$	»

Silver- or gold-doped and undoped PbTe monocrystals were obtained from the gas phase at a growing temperature of 700–800°C in our study [136, 181]. The Ag was introduced as AgTe and the Au was introduced as AuTe during the crystallization processes whose duration was 100 hours in each case. Tables 14 and 15 give the impurity quantities. The grown bulk crystals up to 20 mm in length and 12 mm in diameter consisted of one or several monocrystal blocks. In order to improve the reliability of the results all monocrystal growing processes were duplicated.

Unlike studies [121, 183] we determined the silver and gold concentration in the derived monocrystals (see Tables 14 and 15). We initially employed a local X-ray spectral analysis whose sensitivity to the test element was 0.1 at.%. Successive experiments with a large statistical set of pulses at each test point revealed that the quantity of silver contributed to the monocrystals in all cases was below the sensitivity of this method. A quantitative spectral analysis technique was then employed; the silver sensitivity of this technique was $5 \cdot 10^{-5}$ at.%, and the gold sensitivity was $7 \cdot 10^{-4}$ at.% with an accuracy of 10-20% of the measured quantity. The results from using this method to determine the Ag and Au concentrations in PbTe monocrystals are given in Tables 14, 15. The drop of several orders of magnitude in the silver concentration ($1.5 \cdot 10^{-4}$-$1.9 \cdot 10^{-3}$ at.%) in PbTe monocrystals compared to the concentration of this impurity in the initial blend (see Table 14) is due to the low pressures of the Ag vapors in the gas phase during the crystal growing process. The effective Ag and Au distribution coefficients were 10^{-3} and 10^{-2}, respectively.

The cubic crystal lattice parameter a of the host (undoped) and silver-doped PbTe monocrystals was measured on the {100} cleavage faces by the standard diffractometry technique on the DRON-2 assembly accurate to ±0.0005 Å. The refined value of the parameter a for each test crystal was determined by extrapolating the a quantities calculated by the X-ray interference beams (200), (400), (600), and (800) by the function $0.5 \cdot (\cos^2\theta/\sin\theta + \cos^2\theta/\theta)$, where $\theta \to 90°$, where θ is the diffraction angle. The derived results are given in Fig. 42.

The electrical properties of the crystals were investigated by the Hall technique at 300 and 77 K. The dependencies of the calculated carrier concentrations and carrier mobilities on the silver and gold concentration in the crystals are given in Tables 14, 15 and in Fig. 43.

As is clear from the experimental results given in Table 14 and Fig. 43 at a silver concentration $\sim 2 \cdot 10^{-3}$ the PbTe crystals have stable n-type conductivity with a donor concentration of $(3\text{-}5) \cdot 10^{17}$ cm^{-3} regardless of the type of conductivity of undoped PbTe crystals grown in these conditions. It is important to note that in this doped material the carrier concentration is an order of magnitude lower while the mobility is one and a half orders of magnitude greater compared to the same characteristics of the initial (undoped) material. All crystals with a high Ag concentration have p-type conductivity; when the Ag concentration is increased from $3 \cdot 10^{16}$ cm^{-3} to $3 \cdot 10^{17}$ cm^{-3} the hole concentration and mobility increase from $2 \cdot 10^{18}$ to $4 \cdot 10^{18}$ cm^{-3} (77 K) and from $1 \cdot 10^3$ to $8 \cdot 10^3$ cm^2/(v·s), respectively.

These changes in the electrical properties of PbTe with Ag are not accompanied (within experimental accuracy) by changes in the crystal lattice parameter across the entire test doping range (see Fig. 42), due to the low doping level ($2 \cdot 10^{16}$–$3 \cdot 10^{17}$ cm^{-3}) as well as the proximity of the ionic radii of Ag^{+} (1.13 Å) and Pb^{2+} (1.26 Å).

We carried out an investigation of the photoluminescence of PbTe as a factor of the concentration of the added Ag [184]. A neodymium pulsed laser (λ = 1.06 μm) was used for excitation. Fresh {100} cleavage faces of monocrystal samples were used as the specimens. It was discovered that introducing Ag increases the quantum efficiency of luminescence by a factor of several times; the greatest increase (by a factor of 4) is observed in PbTe samples with a silver concentration of $5 \cdot 10^{16}$–$5 \cdot 20^{17}$ cm^{-3}. In this range the quantum efficiency is independent of the carrier concentration.

As we see from the experimental data in Table 15 in all *p*-PbTe(Au) test specimens the gold has a weak influence on the electrical properties of lead telluride at a concentration $(1-3) \cdot 10^{-3}$ at.%, although, as in the case of silver, it increases the quantum efficiency of luminescence. The introduction of gold, beginning at 10^{-2} at.% (10^{18} cm^{-3}) serves to increase the hole concentration at 10^{19} cm^{-3}.

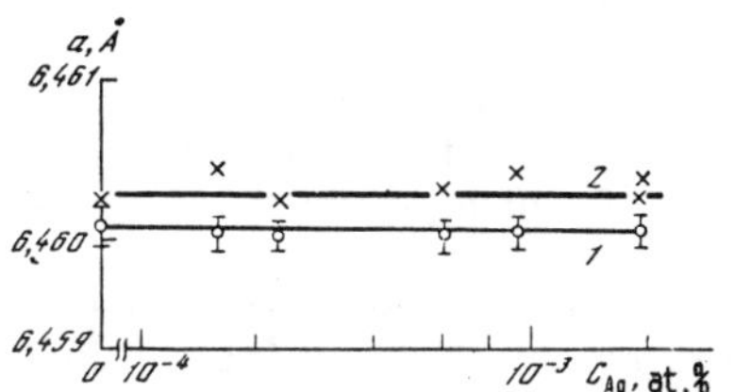

Fig. 42. The crystal lattice parameter a plotted as a function of the Ag concentration in PbTe crystals determined by the (800) X-ray interference beam (1) and by the extrapolation $\theta \rightarrow 90°$ (2)

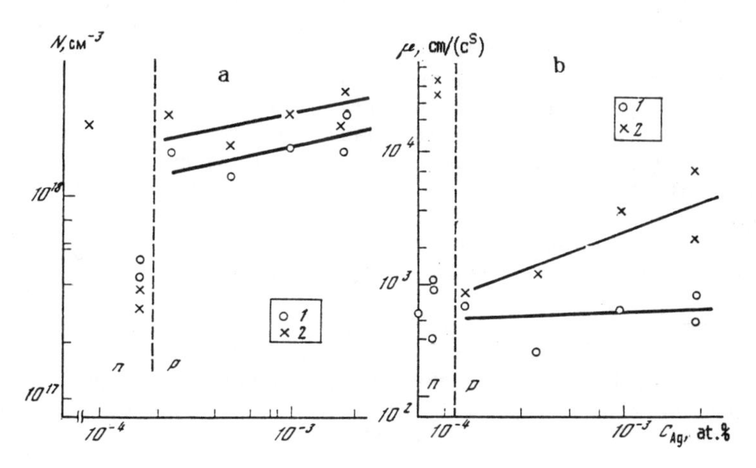

Fig. 43. The Hall carrier concentration (a) and carrier mobility (b) plotted as a function of Ag concentration in PbTe crystals at 300 K (1) and 77 K (2)

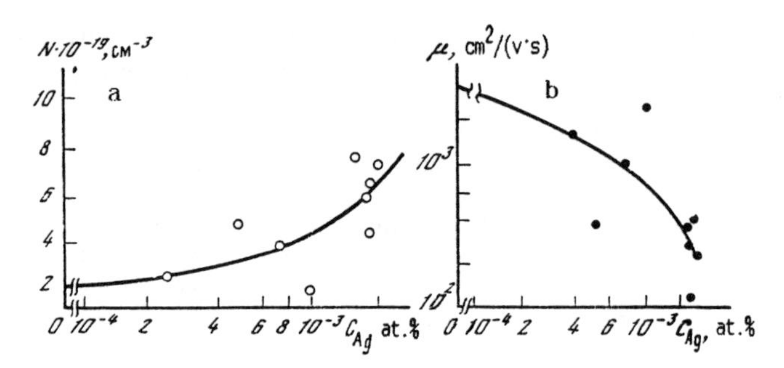

Fig. 44. The Hall carrier concentration (a) and carrier mobility (b) plotted as a function of Ag concentration in $Pb_{0.8}Sn_{0.2}Te$ crystals at 77 K

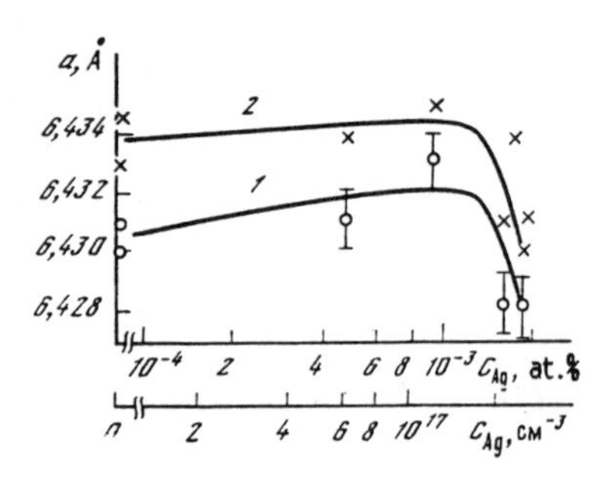

Fig. 45. The lattice parameter a plotted as a function of Ag concentration in $Pb_{0.8}Sn_{0.2}Te$ crystals determined by the (600) X-ray interference beam (curve 1) and by extrapolation (curve 2)

Our investigations did not reveal any degradations in the electrical or luminescence properties of samples stored in air for a period of three years.

Studies [136, 181] were the first to establish that the level of variation in the concentration of charge carriers/holes in silver- and gold-doped lead telluride is at least an order of magnitude greater than the level of variation in the concentration of the doping impurity itself. Accounting for the fact that the hole conductivity of PbTe is determined by the vacancies at the Pb sites this interesting aspect can be attributed to the fact that each added Ag or Au atom results in the creation of at least 10 electrically active metal vacancies/holes. In this case the Ag and Au impurities cannot be considered simple acceptor impurities and in order to explain the derived experimental results it is necessary to posit the existence of a more complex interaction mechanism between the silver and gold atoms and the vacancy subsystem of the PbTe crystal lattice.

Unlike PbTe, silver-doped $Pb_{0.8}Sn_{0.2}Te$ crystals produced stable p-type conductivity across the entire range of impurity concentrations (Table 16). With growth of the Ag concentration the hole concentration increases from $2 \cdot 10^{19}$ to $8 \cdot 10^{19}$ cm^{-3}, while the mobility drops to $3 \cdot 10^3$ to 10λ2 cm^2/(v·s), see Fig. 44 [79].

Unlike PbTe the silver-doping of $Pb_{0.8}Sn_{0.2}Te$ crystals on an impurity level of $\geqslant 10^{17}$ cm^{-3} will cause a significant decrease in the crystal lattice parameter (Fig. 45). This result accounting for the increase in the hold concentration and the decrease in the hole moility detected at identical concentrations of Ag indicates avalanche formation of single vacancies in the metallic sublattice of this material.

Table 16

Certain parameters of $Pb_{0.8}Sn_{0.2}Te(Ag)$ at 77 and at 300 K

Ag concentration, at.%		Type of conductivity		Resistivity, ohm·cm		Carrier concentration, cm^{-3}		Carrier mobility, $cm^2/(v \cdot s)$	
blend	crystal	300 K	77 K	300 K	77 K	300 K	77 K	300 K	77 K
0	0	p	p	$4.53 \cdot 10^{-4}$	$1,29 \cdot 10^{-4}$	$3,65 \cdot 10^{19}$	$1.98 \cdot 10^{19}$	$3.7 \cdot 10^{2}$	$3.44 \cdot 10^{3}$
0.8	$4,9 \cdot 10^{-4}$	p	p	$8.0 \cdot 10^{-4}$	$3.50 \cdot 10^{-4}$	$3.07 \cdot 10^{19}$	$4.80 \cdot 10^{19}$	$2,5 \cdot 10^{2}$	$3.7 \cdot 10^{2}$
0.1	$7,0 \cdot 10^{-4}$	p	p	$5.57 \cdot 10^{-4}$	$1.76 \cdot 10^{-4}$	$1,61 \cdot 10^{19}$	$3.72 \cdot 10^{19}$	$3,72 \cdot 10^{2}$	$9.54 \cdot 10^{2}$
0,1	$9,5 \cdot 10^{-4}$	p	p	$1.38 \cdot 10^{-3}$	$4,0 \cdot 10^{-4}$	$2,08 \cdot 10^{19}$	$1.60 \cdot 10^{19}$	$2,10 \cdot 10^{2}$	$2,40 \cdot 10^{3}$
1	$1.6 \cdot 10^{-3}$	p	*p	$6.13 \cdot 10^{-4}$	$6.57 \cdot 10^{-4}$	$2.14 \cdot 10^{19}$	$7.81 \cdot 10^{19}$	$4,76 \cdot 10^{2}$	$1.21 \cdot 10^{2}$
0,6	$1.7 \cdot 10^{-3}$	p	p	$7.88 \cdot 10^{-4}$	$2.87 \cdot 10^{-4}$	$2,55 \cdot 10^{19}$	$5.95 \cdot 10^{19}$	$3,1 \cdot 10^{2}$	$3.65 \cdot 10^{2}$
0.6	$1.8 \cdot 10^{-3}$	p	p	$8,2 \cdot 10^{-4}$	$3,37 \cdot 10^{-4}$	$4,46 \cdot 10^{19}$	$4,46 \cdot 10^{19}$	$1.7 \cdot 10^{2}$	$4,15 \cdot 10^{2}$
3	$1,8 \cdot 10^{-3}$	p	p	$9.70 \cdot 10^{-4}$	$3.6 \cdot 10^{-4}$	$2,50 \cdot 10^{19}$	$6,46 \cdot 10^{19}$	$2,50 \cdot 10^{2}$	$2,60 \cdot 10^{2}$
1	$1.9 \cdot 10^{-3}$	p	p	$7.59 \cdot 10^{-1}$	$3.86 \cdot 10^{-4}$	$5.78 \cdot 10^{19}$	$6,72 \cdot 10^{19}$	$1,42 \cdot 10^{2}$	$2.4 \cdot 10^{2}$

***The materials *p*-type material contains *n*-type regions.**

It was demonstrated previously in Section 2 of the present survey using CdTe and PbTe as examples that the concentration of thermodynamically-equilibrium vacancies can exceed the charge carrier/hole concentration by two orders of magnitude. This discrepancy is due to the formation of electrically neutral vacancy complexes. Within the framework of this crystallochemical model the Ag and Au atoms will prevent the formation of neutral vacancy complexes during the growing of the PbTe and $Pb_{0.8}Sn_{0.2}$ Te crystals thereby causing an increase in the concentration of single vacancies in the metal sublattice and therefore an increase in hole concentration.

A comparison of the concentration levels of the doping impurity and the carrier concentration in $Pb_{0.8}Sn_{0.2}Te$ reveals that each Ag atom at a concentration of $\geqslant 10^{17}$ cm^{-3} forms at least 10^2 electrically active vacancies. It follows that in the case of $Pb_{0.8}Sn_{0.2}Te$ the "gain" in hole concentration from silver is an order of magnitude higher than in PbTe. The more significant influence of Ag on the vacancy system of the $Pb_{0.8}Sn_{0.2}Te$ metallic sublattice is due to the order of magnitude greater large composition stability range on the tellurium side of this material compared to PbTe (at a growing temperature of 700-800°C).

Generalizing the experimental results from the doping of $Pb_{1-x}Sn_xTe$ with different impurities (Bi, In, Tl, Ag, Au) having a variety of valence properties makes it possible to suggest a few comments regarding the features of impurity dissolution in materials with a high intrinsic defect concentration.

1. The donor or acceptor action of the impurity is determined not by the valency of the added atom but rather by its interaction with the prevalent intrinsic defect. If such defects (vacancies) in A^4B^6 are characterized by a low formation energy (see Table 1) they are actively involved in the electronic processes occurring in the doped materials. For example, in an undoped PbTe crystal where vacancies at the Pb sites are predominant, the impurity atoms (In, Ag) interact with the cation sublattice vacancies, giving their valance electrons (either partially or totally) to the tellurium, and the hole concentration drops.

2. The solubility of the doping impurities in compounds of variable composition grows with growth of the vacancy concentration in the solid solutions, i.e., with an increase in the composition stability range size. For example, the solubility of In and Bi in PbTe is approximately 10^{18} atoms/cm^3, while in $Pb_{0.8}Sn_{0.2}Te$ it is 10^{20} atoms/cm^3.

3. When doping a bilateral phase of variable composition such as PbTe the impurities will interact with the vacancies more efficiency if they are introduced in an active form such as InTe, Ag_2Te, TlTe, and not as element atoms. Moreover, as we know if the doping process occurs in the presence of an impurity of opposite sign in an equiatomic quantity this will eliminate the decay of the impurity/ vacancy solid solution [185]. From this viewpoint introducing Ag^+ as Ag_2Te increases the solubility and reduces the retrograde nature of the composition stability range.

4. The nontrivial mechanism of Ag and Au impurity action on the properties of PbTe resulting in carrier "multiplication" detected for the first time is experimental proof of the interaction of the impurity with the entire vacancy subsystem of the crystal lattice. This indicates a change in the width of the composition stability range of PbTe and in its configuration in the presence of a third (doping) component. It seems that further achievements in controlling the properties of superconductors of variable composition by means of doping will involve investigations of their microdiagrams in the composition stability range.

4.3. Controlling the Luminescence Properties of Lead-Tin Tellurides

We know that interest in lasers based on narrowband superconductors can be traced to their promise for applications in high resolution laser IR spectroscopy.

Lead and tin chalcogenides and solid solutions based on these components are used in research in semiconductor lasers operating in the near and far IR (up to 32 μm). Spontaneous recombination radiation [186, 188], laser radiation from *p-n*-junctions [189, 190] and heterostructures [191, 192] were investigated in the $Pb_{1-x}Sn_x$ and $Pb_{1-x}Sn_xTe$ solid solutions.

Using $Pb_{1-x}Sn_xTe$ solid solutions and the capabilities of gas-phase epitaxy that make it possible to apply a more wideband layer on a $Pb_{1-x}Sn_xTe$ solid solution substrate of variable composition the authors of study [193] fabricated a set of diodes and by altering their temperature it was possible to obtain laser radiation of any wavelength in the 5-15 μm range. The diodes were fabricated from *p*-$Pb_{1-x}Sn_xTe$ substrates with applied *n*-PbTe layers (photostimulated epitaxy technique). The material for the substrates was fabricated using a variety of techniques, including the Czocharlski, Bridgman, and gas and liquid phase growing techniques.

Previous experiments have revealed that $Pb_{1-x}Sn_xTe$ and $Pb_{1-x}Sn_xSe$ substrates grown from the melt and the gas phase require long-term homogenizing annealing in order to be suitable for fabricating laser structures, since the cathodoluminescent properties are manifest only after long-term annealing in these samples. On this basis study [104] concluded that in $Pb_{1-x}Sn_xTe$ and $Pb_{1-x}Sn_xSe$ materials an additional reason for the lack of luminescence are the quenching centers that can be eliminated by annealing.

Zh. Pankov [194] examined possible causes of nonradiative recombination of the electron-hole pairs in semiconductors and described, in particular, the nonradiative recombination mechanism through defects and inclusions. Following study [194] we will consider a localized defect which, as a microscopic internal surface or metallic inclusion, creates a continuous set of electron states (Fig. 46). The electrons and holes at a distance less than the diffusion length *l* from the boundary of the given defect with an effective radius pr will merge at this defect and will nonradiatively recombine through the continuous set of states. Such a model explains values of the quantum efficiency significantly less than unity observed even at a low temperature in "pure" materials. It has been demonstrated in study [195] that in *n*-type GaAs heavily doped by Se or Te and therefore containing Ga_2Se_3 and Ga_2Te_3 precipitates, the radiation efficiency drops by an order of magnitude. According to Pankov [194] the homogeneous distribution of minor precipitates will influence radiation efficiency to a much greater extent than a small concentration of large precipitates. Therefore if a small number of minority phase particles are accumulated in a few large groups, thereby eliminating the small precipitates in the surrounding material, the radiation efficiency will be greater in the vicinity in each of these accumulations than in regions with a dispersive microprecipitate distribution.

Accounting for these results we proposed [193] that the nonradiative recombination in lead and tin chalcogenide monocrystals is related to the precipitation of excess component precipitates (the metal or the chalcogen) as a result of the retrograde nature of the solidi of these materials. For example, in the case of PbTe-based solid solutions (Fig. 47) the excess component will be released as tellurium precipitates in the *p*-region and metallic precipitates in the *n*-region upon cooling in monocrystals with compositions corresponding to *A-B*, *C-D* and *C'-D'*. According to our data [23] the total concentration of the component precipitated from cooling crystals is 10^{19}-10^{20} cm^{-3}. Such precipitates having dimensions of the order of a few nanometers [193] and separated by several tens of nanometers will function as efficient nonradiative recombination centers and therefore will significantly degrade the luminescence and photoelectric properties of the material. The validity of our hypothesis is confirmed by the fact that we have indeed discovered high quality luminescence without preliminary annealing in PbSnTe crystals obtained from the liquid phase at low temperatures. According to the phase diagram (see Fig. 47) the quantity of precipitates is insignificant in such crystals.

Long-term (several days) annealing of samples in the presence of a blend of appropriate composition is the normal technique used to reduce the number of nonradiative recombination centers in a material. One drawback of this method is degraded structural characteristics of the material. Investigations of the diffraction and oscillation curves in the two-crystal spectrometer mode of $Pb_{0.8}Sn_{0.2}Te$ monocrystals grown at high temperatures (initially nonluminescing) revealed that annealing produces a significant increase in the (600) X-ray diffraction interference beams from 0.09° to 0.15° and in the (200) oscillation curves from 0.2° to 1°. These expansions in the half-widths reveal that annealing causes a significant change in the monocrystal nature of the material: the average size of the monocrystal microblocks drops (from 3500 Å to 1500 Å) and their disorientation increases, i.e., the structure undergoes polygonization and forms multiple small-angle boundaries that function as precipitate drains. As a result the annealing process is accompanied by a significant increase in the microinhomogeneity of the blocks themselves, resulting in the luminescence properties of the monocrystal [104].

It is possible within the framework of the proposed model of the luminescence quenching centers that if the crystallization process of the material (and its subsequent cooling from the growing temperature to room temperature) were to be carried out in conditions that eliminate the precipitation process, the derived crystals could luminesce immediately following the growing process. Since the primary cause behind tellurium precipitate formation in *p*-$Pb_{1-x}Sn_xTe$ crystals is the existence of thermodynamically equilibrium vacancies in the metallic sublattice, a drop in the number of these vacancies in the crystallizing material will eliminate precipitation and will increase the quantum efficiency of luminescence.

Clearly it is possible to reduce the concentration of thermodynamically equilibrium vacancies either by crystallization of the material by altering its composition to be more stoichiometric (for example, by reducing the growing temperature) or by reducing the size of the composition stability range of the material by irradiating the crystal growth zone (see Section 3.2 of this survey).

Today the issue of the causes of luminescence quenching in epitaxial lead and tin chalcogenide layers has been the subject of increasing interest since twin epitaxial structures of these compounds have been used to fabricate CW IR lasers [196]. The epitaxial layers used as the active medium in a laser must have luminescent properties immediately following the growing process since annealing of such structures would cause mutual diffusion of the layers and would blur out their *p-n*-junctions. We therefore carried out research [197] whose purpose was to investigate the influence of the degree of deviation of lead selenide epitaxial layers from stoichiometry on their luminescence properties and to determine the growing conditions that would guarantee such properties in the grown layers. Indeed the phase diagram of the Pb-Se system (Fig. 48) reveals that as the composition of the grown PbSe film approaches stoichiometry the concentration of the excess component precipitated upon cooling will drop and if the composition stability range of the PbSe compound is sufficiently broad at low temperatures (where there is virtually no diffusion of components in the solid phase) it is possible to completely avoid precipitation which will enhance the luminescence of the synthesized material.

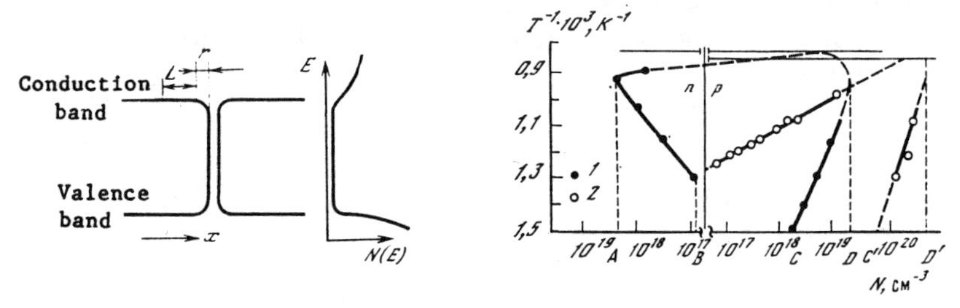

Fig. 46. Model of a defect or precipitate producing a localized continuum of states

Fig. 47. The composition stability ranges of PbTe and the $Pb_{0.8}Sn_{0.2}Te$ solid solution
1 - PbTe, 2 - $Pb_{0.8}Sn_{0.2}Te$

The lead selenide layers of *p*- and *n*-type conductivity with controlled deviation from stoichiometry were grown by the photostimulated epitaxy technique (see Section 3.2). An additional lead source with an independently controlled temperature was used to manipulate the composition of the gas phase over the growing film surface. Since the degree of dissociation of the gaseous lead selenide depends on the temperature and composition of the subliming material [198] in order to avoid uncontrollable variations in the composition of the growing film, crystals.with a maximal selenium concentration (an acceptor concentration of $7.3 \cdot 10^{18}$ cm^{-3}) were employed as the blend, while the substrate and blend temperatures were maintained at a constant level. The layer width in all cases was 50-60 μm. The $Pb_{0.8}Sn_{0.2}Se$ and KCl monocrystal substrates were oriented in {100} planes accurate to better than 1-2°. In these technological conditions the epitaxial layers

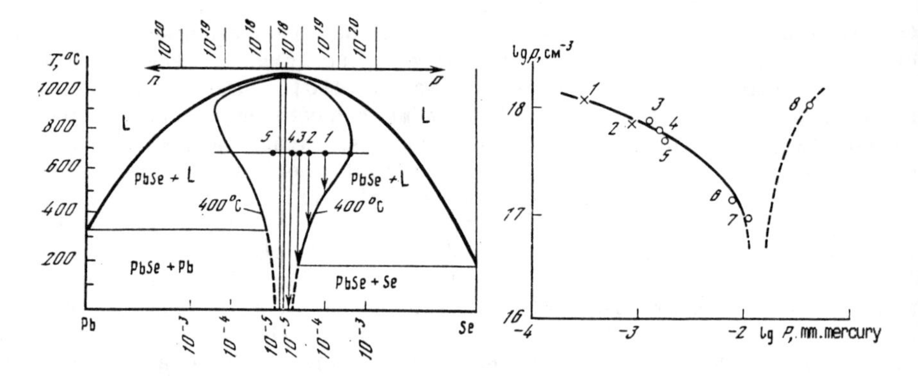

Fig. 48. Phase diagram of the Pb-Se system
1, 2, 3 - compositions of the lead selenide with a diminishing quantity of precipitates; 4 - the composition at which there is no precipitation

Fig. 49. The carrier concentration *p* in PbSe films plotted as a function of lead vapor pressure *P*
1, 2 - nonluminescing *p*-layers; 3-7 - luminescing *p*-layers; 8-luminescing *n*-layer

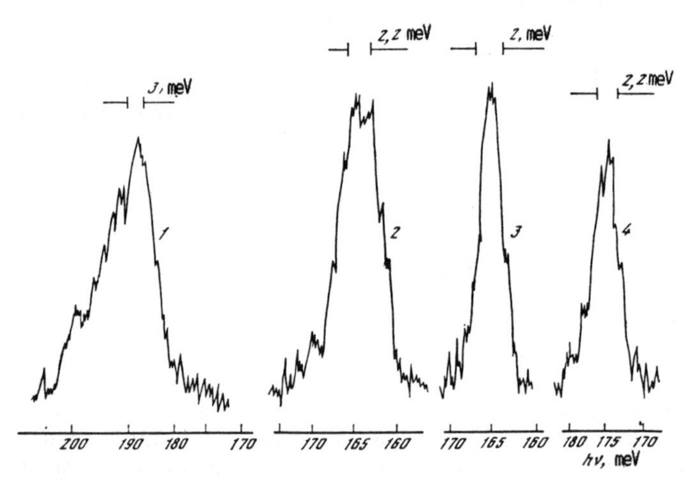

Fig. 50. The luminescence spectra of PbSe films with a variable carrier concentration
1 - $p = 6.2 \cdot 10^{17}$; 2 - $1.3 \cdot 10^{17}$; 3 - $9 \cdot 10^{16}$; 4 - n - $1.04 \cdot 10^{18}$ cm^{-3}

were monocrystalline and had a crystallographic orientation identical to that of the substrate. The discrepancy of the lattice parameters of the $Pb_{0.8}Sn_{0.2}Se$ and KCl grown layers and substrates were 0.7 and 2.7%, respectively. Primary research was carried out on PbSe/KCl structures. KCl dielectric substrates were necessary for the experiment due to the need to control the free carrier concentration in the luminescing and nonluminescing PbSe films. The carrier concentration was measured in all test films by the Van der Paue method.

Fig. 39 shows the carrier concentration in the films plotted as a function of lead vapor pressure. The graph clearly indicates that as the lead vapor pressure jumps from $3 \cdot 10^{-4}$ mm/mercury to $9 \cdot 10^{-3}$ mm/mercury the hole concentration in the films dropped from $1.3 \cdot 10^{18}$ cm^{-3} to $9 \cdot 10^{16}$ cm^{-3}. The range of lead vapor pressures of $9 \cdot 10^{-3}$-$3 \cdot 10^{-2}$ mm/mercury represents the range of unstable composition of lead selenide. The film conductivity changes in this range. At a lead vapor pressure of $3 \cdot 10^{-2}$ mm/mercury the electron concentration in the films was $1.0 \cdot 10^{18}$ cm^{-3}.

The photoluminescence properties were investigated using these films. Photoluminescence was excited by means of YAG:Nd3+ Q-switched laser [199]. A Ge-Au photocell with a sensitivity of 10^{-4} v/w was used to record the radiation. Stimulated luminescence was observed on films with hole concentrations of less than $6.5 \cdot 10^{17}$ cm^{-3} at 78 K. An analogous luminescence spectrum was obtained on films with an electron concentration of $1.0 \cdot 10^{18}$ cm^{-3}. Fig. 50 shows characteristic luminescence spectra for different carrier concentration levels. The luminescence properties of the films were sensitive to the cooling conditions. Films grown at different lead vapor pressures manifested luminescent properties only at cooling durations of less than 30 min.

Since it was previously established that the annihilation of the nonradiative recombination centers that occurs in $Pb_{0.8}Sn_{0.2}Te$ monocrystals takes place against a background of degrading structural characteristics of the material [104] we investigated the structural characteristics of luminescing and nonluminescing lead selenide films and then performed a comparative analysis of results obtained for luminescing and nonluminescing layers with an identical carrier concentration.

The investigations were carried out on a DRON-2 X-ray diffractometer using a BSV-11 tube and monochromatic Cu K_{α}-radiation. The vertical divergence of the primary beam was less than $1.7 \cdot 10^{-2}$ rad. A 50 μm (in width) slit was used in front of the counter. Diffraction reflection curves were obtained in these conditions with the sample and the counter rotated simultaneously ($\theta/2\theta$-curves) and with rotation of the sample only (fixed counter) (ω-curves). Fig. 51 gives standard ω-oscillation curves for luminescing and nonluminescing lead selenide layers on BaF_2 substrates.

The physical expansion quantities β of $\theta/2\theta$- and ω-curves were selected, consistent with Section 2.2 of this survey, as the structural perfection criteria. The corresponding physical expansions β_{θ} and β_{ω} were determined from the measured integrated half-widths B of these and other curves for the (200), (400) and (600) reflections by approximating the experimental profile with a Gaussian function. The instrument expansion was estimated by near-perfect nonluminescing lead selenide films applied to $Pb_{0.8}Sn_{0.2}Se$ substrates. The substructural elements of the crystal lattice of the epitaxial layers were determined by formulae (19-23) given in Section 2.2 from the values of β_θ and β_ω: the dimensions of the coherent scattering regions on the planes reflecting the X-ray radiation and on the normal to them (L_t and N_n, respectively); the microdeformation of the crystal lattice $\varepsilon = \pm\Delta d/d$ (d is the interplanar spacing for the reflecting family of planes) on the reflecting planes and on the normal to these planes (ε_t and ε_n, respectively); the angular disorientation α of the coherent scattering regions with respect to one another.

Calculation results for films with a hole concentration of $1.3 \cdot 10^{17}$ cm^{-3} are given in Table 17. This same table gives the dislocation densities N calculated by existing formulae

Table 17

Substructural characteristics of lead selenide epitaxial layers

Layer	L_n, Å	L_l, Å	ε_n, rel. units	ε_l, rel. units	α angular units	N, cm^{-1}
Nonluminescing	3500	2000	$0.25 \cdot 10^{-3}$	0	0.4	$5 \cdot 10^{9}$
Luminescing	2500	600	$0.25 \cdot 10^{-3}$	$6 \cdot 10^{-3}$	1.6	10^{11}

Table 17 reveals that the nonluminescing films consist of large regions with a perfect crystal structure with dimensions $\sim 3 \cdot 10^{3}$/Å along the epitaxial layer and on the normal to the layer. There is no lattice deformation along the layer within the epitaxial layer ($\varepsilon_t = 0$). Only insignificant lattice stresses are observed on the normal to the epitaxial layer surface ($\varepsilon_n = 2.5 \cdot 10^{-4}$). The angular disorientation α of the separate substructural elements does not exceed a few tenths of a degree. In the luminescing layers the structure was less uniform. While the quantities L_n and ε_n were conserved, the value of L_t was reduced somewhat and a deformation ε_t an order of magnitude greater than ε_n arose. Moreover the angular disorientation α of the block increased by a factor of several times and the dislocation density N grew by an order of magnitude. Both types of layer structures are shown schematically in Fig. 52.

The derived data make it possible to conclude that the imperfections arising in the luminescing layers (where the distances between these imperfections are comparable to the distances between the precipitates [193]) are efficient escape channels for the nonradiative recombination centers (precipitates) formed in the film due to retrograde solubility of the excess component. Since the disorientation of the substructural blocks with respect to the normal n increased and a strong microdeformation ε_t arose along the layer, we can expect that the small-angle boundaries of the blocks perpendicular to the layer surface serve as the precipitate drains.

We should note that the correlation between the luminescence properties of the epitaxial lead selenide films and the polygonization of their microstructure was observed across the entire range of test carrier concentrations from $p = 6.5 \cdot 10^{17}$ to $n = 1.0 \cdot 10^{18}$ cm^{-3}.

Study [17] has demonstrated that luminescence is achieved only in polygonized PbSe films across the entire test range of both types of carrier concentrations, where the multiple small-angle boundaries function as effective drains for the nonradiative recombination centers (precipitates); as indicated by experiment the minimum drainage time for such centers at the block boundaries (with the given block dimensions; see Table 18) is 30 minutes.

The need for polygonization of the film structure in order to achieve luminescent properties even at a hole concentration of $9 \cdot 10^{16}$ cm^{-3} reveals a significant narrowing of the composition stability range of PbSe below 400°C which makes it impossible to avoid precipitation or a significant reduction in precipitate quantity (see Fig. 48)

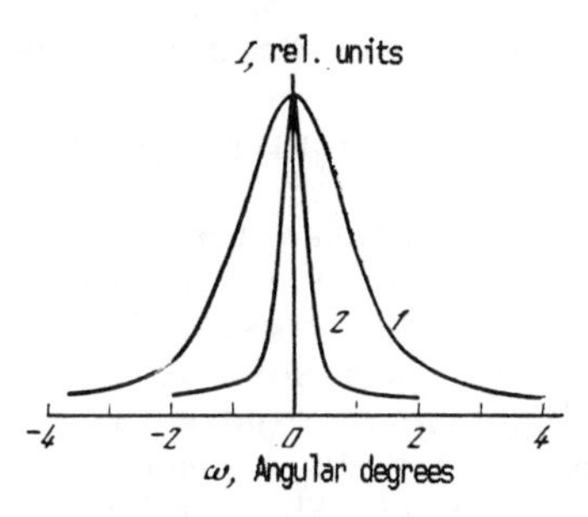

Fig. 51. Rocking curves for luminescing (1) and nonluminescing (2) PbSe layers

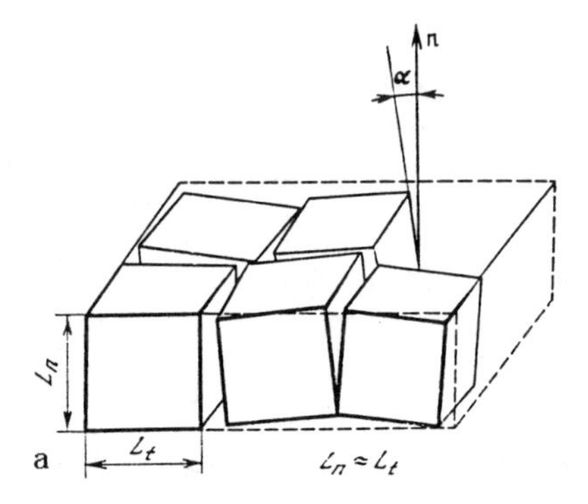

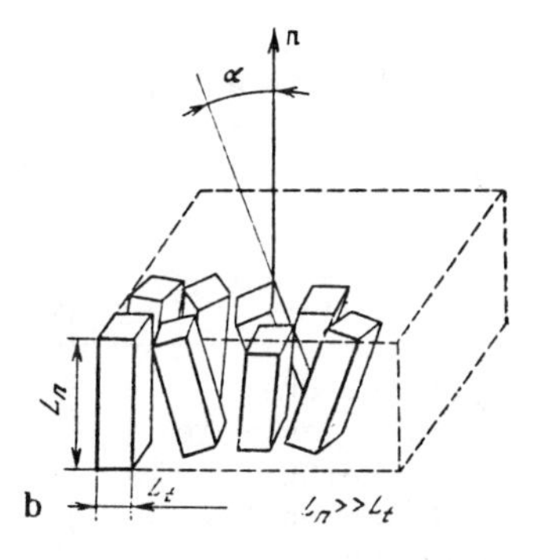

Fig. 52. Block structure of nonluminescing (a) and luminescing (b) lead selenide films

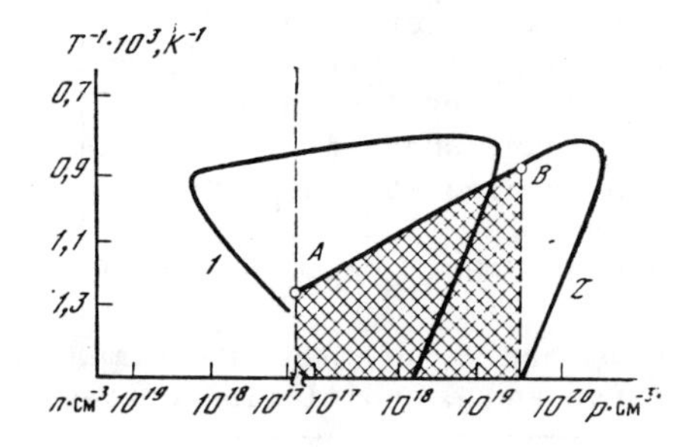

Fig. 53. Range of $Pb_{1-x}Sn_xTe$ with the *AB* region (hatched) indicating where zero precipitates in the crystallized material can be expected

We solved an analogous problem of growing luminescing epitaxial layers by "doping" with intrinsic components for a $Pb_{0.8}Sn_{0.2}Te$ solid solution. A gas-phase photostimulated epitaxy technique for PbSnTe layers using an Me-rich source was developed [116, 200]; layer growing in the presence of excess metal will reduce the metal vacancy concentration and thereby eliminate precipitation of the tellurium precipitates.

This method of doping layers with intrinsic components or a foreign impurity during their growth process in photostimulated epitaxy makes it possible to dope either from a temperature-independent Me-rich source or from the $Pb_{1-x}Sn_xTe$ source blend with the appropriate added impurity [200]. The technological parameters of the process are selected so that the layer is grown in quasi-equilibrium conditions following the vapor-liquid-solid mechanism.

In our case the epitaxial growing of $Pb_{1-x}Sn_xTe$ layers from an Me-rich source followed this same mechanism: the metallic film on the surface of the growing layer functioned as the solvent and any excess amount of this solvent in the form of droplets was removed from the growing layers. The metal would then enter the lattice of the growing crystal only in a quantity determined by the composition stability range of solid $Pb_{1-x}Sn_xTe$. The layers of the compounds corresponding to the *AB* segment of the solidus of the phase microdiagram for $Pb_{1-x}Sn_xTe$ (Fig. 53) would not withstand retrograde precipitation of the precipitates from cooling down from the growth temperature at any cooling rate.

Table 18
X-ray structural and luminescence properties of PbSnTe samples

Samples	B_θ, angular deg. [(600) X-ray interference beam]*	B_ω, angular deg. [(200) X-ray interference beam]*	Luminescence
$Pb_{0.8}Sn_{0.2}Te$ (Pb+Sn) epitaxial layer on PbTe	0.07	0.07	Yes
$Pb_{0.8}Sn_{0.2}Te$ (Pb+Sn) epitaxial layer on BaF_2	0.085	0.25	"
In-doped $Pb_{0.8}Sn_{0.2}Te$ epitaxial layer on BaF_2	0.07	0.17	"
$Pb_{0.8}Sn_{0.2}Te$ epitaxial layer obtained by liquid phase epitaxy, undoped	0.05	0.1	"
Nonannealed $Pb_{0.8}Sn_{0.2}Te$ crystal	0.09	0.2	No
Annealed $Pb_{0.8}Sn_{0.2}Te$ crystal	0.15	1.0	Yes

* B_θ and B_ω are the experimental half-widths of the diffraction $\theta/2\theta$-curve and ω-oscillation curves, respectively

X-ray structural investigations have revealed the high level of structural perfection of the grown layers. No expansions of the diffraction curves or oscillation curves were observed; these normally occur in crystals annealed in a metal-rich medium (Table 18). As indicated by the data in Table 18 the epitaxial PbSnTe layers grown by this technique from the Me-rich source indeed have efficient luminescence, which makes it possible to use these layers as an active medium for an IR laser without additional homogenizing annealing.

It was demonstrated previously (see Section 4.2) that indium impurity doping during the growth of p-$Pb_{1-x}Sn_xTe$ bulk crystals in an indium range from 0 to 0.6 at.% reduces the concentration of charged vacancies in the metallic sublattice which, in principle, will eliminate the precipitation of tellurium precipitates and will increase the quantum efficiency of luminescence. We therefore deliberately doped the $Pb_{1-x}Sn_xTe$ epitaxial layers (x = 0.20 and 0.22) with an In impurity during their growth process and then investigated their electrical and luminescence properties.

The photostimulated gas epitaxy technique was used to grow the epitaxial layers. PbTe, $Pb_{1-x}Sn_xTe$, and BaF_2 monocrystals with a {100} and {111} crystallographic orientation were employed as the substrates. The insulating BaF_2 substrates were needed to investigate the electrical properties of the epitaxial layers. The derived monocrystal layers between 5 and 200 μm in thickness from $Pb_{1-x}Sn_xTe$ solid solutions (x = 0.20, 0.22) followed the crystallographic orientation of the substrate. The indium concentration in these layers was determined by means of local X-ray spectral analysis on the "Camebax" and MAR-2 units.

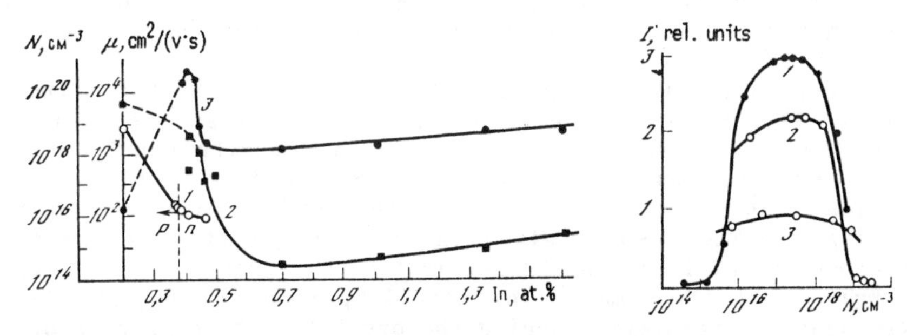

Fig. 54. The Hall concentration (1, 2) and carrier mobility (3) plotted as a function of indium concentration in $Pb_{1-x}Sn_xTe$ solid solution layers (1 - x = 0.20, 2, 3 - x = 0.22)

Fig. 55. The luminescence intensity of indium-doped (1, 2) and undoped (3) $Pb_{1-x}Sn_xTe$ solid solution layers plotted as a function of carrier concentration [76]
1 - x = 0.22; 2, 3 - x = 0.20

Hall variations in the carrier concentration in the epitaxial layers at 77 K (Fig. 54) have revealed that indium introduced into the

$Pb_{0.8}Sn_{0.22}Te$ layers in a concentration range from 0 to 0.4 at.% is a donor impurity and alters the carrier concentration from $p = 5 \cdot 10^{19}$ to $n = 10^{18}$ cm^{-3}. Therefore the position of the *p-n*-junction corresponds to an indium concentration of 0.4 at.% for this composition. With a further increase in the concentration the indium functions as an acceptor, reducing the donor concentration to $2 \cdot 10^{14}$ cm^{-3}.

Fig. 55 gives the integral luminescence intensity of indium-doped $Pb_{1-x}Sn_xTe$ solid solution layers (x = 0.20, 0.22) plotted as a function of carrier concentration. For comparison the graph gives an analogous relation for undoped solid solution layers for x = 0.20. It follows from these graphs that the introduction of indium in a concentration range from 0.35 to 0.5 at.% increases the quantum efficiency of luminescence by a factor of 3-5 [76].

Measurements of the half-width of the X-ray diffraction $\theta/2\theta$-curves and ω-oscillation curves have revealed that good luminescence properties of the doped layers are achieved against a background of good structural perfection of the crystal lattice.

The data in Table 18 also reveal that in terms of the half-widths of the diffraction curves and the oscillation curves these layers are comparable to luminescing liquid-phase epitaxial layers grown in equilibrium conditions. For comparison purposes Table 18 also gives the structural characteristics of luminescing annealed bulk $Pb_{0.8}Sn_{0.2}Te$ crystals grown from the gas phase.

Photostimulated indium doping in $Pb_{1-x}Sn_xTe$ solid solutions not only eliminates precipitation of precipitates but also significantly increases the quantum efficiency of luminescence due to the reduction in the concentration of thermodynamically equilibrium vacancies.

In conclusion it is important to note that with photostimulated doping of $Pb_{1-x}Sn_xTe$ epitaxial layers the intrinsic metallic components (Pb + Sn) and the metallic indium impurity can be used to achieve a $Pb_{1-x}Sn_x$ composition with a high ($\sim 10^{18}$ cm^{-3}) electron concentration that cannot be obtained with regular gas-phase epitaxy. Accounting for the results given in Section 3.2 it would seem that these experimental facts can be attributed to a change in the size and configuration of the composition stability range of $Pb_{1-x}Sn_xTe$ with electromagnetic irradiation during the crystallization process of this material.

References

1. "Sb. Fizika i khimiya soedineniy A^2B^6" [Handbook: the physics and chemistry of A^2B^6 compounds] Edited by Prof. S.A. Medvedev, Moscow: Mir, 1970, pp. 140-141.

2. Wagner, C., Shottky, W. "Theorie der geoirdneten Mischphasen" ZTSCHR. PHYS. CHEM., 1930-1931, vol. 11, pp. 163-210.

3. Kröger, F.A., Vink, H.Z. In: Halbleiterproblem, vol. 1, Ed. W. Shottky, Braunschweig: Vieweg, 1954, p. 128.

4. Kröger, F.A., Vink, H.Z. "Relations between the concentrations of imperfections in crystal solids" SOLID STATE PHYS., 1956, vol. 3, pp. 307-435.

5. Bublik, V.T. "X-ray investigation of the atomic structure and the thermodynamics of semiconductor solid solutions": Dissertation for Doctorate of Physics and Mathematics, Moscow, 1979, 379 pp.

6. "Soedineniya peremennogo sostava" [Compounds of variable composition] Edited by B.F. Ormont, Khimiya, 1969, pp. 395-405.

7. Bloem, J., Kröger, F.A. "The p-T-x-phase diagram of the lead-sulphur system" ZTSCHR PHYS. CHEM., Neue Folge, 1956, vol. 7, pp. 1-14.

8. Brebrick, R.F., Strauss, A.J. "Partial pressures in equilibrium with group IV tellurides. I. Optical absorption method and results for PbTe" J. CHEM. PHYS., 1964, vol. 40, pp. 3230-3241.

9. Gas'kov, A.M., Zlomanov, V.P., Novoselova, A.V. "P-T-x diagram of lead telluride" Vestn. MGU. Ser. 2, KHIMIYA, 1970, vol. 11, pp. 49-50.

10. Strauss, A.J., Brebrick, R.F. "Deviations from stoichiometry and lattice defects" J. PHYS. COLLOQ. C4, Suppl. 11-12, 1968, vol. 29, fasc. XI-XII, pp. C4-21.

11. Brebrick, B.F., Allgaier, R.S. "Composition limits of stability of PbTe" J. CHEM. PHYS., 1960, vol. 32, pp. 1826-1831.

12. Miller, E., Komarek, K., Cordoff, I. "Stoichiometry of lead telluride"AIME TRANS. METAL. SOC., 1959, vol. 215, pp. 882-887.

13. Brebrick, R.F. "Deviation from stoichiometry and electrical properties in SnTe" J. PHYS. AND CHEM. SOLIDS, 1963, vol. 24, pp. 27-36.

14. Fujimoto, M., Sato, Y. "P-T-x phase diagram of the lead telluride system" JAP. J. APPL. PHYS., 1966, v. 5, pp. 128-133.

15. Brebrick. R.F., Strauss, A.J. "Partial pressures in equilibrium with group IV tellurides II. Tin telluride" J. CHEM. PHYS., 1964, vol. 41, pp. 197-205.

16. Abrikosov, N.Kh., Shalimova, L.E. "Poluprovodnikovye materialy na osnove soedineniy A^4B^6" [Semiconductor materials based on A^4B^6 compounds] Moscow: Nauka, 1975, 195 pp.

17. Vereshchagin, L.F., Kabalkina, S.S. "Rentgenostrukturnye issledovaniya pri vysokom davlenii" [High-pressure X-ray structural analyses] Moscow: Nauka, 1979, pp. 113-124.

18. Ravich, Yu.I., Efimova, B.A., Smirnov, I.A. "Metody issledovaniya poluprovodnikov v primenenii k khal'kogenidam svintsa PbTe, PbSe, PbS" [Methods of investigating semiconductors with application to PbTe, PbSe, and PbS lead chalcogenides] Moscow: Nauka, 1968, 380 pp.

19. Smith, R.A. "The electronic and optical properties of the lead sulphide group of semiconductors" PHYSICA, 1954, vol. 20, pp. 910-929.

20. Scanlon, W.W. "Polar semiconductors" SOLID STATE PHYS., 1959, vol. 9, pp. 83-138.

21. Palmetschofer, L. "Ion implantation in IV-VI semiconductors" APPL. PHYS. A, 1984, vol. 34, pp. 139-153.

22. Brebrick, R.F. "Composition stability limits for the rocksalt-structure case $(Pb_{1-y}Sn_y)_{1-x}Te$ from lattice parameter measurements" J. PHYS. AND CHEM. SOLIDS, 1971, vol. 32, no. 3, pp. 551-562.

23. Gorina, Yu.I., Kalyuzhnaya, G.A., Kiseleva, K.V., et al. "The nature of defects in undoped PbTe" KRATKIE SOOBSHCH PO FIZIKE FIAN, 1975, no. 11, pp. 24-28.

24. Aleksandrov, O.V., Zaytsev, V.V., Zelikman, I.N., et al. "The influence of pulsed laser radiation on the epitaxy of lead and tin chalcogenides" KRATKIE SOOBSHCH PO FIZIKE FIAN, 1984, no. 2, pp. 45-49.

25. Shelimova, L.E., Abrikosov, N.Kh. "The Sn-Te system in the SnTe compound range" ZhURN. NEORGAN. KHIMII, 1964, vol. 9, no. 8, pp. 1879-1882.

26. Bis, R.F., Dixon, J.R. "Applicability of regards low to the $Pb_{1-x}Sn_xTe$ alloy system" J. APPL. PHYS., 1969, vol. 40, no. 4, pp. 1918-1921.

27. Tairov, S.M., Ormont, B.F., Shostak, N.O. "Investigation of the Pb-Sn-Te system near the PbTe-SnTe pseudobinary plan" IZV. AN SSSR. NEORGAN. MATERIALY, 1970, vol. 6, no. 9, pp. 1584-1588.

28. Kozlovskiy, V.F., Katsnel'son, A.A., Gas'kov, A.M., Zlomanov, V.P. "The near order in $Sn_xPb_{1-x}Te$" DAN SSSR, 1980, vol. 251, no. 5, pp. 1162-1166.

29. Kaukis, A.A., Banishkin, A.M., Bekker, A.A., et al. "The Mössbauer

spectra of $Pb_{1-x}Sn_xTe$ alloys" IZV. AN SSSR. NEORGAN. MATERIALY, 1972, vol. 8, no. 9, pp. 1667-1668.

30. Bekker, A.A., Efremova, E.N., Nesmeyanov, A.M. "Comprehensive investigation of the nature of chemical bonds and the local symmetry of Sn and Te in PbTe- and SnTe-based solid solutions" Proc. 5th Intern. Conf. Mössbauer spectroscopy, Bratislava, 1973, pp. 267-269.

31. Nikolaev, I.N., Potapov, V.P., Shotov, A.P., Yurchakevich, E.E. "Symmetry breaking in the local tin atomic configuration in a $Pb_{1-x}Sn_xTe$ crystal lattice" PIS'MA V ZhETF 1977, vol. 25, no. 4, pp. 185-187.

32. Levanyuk, A.P., Sigov, A.S. "The influence of defects on the properties of ferroelectrics and related materials near the second order phase transition point" IZV. AN SSSR. SER. FIZ., 1981, vol. 45, no. 9, pp. 1640-1645.

33. Novikova, S.I., Shelimova, L.E. "Phase transition in SnTe" FTT, 1965, vol. 7, no. 8, pp. 2544-2545.

34. Novikova, S.I., Shelimova, L.E., "Low-temperature phase transition in tin telluride" FTT, 1967, vol. 9, no. 5, pp. 1336-1338.

35. Muldawer, L. "The low temperature transformation in SnTe" BULL. AMER. PHYS. SOC., 1971, vol. 16, no. 1, p. 84.

36. Muldower, L. "New studies of the low-temperature transformation of SnTe" J. NONMETALS", 1973, vol. 1, no. 2, pp. 177-182.

37. Iizumi, M., Hamaguchi, Y., Komatsubara, K.F., Kato, Y. "Phase transition in SnTe with low-carrier concentration" J. PHYS. SOC. JAP., 1975, vol. 38, no. 2, pp. 443-449.

38. Kobayashi, K.L.I., Katayama, J., Narita, K., et al. "Observation of band structure changes due to structural phase transition in $Pb_{1-x}Sn_xTe$" Proc. 14th Intern. Conf. Phys. Semicond. Edinburg, Sept, 1978, pp. 441-444.

39. Vallasiades, O., Economou, N.A. "On the phase transformation of SnTe" PHYS. STATUS SOLIDI (a), 1975, vol. 30, pp. 287-195.

40. Natori, A. "Displacive phase transition in narrow-gap semiconductors" J. PHYS. SOC. JAP., 1976, vol. 40, no. 1, pp. 163-171.

41. Kristofel, N., Konsin, R. "Displacive vibronic phase transitions in narrow-gap semiconductors" PHYS. STATUS SOLIDI, 1968, vol. 28, no. 1, pp. 331-377.

42. Brillson, L.J., Burstein, E., Muldawer, L. "Raman observation of

the ferroelectric phase transition in SnTe" PHYS. REV. B, 1974, vol. 9, no. 4, pp. 1547-1551.

43. Fano, V. "Phase transition in SnTe by Mössbauer spectroscopy" SOLID. STATE COMMUNS., 1977, vol. 22, pp. 467-470.

44. Foley, G.M., Langenlerg, D.N. "Microwave magnetoplasma study of lattice and electronic properties of PbTe" PHYS. REV. B, 1977, vol. 15, no. 10, pp. 4830-4849.

45. Hohnke, D.K., Holloway, H., Kaiser, S. "Phase relations and transformations in the system PbTe-GeTe" J. PHYS. AND CHEM. SOLIDS, 1972, vol. 33, pp. 2053-2062.

46. Kawamura, H. "Phase transition in IV-VI compounds" Proc. Intern. Summer School, France, Nimes, Sept. 1979, pp. 470-494.

47. Littlewood, P.B. "The critical structure of IV-VI compounds: II. A microscopic model for cubic rhombohedral materials" SOLID STATE PHYS., 1980, vol. 13, no. 26, pp. 4875-4892.

48. Aleksandrov, O.V., Gorina, Yu.I., Kiseleva, K.V. "The problem of low-temperature lattice instability in the PbTe-SnTe system" ZhTF, 1980, vol. L, pp. 2473-2475.

49. Erasova, N.A., Lykov, S.N., Chernik, I.A."The superconductivity mechanism of PbTe(Tl) and the phase transition in the PbTe-GeTe system" FTT, 1983, vol. 25, no. 11, p. 269-271.

50. Herrmann, K.H., Möllmann, K.P. "Curie temperature as a critical temperature for dielectric, galvanomagnetic and photoelectric phenomena in strongly doped $Pb_{1-x}Sn_xTe$" PHYS. STATUS SOLIDI (a), 1983, vol. 80, pp. K101-K104.

51. Nasybbulin, R.A., Girshberg, N.N., et al. "The nonmonotonic dependence of the ferroelectric phase transition temperature in $Pb_{1-x}Sn_xTe$ on composition" FTT, 1983, vol. 25, no. 3, pp. 784-788.

52. Nasybbulin, R.A., Kalimullin, R.Kh., Shapkin, V.V., et al. "High-temperature phase transitions in $Pb_{1-x}Sn_xTe$ solid solutions" FTT, 1981, vol. 23, no. 1, pp. 300-302.

53. Baginskiy, V.M., Kikodze, R.O., Lashkarev, G.V., Radchenko, M.V. "Magnetic susceptibility and thermo-EMF of narrowband $Pb_{1-x}Sn_xTe$ semiconductors with a structural phase transition" Preprint. In-ta Fiziki AN USSR, Kiev, 1978.

54. Lashkarev, G.V., Baginski, V.M., Kikodze, R.O., Radchenko, M.V. "Magnetic and kinetic properties of narrow-gap $Pb_{1-x}Sn_xTe$ semiconductors at low temperatures" Proc. 14th Intern. Conf. Phys. Semicond., Edinburgh, Sept. 1978, pp. 597-600.

55. Lashkarev, G.V., Radchenko, M.V., Orletskiy, V.B., et al. "The features of thermoelectric phenomena in a $Pb_{1-x}Sn_xTe$ solid solutions ($0 \leqslant x \leqslant 0.23$) at low temperatures" FTP, 1980, vol. 14, no. 3, pp. 490-495.

56. Fano, V. "Growth and characterization of $A^{IV}B^{VI}$ single crystals for IR technology and thermoelectric energy conversion" PROGR. CRYST. GROWTH CHARACT., 1981, vol. 3, pp. 287-308.

57. Katayama, S. "Anomalous resistivity in structural phase transition of IV-VI compound: p-SnTe" SOLID STATE COMMUNS, 1976, vol. 19, no. 4, pp. 381-383.

58. Kobayashi, K.L.I., Kato, Y., Katayama, Y., Komatsubara, K.F. "Carrier-concentration-dependent phase transition in SnTe" PHYS. REV. LETT., 1976, vol. 37, no. 12, pp. 772-774.

59. Grassie, A.D.C., Agapito, J.A., Gonzales, P. "Anomalous resistivity at the structural phase transition of polycrystalline SnTe" J. PHYS. C: SOLID STATE PHYS., 1979, vol. 12, no. 24, pp. L925-L927.

60. Ortalli, I. "How many phase transitions in SnTe?" FERROELECTRICS, 1984, vol. 54, pp. 325-328.

61. Murase, K., Sugai, S., Higushi, T., et al. "Band-gap and phase transformation in IV-VI semiconductors" Proc. 14th Intern. Conf. Semicond., Edinburg, Sept. 1978, pp. 437-440.

62. Nakanishi, A., Matsubara, T. "Structural phase transition in mixed IV-VI compounds" SOLID STATE COMMUNS., 1979, vol. 32, no. 7, pp. 577-580.

63. Bate, R.T., Caster, D.L., Wrobel, J.S. "Evidence for paraelectric behavior in PbTe and pseudobinaries based on PbTe" Proc. 10th Intern. Conf. Phys. Semicond., Cambridge (Mass.), 1970, pp. 125-128.

64. Shimada, T., Kobayashi, K.L.I., Kato, Y., et al. "Soft-phonon-induced anomalies in resistivity, Hall coefficient and Raman scattering intensity in $Pb_{1-x}Sn_xTe$" Proc. 13th Intern. Conf. Rome, 1976, pp. 314-317.

65. Murase, K., Sugai, S., Takaoka, S., Katayama, S. "Study on the phase transition of IV-VI compound alloy semiconductors" Proc. 13th Intern. Conf., Rome, 1976, pp. 305-308.

66. Suski, T., Katayama, S. "Phonon induced anomalous resistivity in structural phase transition of PbSnTe" J. PHYS. C6, Suppl., 1981, vol. 42, no. 12, pp. 758-760.

67. Suski, T. "Ferroelectric phase transition in electron-irradiated

PbSnTe crystals" J. PHYS. C: SOLID STATE PHYS., 1982, vol. 15, no. 27, pp. L953-L956.

68. Bratashevskiy, Yu.A., Prozorovskiy, V.D., Kharionovskiy, Yu.S. "Phase transition in $Pb_{1-x}Sn_xTe$" FTP, 1975, vol. 9, no. 8, pp. 1612-1613.

69. Nasybbulin, R.A., Kalimullin, R.Kh., Shapkin, V.V., Kukharskiy, A.A. "The behavior of the ferroelectric phase transition in $Pb_{1-x}Sn_xTe$ near the band inversion" IZV. AN SSSR. SER. FIZ., 1983, vol. 47, no. 4, pp. 702-704.

70. Aleksandrov, O.V., Kiseleva, K.V. "Structural phase transitions in $Pb_{0.78}Sn_{0.22}Te$ monocrystals in the 10-300 K temperature range" KRATKIE SOOBSHCH. PO FIZIKE FIAN, 1984, no. 4, pp. 18-21.

71. Aleksandrov, O.V. "The structural instability of $Pb_{0.78}Sn_{0.22}Te$-In solid solutions" Dissertation for Candidate Degree of Physics and Mathematics, Moscow, 1985, 181 pp.

72. Aleksandrov, O.V., Kiseleva, K.V. "Low-temperature polymorphism in the $Pb_{0.78}Sn_{0.22}Te$-In system" KRATKIE SOOBSHCH PO FIZIKE FIAN, 1984, no. 8, pp. 7-10.

73. "Fizika i khimiya soedineniy $A^{II}B^{VI}$" [The physics and chemistry of A^2B^6 compounds] Ed. by. S.A. Medvedev, Moscow: Mir, 1970, 624 pp.

74. Medvedev, S.A., Maksimovskiy, S.N., Kiseleva, K.V., et al. "The nature of point defects in undoped CdTe" IZV. AN SSSR. NEORGAN. MATERIALY, 1973, vol. IX, no. 3, pp. 356-360.

75. Jantsch, W., Lopez-Otero, A. "Influence of lattice defects on the paraelectric behavior of PbTe" Proc. 13th Intern. Conf., Rome, 1976, pp. 487-490.

76. Kalyuzhanya, G.A., Mamedov, T.S., Kiseleva, K.V., Britov, A.D. "The properties of epitaxial layers of indium-doped $Pb_{1-x}Sn_xTe$ solid solutions" IZV. AN SSSR, NEORGAN. MATERIALY, 1979, vol. 15, no. 2, pp. 231-234.

77. Gorina, Yu.I., Kalyuzhnaya, G.A., Kiseleva, K.V., et al. "The influence of a silver impurity on the electrical properties of a tin-lead telluride solid solution" KRATKIE SOOBSHCH PO FIZIKE FIAN, 1981, no. 7, pp. 42-47.

78. Ratcliff, R.T. "The measurement of small density changes in solids" BRIT. J. APPL. PHYS., 1965, vol. 16, no. 8, pp. 1193-1196.

79. Bell, G.A. "AUSTRAL. J. SCI., 1958, vol. 9, pp. 236-244.

80. Kuhlmann-Wilsdorf, D., Sezaki, K. "J. PHYS. SOC. JAP., Suppl. III, 1963, vol. 18, pp. 54-58.

81. Medvedev, S.A., Kopylovskiy, B.D., Sentyurina, N.N. "Investigation of the temperature dependence of the partial pressures of components over CdTe crystals" In: Tellurid kadmiya [Cadmium telluride] Ed. by B.M. Vul, Moscow: Nauka, 1968, pp. 19-24.

82. Palatnik, L.S., Fuks, M.Ya., Alaverdova, O.G., Shpakovskaya, L.P. "X-ray investigation of the structural perfection of epitaxial PbS films by two-crystal spectrometry" KRISTALLOGRAFIYA, 1977, vol. 22, no. 3, pp. 608-614.

83. Hordon, M.Z., Averbach, B.L. "X-ray measurements of dislocation density in deformed copper and aluminum single crystals" ACTA METALLURGICA, 1961, vol. 9, pp. 237-246.

84. Medvedev, S.A., Kiseleva, K.V., Lykhin, V.A. "The influence of deviations from stoichiometry on select properties of the Nb_3Sn phase" FMM, 1967, vol. 24, no. 6, pp. 1050-1055.

85. Antonova, E.A., Kiseleva, K.V., Medvedev, S.A. "Superconductivity in $NbSe_2$-type layered structures" FMM, 1969, vol. 27, no. 3, pp. 441-445.

86. DeNobel, D. "Phase equilibria and semiconducting properties of cadmium telluride" PHILIPS RES. REP., 1959, vol. 14, pp. 430-492.

87. Brebrick, R.F., Strauss, A.S. "Partial pressures and Gibbs free energy of formation for congruently subliming CdTe" J. PHYS. and CHEM. SOLIDS, 1964, vol. 25, pp. 1441-1145.

88. Medvedev, S.A., Maksimovskiy, S.N., Klevkov, Yu.V., Shapkin, P.V. "The synthesis and crystallization of cadmium telluride with controlled deviation from stoichiometry" In: Tellurid kadmiya [Cadmium telluride], Moscow: Nauka, 1968, pp. 7-12.

89. Medvedev, S.A. "Vvedenie v tekhnologiyu poluprovodnikovykh materialev" [Introduction to the technology of semiconductor materials] Moscow: Vyssh. shk., 1970, 503 pp.

90. Medvedev, S.A., Klevkov, Yu.V., Kiseleva, K.V., Sentyurina, N.N. "Growing CdTe crystals from solution-melts and an investigation of their properties" IZV. AN SSSR. NEORGAN. MATERIALY, 1972, vol. 8, no. 7, pp. 1210-1213.

91. Medvedev, S.A., Maksimovskiy, S.N., Sentyurina, N.N., Ivanov, G.A. "Monitoring deviation from stoichiometry in CdTe" IZV. AN SSSR. NEORGAN. MATERIALY, 1970, vol. 6, no. 6, pp. 1081-1085.

92. Medvedev, S.A., Klevkov, Yu.V. "Growing of CdTe crystals from the

vapor phase by vacuum sublimation" IZV. AN SSSR. NEORGAN. MATERIALY, 1971, vol. 7, pp. 753-756.

93. Plotnikov, A.F., Sal'man, V.M., Sokolova, A.A., Chapnin, V.A. "Light absorption by free carriers in *n*-CdTe crystals" In: Tellurid kadmiya [Cadmium telluride] Moscow, Nauka, 1968, pp. 51-59.

94. Vodop'yanov, L.K., Abramov, A.A. "Electrical properties of CdTe at low temperatures" In: Tellurid kadmiya [Cadmium telluride] Moscow: Nauka, 1968, pp. 122-130.

95. Kroger, F. "Khimiya nesovershennykh kristallov" [The chemistry of imperfect crystals] Moscow: Mir, 1969, pp. 487-492.

96. Bryant, F.T., Cox, A.T.J., Webster, E. "Atomic displacements and the nature of band edge radiative emission in cadmium telluride" J. PHYS., Ser. 2, 1968, vol. 1, pp. 1737-1745.

97. Agrinskaya, N.V. Dissertation for Candidate of Physics and Mathematics, Leningrad, 1971.

98. Sidel'nikov, N.G., Vanyukov, A.V., Ivanov, Yu.M. "The temperature dependence of the partial pressure of Cd in the Cd-Te system" ZhFKh, 1974, vol. 48, no. 8, pp. 2103-2104.

99. Ivanov, Yu.M., Leybov, V.A., Vanyukhov, A.V. "The composition stability range of CdTe" In: Referaty dokl. i soobshch. XII Mendeleevskogo s'ezda po obshch i prikl. khimii" [Abstracts of papers and reports at the Twelfth Mendeleev Conference on General and Applied Chemistry] Moscow: Nauka, no. 1, 1981, p. 50.

100. Kobeleva, S.P., Maksimovskiy, S.N. "The program of dominating point defects in CdTe" IZV. AN SSSR. NEORGAN. MATERIALY, 1968, vol. 22, no. 6, p. 922.

101. Harman, T.C., Melngailis, I. "Narrow gap semiconductors" APPL. SOLID STATE SCI., 1974, vol. 1, pp. 1-94.

102. Preier, H. "Recent advances in lead-chalcogenide diode lasers" J. APPL. PHYS., 1979, vol. 20, pp. 188-206.

103. Vasil'kova, V.V., Kovalev, A.N., Medvedev, S.A. "The influence of repeated sublimation purification on the properties of lead chalcogenide monocrystals" In: Tez. dokl. VI Mezhdunar. konf. po rostu kristallov [Topic papers of the Sixth International Conference on Crystal Growing] Moscow: VINITI, 1980, vol. 1, pp. 216-217.

104. Gorina, Yu.I., Kalyuzhnaya, G.A., Kiseleva, K.V., et al. "Investigation of the crystal characteristics of PbSnTe-based heterostructures fabricated by gas epitaxy" KRATKIE SOOBSHCH PO FIZIKE FIAN, 1976, no. 8, pp. 15-20.

105. Medvedev, S.A., Kovalev, A.N., Perchukova, L.I., Zhuravlev, V.B. "Fabrication of photosensitive lead telluride films by open vaporization" ELEKTRON. TEKHNIKA. MATERIALY, 1976, no. 8, pp. 107-108.

106. Novoselova, A.V., Zlomanov, V.P., Masyakin, E.V., Tananaeva, O.I. "P-T-X phase diagrams and the vapor phase monocrystal growing mechanism for group 4-6 compounds" In: Rost i legirovanie poluprovodnikovykh kristallov i plenok [Growth and doping of semiconductor crystals and films] Novosibirsk: Nauka, 1977, Ch. 1, pp. 30-40.

107. Medvedev, S.A., Ivannikova, G.E. "The features of growing autoepitaxial Si films by radiant heating of substrates" ELEKTRON. TEKHNIKA, Ser. 2, 1970, no. 4, pp. 29-35.

108. Kiseleva, K.V., Ivannikova, G.E., Ershova, L.M. "X-ray diffraction investigations of silicon film structure" ELEKTRON. PROM-ST', 1972, no. 7, pp. 90-92.

109. Vul, B.M., Ivannikova, G.E., Kalyuzhnaya, G.A., et al. Inventor's certificate 447108 (USSR) "Contact fabrication technique" Published in Bulletin of Inventions, 1978, no. 29.

110. Gorina, Yu.I., Kalyuzhnaya, G.A., Kuznetsov, A.V., Maksimovskiy, S.N., et al. Inventor's certificate 466816 (USSR) "Film growing technique" Published in Bulletin of Inventions, 1978, no. 29.

111. Pat. 4115163 US. "Method of growing epitaxial semiconductor films utilizing radiant heating" Ed. J.I. Gorina, G.A. Kaljuzhnaja, A.V. Kuznetsov, et al.

112. Guro, G.M., Kalyuzhnaya, G.A., Mamedov, T.S., Shelepin, L.A. "Activation of narrowband semiconductor growing processes by light irradiation" KRATKIE SOOBSHCH. PO FIZIKE FIAN, 1978, no. 11, pp. 27-32.

113. Guro, G.M., Kalyuzhnaya, G.A., Mamedov, T.S., Shelepin, L.A. "Controlling crystal growth by means of electromagnetic radiation" ZhETF, 1979, vol. 77, pp. 2366-2374.

114. Kalyuzhnaya, G.A., Kozhukhov, V.G., Mamedov, T.S., Sentyurina, N.N. "Investigation of the influence of luminous irradiation on the gas phase in lead telluride epitaxy" KRATKIE SOOBSHCH. PO FIZIKE FIAN, 1982, no. 2, pp. 40-45.

115. Guro, G.M., Kalyuzhnaya, G.A., Mamedov, T.S., Shelepin, L.A. "The influence of radiation on the kinetics of crystal growing" Tr. FIAN, 1980, vol. 124, pp. 127-140.

116. Gorina, Yu.I., Kalyuzhnaya, G.A., Kiseleva, K.V., et al. "Investigation of PbTe(Bi)-PbSnTe-TbTe laser epitaxial structures" FTP, 1979, vol. 13, pp. 305-310.

117. Revokatova, I.P. "Investigation of the dependence of the electrical properties on deviation from stoichiometry in a^2B^6 and A^4B^6 epitaxial films with electromagnetic action on the growing system" Dissertation for Candidates Degree of Technical Sciences, Moscow, 1982, 152 pp.

118. Dvurechenskiy, A.V., Kachurin, G.A., et al. "Impul'snyy otzhig poluprovodnikovykh materialov" [Pulsed annealing of semiconductor materials] Moscow: Nauka, 1982, 208 pp.

119. Wooley, J.C., Berolo, O. "Phase studies of the $Pb_{1-x}Sn_xSe$ alloys MAT. RES. BULL., 1968, vol. 3, no. 5, pp. 445-455.

120. Kroger, F. "Khimiya nesovershennykh kristallov" [The chemistry of imperfect crystals] Moscow: Mir, 1969, 428 pp.

121. Koval'chik, T.L., Maslakovets, Yu.P. "The influence of impurities on the electrical properties of lead telluride" ZhETF, 1956, vol. 26, no. 11, pp. 2117-2131.

122. Nemov, S.A., Kaydanov, V.I., Leonov, A.S., et al. "The influence of double Tl and Na doping on the electrical properties of PbTe" In: Materialy 5-i Vsesoyuz. konf. po fiz.-khim. osnovam legirovaniya poluprovodnikov [Materials of the Fifth All Union Conference on the Physiochemical Principles of Semiconductor Doping] Moscow, 1982, p. 27.

123. Belokon', S.A., Gromovoy, Yu.S., Lakeenkov, V.M., et al. "The current state of MnTeCr transition metal impurities in PbTe monocrystals" In: Materialy 5-i Vsesoyuz. konf. po fiz.-khim. osnovam legirovaniya poluprovodnikov [Papers of the Fifth All Union Conference on the Physiochemical Principles of Semiconductor Doping] Moscow, 1982, p.29.

124. Rustamov,P.G., Alidzhanov, M.A., Abilov, Ch.I. "Investigation of lead telluride doped by group III elements" In: Materialy 5-i Vsesoyuz. konf. po fiz.-khim. osnovam legirovaniya poluprovodnikov [Papers of the Fifth All Union Conference on the Physiochemical Principles of Semiconductor Doping] Moscow, 1982, p.30.

125. Aver'yanov, I.S., Pyregov, B.P., Petrova, O.A., et al. "The properties of doped $Pb_{1-x}Sn_8Se$ crystals" In: Materialy 5-i Vsesoyuz. konf. po fiz.-khim. osnovam legirovaniya poluprovodnikov [Papers of the Fifth All Union Conference on the Physiochemical Principles of Semiconductor Doping] Moscow, 1982, p. 33.

126. Sizova, L.A., Berezhnaya, I.A., Generalova, D.A., et al. "Conductivity inversion in PbTe and $Pb_{1-x}Sn_xTe$ epitaxial films by intro-

ducing Zn ions" In: Materialy 5-i Vsesoyuz. simpoz. "Poluprovodniki s uzkoy zapreshchennoy zonoy" [Papers of the Fifth All Union Conference on the Physiochemical Principles of Semiconductor Doping] L'vov, 1980, p. 75.

127. Kondratenko, M.M., Omanchukovskaya, I.V., Orletskiy, V.B. "Diffusion, solubility and the phase of Cd, Ge, and Symmetrical basis impurities in $Pb_{1-x}Sn_xTe$ solid solution monocrystals" In: Materialy 5-i Vsesoyuz. simpoz. "Poluprovodniki s uzkoy zapreshchennoy zonoy" [Papers of the Fifth All Union Conference on the Physiochemical Principles of Semiconductor Doping] L'vov, 1980, p. 72.

128. Kaidanov, V.I., Mel'nik, R.E., Nemov, S.A. "The electrical properties of thallium-doped lead chalcogenides" In: Materialy I-i Vsesoyuz. nauch.-tekhn. konf. "Poluchenie i svoystva poluprovodnikovykh soedineniy A^2B^6 and A^4B^6" [Materials of the First All-Union Scientific and Technical Conference "The Manufacture and Properties of A^2B^6 and A^4B^6 Semiconductor Compounds"] Moscow, 1977, p. 172.

129. Andramonov, V.S., Ezhova, L.N., Kharionovskiy, Yu.S., et al. "The electrical properties of cadmium-doped $Pb_{1-x}Sn_x$ and $Pb_{1-x}Sn_xSe$ monocrystals" In: Materialy I-i Vsesoyuz. nauch.-tekhn. konf. "Poluchenie i svoystva poluprovodnikovykh soedineniy A^2B^6 and A^4B^6" [Materials of the First All-Union Scientific and Technical Conference "The Manufacture and Properties of A^2B^6 and A^4B^6 Semiconductor Compounds] Moscow, 1977, p. 325.

130. Aleksandrov, O.V., Kalyuzhnaya, G.A., Kiseleva, K.V., Strogankova, N.I. "Investigation of narrowband material based on PbSnTe with a low carrier concentration" In: Materialy I-i Vsesoyuz. nauch.-tekhn. konf. "Poluchenie i svoystva poluprovodnikovykh soedineniy A^2B^6 and A^4B^6" [Materials of the First All-Union Scientific and Technical Conference "The Manufacture and Properties of A^2B^6 and A^4B^6 Semiconductor Compounds"] Moscow, 1977, p. 293.

131. Plyatskoi, S.V., Sizov, F.F., Teperkin, V.V. "The influence of Ga impurities on the $Pb_{1-x}Sn_xTe$ solid solution monocrystals ($0 < x < 0.20$) In: Materialy 6-go Vsesoyuz. simpoz. "Poluprovodniki s uzkoy zapreshchennoy zonoy" [Materials of the Sixth All Union Symposium "Narrow bandgap semiconductors"] L'vov, 1983, pp. 191-192.

132. Takaoka, S., Hamaguchi, T., Shimomura, S., Murase, K. "Investigation of roles of atmospheric impurity in $Pb_{1-x}Sn_xTe$" Proc. 17th Intern. Conf. Phys. Semicond. (ICPS), Springer, 1984, pp. 663-666.

133. Dubkin, A.D., Erasova, N.A., Kaydanov, V.I., et al. "The in-

fluence of indium doping on the electrical properties of tin telluride" FTP, 1972, vol. 6, no. 8, pp. 2294-2297.

134. Strauss, A.J. "Effect of Pb and Te-saturation on carrier concentrations in impurity-doped PbTe" J. ELECTRON. MATER., vol. 2, no. 4, pp. 553-570.

135. Bushmarina, G.S., Gruzinov, B.F., Lev. E.Ya., et al. "The features of impurity dissolution in $Ge_{1-x}Te_x$ and $Sn_{1-x}Te$ compounds of variable composition" In: Rost I legirovanie poluprovodnikovykh kristallov I plenok. [Growth and doping of semiconductor crystals and films] Novosibirsk: Nauka, Sib. otd-nie, 1977, Ch. 1, pp. 286-290.

136. Aleksandrov, O.V., Gorina, Yu.I., Kalyuzhnaya, G.A., et al. "The influence of silver doping on the electrical properties of lead telluride" KRATKIE SOOBSHCH PO FIZIKE FIAN, 1981, vol. 20, pp. 1112-1116.

137. Gorina, Yu.I., Zaynudinov, S., Kalyuzhnaya, G.A., Kiseleva, K.V. "Gas phase growing and doping of lead-tin telluride" IZV. AN SSSR. NEORGAN. MATERIALY, 1984, vol. 20, pp. 1112-1116.

138. Aleksandrov, O.V., Kalyuzhnaya, G.A., Kiseleva, K.V., Strogankova, N.I. "Investigation of indium-doped $Pb_{0.78}Sn_{0.22}Te$ solid solution with a low carrier concentration" IZV. AN SSSR. NEORGAN. MATERIALY, 1978, vol. 14, no. 7, pp. 1277-1279.

139. Aleksandrov, O.V., "Structural instability of $Pb_{0.78}Sn_{0.22}Te$ solid solutions" In: Dissertation for Candidate of Physics and Mathematics, Moscow, 1985.

140. Zaslavskiy, A.I., Sergeeva, V.M. "Polymorphism of In_2Te_3" FTT, 1960, vol. 2, no. 11, pp. 2872-2880.

141. Valatska, K., Lideykis, T., Shil'kene, I., et al. "Liquid-phase epitaxy of indium-doped lead-tin telluride" IZV. AN SSSR. NEORGAN. MATERIALY., 1983. vol. 19, no. 1, pp. 51-44.

142. Akimov, B.A., Brandt, N.B., Zhukov, A.A., Ryabova, L.I., Khotlov, D.R. "The features of the band structure of $Pb_{1-x}Sn_xTe$ <In> alloys with a high indium concentration" FTP, 1981, vol. 15, no. 11, pp. 2232-2234.

143. Averkin, A.A., Kaydanov, V.I., Mel'nik, R.B., et al. "The nature of impurity indium states in lead telluride" FTP, 1975, vol. 9, no. 10, pp. 1873-1878.

144. Drabkin, I.A., Moyzhes V.Ya. "Spontaneous dissociation of neutral impurity states into positively and negatively charged states" FTP, 1981, vol. 15, no. 4, pp. 625-648.

145. Andreev, Yu.V., Geyman, K.I., Drabkin, I.A., et al. "The electrical properties of indium-doped $Pb_{1-x}Sn_xTe$" FTP, 1975, no. 9, no. 10, pp. 1873-1878.

146. Drabkin, I.A., Kvantov, M.A., Kompaniets, V.V., Kostikov, Yu.P. "Charged In states in PbTe" FTP, 1982, vol. 16, no. 7, pp. 1276-1277.

147. Dixon, J.R., Riedl, H.R. "Electric susceptibility hole mass of lead telluride" PHYS. REV. A, 1965, vol. 138, no. 2, pp. 873-879.

148. Burkhard, H., Bauer, G., Lopez-Otero, A. "Far-infrared reflectivity of PbTe films on NaCl substrates" SOLID. STATE COMMUNS, 1976, vol. 18, no. 6, pp. 773-778.

149. Parada, N.J., Pratt, G.W. "New model for vacancy states in PbTe" PHYS. REV. LETT., 1969, vol. 22, no. 5, pp. 180-182.

150. Parada, N.J. "Localized defects in PbTe" PHYS. REV. B., 1971, vol. 3, no. 6, pp. 2042-2055.

151. Brus, A., Kauli, R. "Strukturnye fazovye perekhody" [Structural phase transitions] Moscow: Mir, 1984, pp. 394-395.

152. Levanyuk, A.P., Osipov, V.V., Sigov, A.S., Sobyanin, A.A. "Changes in defect structure and related anomalies in substances near phase transition points" ZhETF, 1979, vol. 76, no. 1, pp. 345-368.

153. Herrmann, K.H., Kalyuzhnaya, G.A., Möllmann, K.P., Wendt, M. "Photoelectric behavior in $Pb_{0.78}Sn_{0.22}Te$, In: Phys. status solidi (a), 1982, vol. 71, pp. K21-K24.

154. Gufan, Yu.M., Larin, E.S. "A phenomenological examination of isostructural phase transitions" DAN SSSR, 1978, vol. 242, no. 6, pp. 1311-1313.

155. Tonkov, E.Yu., Aptekar' I.L. "The critical point on the isomorphic transition curve of SnS" FTT, 1974, vol. 16, no. 5, pp. 1507-1508.

156. Jayaraman, A., Bucker, E., Dernier, P.D., Longinotti, L.D. "Temperature-induced explosive first-order electronic phase transition in Gd-doped SmS, PHYS. REV. LETT., 1973, vol. 31, no. 11, pp. 700-703.

157. Vereshchagin, L.F., Kabalkina, S.S. "Rentgenostrukturnye issledovaniya pri vysokom davlenii" [High-pressure X-ray structural investigations] Moscow: Nauka, 1979, pp. 113-124.

158. Melngailis, I. "Photovoltaic effect in $Pb_{1-x}Sn_xTe$ diodes" APPL. PHYS. LETT., 1966, vol. 9, no. 8, pp. 304-306.

159. Vul, B.M., Grushechkina, S.P., Ragimova, T.Sh., Shotov, A.P. "Photoelectric phenomena in $Pb_{0.78}Sn_{0.22}Te$" In: Transition radiation. Vsesoyuz. konf. po fizike poluprovodñikov [Papers of the All-Union Conference on Semiconductor Physics] Baku, 1982, vol. 2, pp. 126-127.

160. Karlson, T.A. "Fotoelektronnaya i Ozhe-spektroskopiya" [Photoelectron and Auger spectroscopy] Leningrad: Mashinostroenie, 1981, 431 pp.

161. Wagner, C.D, Biloen, P. "X-ray excited auger and photoelectron spectra of partially oxidized magnesium surfaces: the observation of anormal chemical shifts" SURF. SCI., 1973, vol. 35, pp. 82-95.

162. Wilmsen, C.W., Kee, R.W. "Auger analysis of the anodic oxide/InP interface" J. VAC. SCI. AND TECHNOL., 1977, vol. 14, no. 4, pp. 953-956.

163. Grishina, T.A., Drabkin, I.A., Kostikov, Yu.N., et al. "An Auger spectroscopic investigation of processes at the metal/semiconductor boundary in the In-$Pb_{1-x}Sn_xTe$ system" IZV. AN SSSR. NEORGAN. MATERIALY., 1982, vol. 18, no. 10, pp. 1709-1713.

164. Smorodina, T.A., Sheftal' A.N., "Introduction of impurities in lead chalcogenide" 1983, vol. 19, no. 1, pp. 36-40.

165. Muitenberg, G.E. "Handbook of X-ray photoelectron spectroscopy" Minnesota, 1979.

166. Aleksandrov, O.V., Kiseleva, K.V., Kuchaev, S.V. "The In-Te chemical bond in the $Pb_{0.78}Sn_{0.22}Te$ (In) lattice" KRATKIE SOOBSHCH PO FIZIKE FIAN, 1985, no. 1, pp. 60-63.

167. Gin'e, A. "Neodnorodnye metallicheskie tverdye rastvory" [Inhomogeneous metallic solid solutions] Moscow: Izd-vo inostr. lit., 1962, pp. 51-55.

168. Panin, V.E., Khoy, Yu.A., Naumov, I.I. "Teoriya faz v splavakh" [Phase theory in alloys] Novosibirsk: Nauka, 1984, pp. 105-117.

169. Gorelik, S.S., Dashevskiy, M.Ya. "Materialovedenie poluprovodnikov i metallovedenie" [Semiconductor materials processing and metal processing] Moscow: Metallurgiya, 1973, p. 259.

170. Fistul', V.I., "The interatomic interaction of impurities in heavily-doped semiconductors" In: Khimicheskaya svyaz' v kristallakh [Chemical bonds in crystals] Minsk: Nauka i tekhnika, 1977, p. 447.

171. Aleksandrov, O.V., Kiseleva, K.V. "The Fermi level stabilization mechanism in $Pb_{0.78}Sn_{0.22}Te(In)$" KRATKIE SOOBSHCH PO FIZIKE FIAN, 1985, no. 1, pp. 64-67.

172. Zhuze, V.P., Sergeeva, V.M., Shelykh, A.I. "The electrical properties of In_2Te_3: a semiconductor with a defect structure" FTT, 1960, vol. 2, no. 11, pp. 2858-2871.

173. Ormont, B.F., "Vvedenie v fizicheskuyu khimiyu i kristallokhimiyu poluprovodnikov" [Introduction to physical chemistry and the crystal chemistry of semiconductors] Moscow: Vyssh. shk., 1982, 528 pp.

174. Akimov, B.A., Elizarov, A.I., Kurmanov, K.R., et al. "The features of the electrical properties of combination-doped $Pb_{1-x}Sn_xTe$ alloys" FTP, 1982, vol. 17, no. 6, pp. 1003-1008.

175. Vul, B.M., Voronova, I.D., Kalyuzhnaya, G.A., et al. "The features of transport phenomena in $Pb_{0.78}Sn_{0.22}Te$ with a high indium concentration" PIS'MA V ZhETF, 1979, vol. 29, no. 1.

176. Vul, B.M., Voronova, I.D., Grushechkina, S.P., Ragimova, T.Sh. "Electron accumulation and relaxation time with photoresponse in $Pb_{0.78}Sn_{0.22}Te$" PIS'MA V ZhETF, 1981, vol. 33, no. 6, pp. 346-350.

177. Akimov, V.A., Brandt, N.B., Klimonskiy, S.O., et al. "Dynamics in the semiconductor-metal transition induced by infrared illumination in $PB_{1-x}Sn_xTe(In)$ alloys" PHYS. LETT. A., 1982, vol. 88, no. 9, pp. 483-486.

178. Vinogradov, V.S., Voronova, I.D., Ragimova, T.Sh., Shotov, A.P. "Relaxation of photoconductivity in $Pb_{1-x}Sn_xTe$: Experiment and theory" Preprint FIAN, no. 218, Moscow, 1983.

179. Volkov, B.A., Pankratov, O.A. "The Yahn-Teller instability of the crystal configuration of point defects in A^4B^6 semiconductors" DAN SSSR, 1980, vol. 255, no. 1, pp. 93-97.

180. Volkov, B.A., Voronova, I.D., Shotov, A.P. "Logarithmic relaxation of long-term photoconductivity and negative photoconductivity in $Pb_{0.78}Sn_{0.22}Te$" DAN SSSR (this volume).

181. Gorina, Yu.I., Pashunin, Yu.M., Pruchsenkov, S.I. "The influence of a gold impurity on the electrical and optical properties of lead telluride" In: Tez. dikl. IV Vsesoyuz. konf. "Troynye poluprovodniki i ikh primenenie" [Topic papers of the Fourth All Union Conference "Ternary Semiconductors and Their Application] Kishinev, 1983, p. 250.

182. Fano, V., Mignoni, G., Pergolari, B. "Correlation of electrical

measurements with chemical analysis in silver-doped lead telluride" MATER. CHEM. AND PHYS., 1979, vol. 4, pp. 507-513.

183. Borisova, L.D., Dimitrova, S.K. "Thermoelectric properties of silver doped PbTe" PHYS. STATUS SOLIDI, 1980, vol. 61, no. 1, pp. K25-K29.

184. Gorina, Yu.I., Zaynudinov, S., Kalyuzhnaya, G.A., et al. "The influence of silver on the electrical properties and photoluminescence of lead telluride" IZV. AN SSSR. NEORGAN. MATERIALY., 1986, no. 7, pp. 1105-1108.

185. Glazov, V.M., Zemskov, V.S. "Fiziko-khimicheskie osnovy legirovaniya poluprovodnikov" [The physiochemical principles of semiconductor doping] Moscow: Nauka, 1967, 213 pp.

186. Salawa, A.R. "Small bandgap lasers and their uses in spectroscopy" J. LUMINESCENCE, 1973, vol. 7, pp. 477-500.

187. Colles, M.J., Pidgeon, C.R. "Tunable lasers" REP. PROG. PHYS., 1975, vol. 38, no. 3, pp. 331-452.

188. Cuff, K.F., Ellett, M.R., Kuglin, C.D., Williams, L.R. "The band structure of PbTe, PbSe, and PbS" Proc. VIIth Intern. Conf. Phys. Semicond., Paris, 1964, pp. 677-684.

189. Butler, J.F., Calawa, A.R., Phelan, R.J., et al. "PbTe diode laser" APPL. PHYS. LETT., 1964, vol. 5, no. 4, pp. 75-77.

190. Dimmok, J.O., Melngailis, I., Strauss, A.J. "Band structure and laser action in $Pb_xSn_{1-x}Te$" PHYS. REV. LETT., 1966, vol. 16, pp. 1193-1196.

191. Tomasetta, L.R., Fonstad, C.G. "Liquid epitaxial growth of laser heterostructures in $PB_{1-x}Sn_xTe$" APPL. PHYS. LETT., 1974, vol. 24, pp. 567-570.

192. Kurbatov, L.N., Brutov, A.D., Kalyuzhnaya, G.A., et al. "A PbSnTe-based heterolaser operating at 10 μm" KVANTOVAYA ELEKTRON., 1975, vol. 2, no. 9, pp. 2084-2086.

193. Britov, A.D., Karavaev, S.M., Kalyuzhnaya, G.A., et al. "Laser diodes based on PbSnTe in the 5-15 μm range" KVANTOVAYA ELEKTRON., 1976, vol. 3, no. 10, pp. 2238-2242.

194. Pankov, Zh. "Opticheskie protsessy v poluprovodnikakh" [Optical processes in semiconductors] Moscow: Mir, 1973, 457 pp.

195. Kressel, H., Hawrylo, F.Z., Abrahams, M.S., Buiocchi, C.J. "Observations concerning radiative efficiency and deep-level luminescence in *n*-type GaAs prepared by liquid phase epitaxy" J. APPL. PHYS., 1968, vol. 39, no. 11, pp. 5139-5144.

196. Girich, B.G., Gureev, D.M., Zasavitskiy, I.I., Shotov, A.P. "Laser heterostructures based on a $Pb_{1-x}Sn_xTe$ solid solution" FTP, 1978, V, 12, no. 1, pp. 124-128.

197. Averyushkin, A.S., Aleksandrov, O.V., Kiseleva, K.V., et al. "Investigation of the X-ray structural and luminescence properties of expitaxial lead selenide films" IZV. AN SSSR. NEORGAN. MATERIALY., 1979, vol. 15, no. 3, pp. 380-385.

198. Handsen, E.E., Miner, Z.A., Mitchell, M.J. "Sublimation pressure and sublimation coefficient of single-crystal lead selenide" J. AMER. CERAM. SOC., 1969, vol. 52, pp. 610-612.

199. Ten, A.S., Yunovich, A.E. "The temperature dependence of the photoluminescence spectra of lead selenide monocrystals" FTT, 1976, vol. 10, no. 5, pp. 866-870.

200. Pat. 4115163 US. "Method of growing epitaxial semiconductor films utilizing radiant heating" Ed. J.I. Gorina, G.A. Kaljuzhnaja, A.V. Kuznetsov, et al.

EXTERNAL STIMULATION MECHANISMS FOR CRYSTAL GROWING PROCESSES

G.M. Guro, G.A. Kalyuzhnaya, F.Kh. Mirzoev, L.A. Shelepin

ABSTRACT

The dependencies of epitaxial crystal growth on external factors are investigated. The mechanisms for influencing the gas phase, the growing surface and the surface layers are discussed. Equations of growing kinetics and cluster formation are examined including the case of precipitate formation with deviation from stoichiometry. The possibilities for applying external influences in growing planar and nonplanar structures and for controlling deviation from stoichiometry are discussed.

Introduction

Crystal growing processes and properties are dependent on external conditions. Electromagnetic radiation, acoustic waves, charged particle beams, electrical fields, and other factors may have a significant influence on these processes. They can accelerate and decelerate growth, stimulate monocrystal, polycrystal, or amorphous structures, implement various degrees of doping or crystal defect levels, influence the stoichiometry, i.e., it is possible to control the growth processes by altering the external factors. The most precise instrument for influencing the properties of a solid is the laser. Specific power, frequency, and pulse duration ranges are required for effective laser application in each operation. Much that has been done in this area so far has been empirical and much remains unclear. Therefore the problem of establishing and analyzing specific external stimulation mechanisms for influencing crystal structure and growing processes acquires special significance. Due to the complex nature of crystal growing it is possible to identify a complete set of mechanisms for influencing various aspects and stages of the process as well as the material properties. Experiments have revealed that proper dosage and selectivity are required for efficient action, i.e., specific intensities are required for stimulating each sample together with a specific spectral range, pulse duration and power.

Overall the problems of developing a theoretical description of laser and other stimulation mechanisms for influencing growing processes as well as material structure are still in the developmental stages. Far from all of the mechanisms are known at present and only a few capabilities of laser technology have been identified. Moreover it is necessary even now to examine available data from a single viewpoint. This is important for technological techniques employed to obtain given structures and given crystal inhomogeneities or in the broader sense to control crystal structure.

The purpose of the present study is also to consider a number of aspects of such a unified approach. The study is largely based on experimental research given in the first chapter of the present volume [1].

The first section considers a collection of specific external stimulation mechanisms including surface, gas-phase, and surface layer actions (where the actual crystal structural formation process takes place) together with the defects and stresses in the crystal. The second section considers kinetic models of the growing processes of binary crystals and cluster formation. In the third section the problems of the practical utilization of the electromagnetic radiation stimulation mechanisms for obtaining given plane structures, for fabricating nonplane structures, and investigating the capabilities of controlling deviation from stoichiometry in growing binary crystals are discussed.

1. External Stimulation Mechanisms

The gas phase. Stimulating the gas phase near the growing surface may have a significant influence on both the ongoing epitaxial growth processes and on the quality of the derived materials. The primary factors here include the formation of free atoms and radicals as well as particles in excited metastable states. Experimental studies of the influence of electromagnetic radiation on gas phase growing of epitaxial films has been ongoing for 15 years [2-5]. Investigations of PbTe, $Pb_{1-x}Sn_xTe$, HgTe crystal growth have been particularly comprehensive [6-8]; a 10 kw high-pressure xenon lamp was used as the source in this case, and this source emits in the 0.2-1.5 μm range with 10% of the power falling in the UV. It was demonstrated that the rate of growth can be significantly increased by radiation and that UV radiation plays the primary role here. Under UV radiation action (filter-selected) an increase in the growth rate of two orders of magnitude was observed (from 0.1-0.2 μm/min without irradiation to 6-15 μm/min with radiation).

The irradiation process also influences epitaxial film quality. Monocrystal layers were obtained in a narrow UV intensity range from 10 to 30 w/cm^2; polycrystal layers were grown at somewhat greater or lesser intensities. At intensities above 40 w/cm^2 polycrystal growing is replaced by flaking, i.e., an uncontrolled rapid growth of crystals in the bulk near the surface in the vicinity of chaotic crystallization centers. A drop in the epitaxy temperature from 800°C to 500°C was also observed in the radiation conditions. A detailed analysis of the experiments is given in Chapter 1 of the present volume.

The primary mechanism responsible for the observed effects is free atom formation [7]. UV irradiation causes dissociation of Te_2 molecules (E_{diss} = 2.8 eV), and PbTe (E_{diss} = 2.7 eV) which then travel to the growing surface. In normal conditions without irradiation the molecular bonds are broken on the surface (necessary for the atom to fit in the lattice sites). The probability of such bond breaking increases with growth of the substrate temperature. In order to introduce a single atom into the lattice 10^3-10^4 molecular collisions with the surface of the growing crystal are required. Unlike intermolecular reactions, reactions involving free atoms and radicals are nonbarrier reactions; the cross-sections are of the order of gas kinetic cross-sections. This is responsible for the increase in the growth rate. The percentage of free atoms already has a significant influence on the epitaxy rate. The reduction in epitaxial temperature from irradiation is related to the fact that in the presence of free atoms the substrate temperature drops and this means that the probabilities of bond breaking in molecules interacting with the surface do not play a significant role.

Results from study [7] have revealed the possibility for significant improvements in crystal structure from UV irradiation of the gas phase near the surface and the formation of free atoms. It is, however, important to emphasize that such stimulation has its own optimum level. At the optimum point the quantity of dissociated molecules forming free atoms accounts for fractions of a percent of the total number of Te_2 molecules. At high free atom concentrations when the growth rate rises multicenter growing is initiated on the surface. A polycrystal is then formed rather than a monocrystal. When the cluster growth rate from free atoms is comparable to the epitaxy rate, chaotic uncontrolled crystal growing ensues in the gas medium near the surface (flaking).

In sum the characteristic features of the gas phase stimulation mechanism include UV irradiation for forming free atoms and a carefully proportioned level of such radiation necessary to obtain enhanced materials at a significantly more rapid growth rate.

Surface. The gas-phase epitaxy process is highly dependent on the surface state. In normal conditions the epitaxy process follows the film growth mechanism [9-12] and is determined by the surface jogs (i.e., boundaries with different levels equal to the intermolecular spacing). Surface crystal growing can be divided into the following stages: transition of the molecules from the vapor phase to an adsorption surface layer; diffusion of the adsorbed molecules along the surface up to the jog; diffusion along the jog boundary to its fissure and embedding in the lattice. The fissure travels gradually along the jog while the jogs progress towards the crystal edge and vanish, forming a finished surface with a specific number of adsorbed molecules and vacancies. Further layer growth is possible when new jogs are formed. For an ideal crystal this occurs when the fluctuation mechanism produces surface (two-dimensional) clusters. However small clusters are unstable due to surface tension forces and will reach critical values only with significant supersaturation. The crystal growing process at low supersaturation is determined entirely by the defect structure of the actual crystal (primarily the existence of screw dislocations). The theory of film growth initially proposed by Barton, Cabrera, and Frank [12] has been fully developed and confirmed experimentally.

External stimulation can have an influence on all stages of the crystal film growing process. Surface charge formation, changes in adhesive properties and atomic excitation are among the factors requiring special attention in the case of UV surface irradiation.

The surface charge resulting from irradiation [13] has a significant influence on the growing processes. Polarization forces act on the molecules adjacent to the charged surface at distances exceeding the reach of the chemical forces. This can cause a redistribution of particles arriving at the surface if there is a nonuniform charge distribution. Significant fields arise at surface fissures, pits, corners, and jogs. Both the gas flow to the jogs and the fissure sites and the surface diffusion rate increase, i.e,. the growing rate increases.

Electromagnetic radiation influences the adsorption properties of the surface. The adsorption properties for a number of semiconductors are determined by the donor-acceptor mechanism [14]. The slow states of these complexes are highly dependent on the ligand configuration of the adsorption center [15]. The control capabilities associated with UV irradiation derive from this.

In addition to the activation and deactivation of existing molecular adsorption centers on the surface, electromagnetic radiation can stimulate their formation by transforming the surface atoms into the excited metastable states with altered valence properties. The electromagnetic radiation can then stimulate fabrication of films on substrates which cannot be used in regular conditions. Study [16] was the first to employ UV radiation for this application.

Therefore irradiation causes an increase in fissure and jog velocity across the surface and improves adsorbability. Accounting for the UV stimulation mechanisms outlined above for application to a defect-free surface it turns out that two-dimensional surface nuclei can form at significantly lower supersaturation than in normal conditions, i.e., under laser action (unlike regular conditions where the defect structure determines the growth of surface clusters and therefore crystal growth) it is possible to examine the growth process independently and to break its interaction with the defect structure. In order to obtain reliable quantitative aspects of these conclusions it is necessary to carry out further experimental investigations.

Crystal surface layers. The actual crystal structure and its defect structures take form in the surface zone during the growth process. The surface layers whose structure differs from the internal layers (due specifically to stresses) establish the configuration of the added material and its defects. In order to influence the properties of the structure the external stimulation must penetrate into the growing crystal. This is possible by pulsed stimulation of the epitaxial film growing process. The significant influence of pulsed radiation on sample structure during the growth process has been established experimentally. Effects have been observed that do not arise from CW electromagnetic irradiation. This was first established in study [17] which was the first study to carry out a detailed experimental investigation of the influence of pulsed UV lasers on the epitaxial growing of PbTe and $Pb_{1-x}Sn_xTe$ crystals. The experiment employed a neodymium laser (λ = 0.35 μm) and a nitrogen laser (λ = 0.337 μm). The pulse duration was $\sim 10^{-8}$ sec with a pulse power of 10^4-10^5 w/cm^2 with a spot diameter of 3 mm. The growing film surface was irradiated by a xenon lamp in conjunction with the laser radiation. The surface was mirror smooth in the laser spot region without significant growth traces and accelerated growth was observed in this area together with an increase in the lattice parameter. Hillocks were observed at the sites exposed to laser irradiation. Overall the results from study [17] revealed that pulsed UV laser action is not equivalent to xenon lamp UV irradiation and does not produce free atoms.

In addition to the formation of free atoms the pulsed UV laser mechanism influencing the epitaxial growth process can be attributed to two additional factors. The first factor is the excitation of an acoustic field in the condensed medium as a result of pulsed laser radiation. The light absorption region is the acoustical wave source. A pressure pulse $p(t)$ arises due to a change in its density. This process can occur with a fixed aggregate state of the material (the thermooptic effect) or can be accompanied by phase transformations that have a significant influence on the pressure level.

Laser sources make it possible to control the parameters of the generated acoustical field. A comprehensive analysis of the laser excitation of a high-power acoustical field can be found in study [18]. The pressure resulting from the thermooptic effect is given by the relation [19]

$$p(t) = (\beta\varkappa/C_p)\dot{T}(t) + (\beta/C_P\alpha_p)I(t),$$

where β is the coefficient of thermal expansion, $\varkappa$ is the thermal conductivity coefficient; α_p is the absorption coefficient, C_p is the specific thermal capacity; T is temperature; I is radiation intensity. Estimates within an order of magnitude for standard values of the test substances ($\beta \sim 10^{-4}$ deg^{-1}, $\varkappa \sim 0.3$ cal/cm·s·deg., $C_p = 0.1$ cal/g·deg., $\alpha_p \sim 10^5$ cm^{-1}) and the laser radiation parameters given above yield a value of $p \sim 10^7$ dynes/cm^2. The second factor is that sufficient lattice atom mobility in the surface layer of the sample is needed for efficient pulsed action. In the experiment from study [17] this was achieved by simultaneous irradiation of the surface by a pulsed UV laser and UV radiation from a xenon lamp. Under xenon-lamp irradiation (as used in study [17]) the material in the surface layer, as indicated by calculation, is near the melting point. Additional local energy release at the defect sites results from pulsed laser action, forming states with a high degree of plasticity.

The action of both these factors (the existence of pressure and sufficient mobility of the lattice atoms) serves to make the medium more dense and reduce the total number of vacancies, thereby increasing the lattice parameter [17]. This process will also be accompanied by a significant reduction in the number of structural defects (dislocations, microvoids, and microcracks).

Study [20] carried out an experimental comparison of a pulsed and CW helium-cadmium laser. An experimental assembly analogous to that of study [17] jointly employed a DKSR-3000 xenon lamp ($\lambda \sim 0.25$-1.4 μm) and an LPM-11 CW laser ($\lambda \sim 0.43$ μm). Laser power was 100 mw. $BaF_5(111)$ freshly cleaved before the epitaxy process was used as the substrate. PbTe and $Pb_{0.78}Sn_{0.22}Te$ films containing an indium impurity were grown in this case. The temperatures of the substrate and the initial material, measured directly during the experiment, were held constant. The temperature of the initial material was 550°C while the substrate temperature was 500°C in all tests. An increase in the growth rate was observed with irradiation. When the lamp alone was employed the growth rate was 2.5 μm/min and the joint lamp/laser action yielded a growth rate of 3.5μm/min. The epitaxial layer structure at the laser beam spot was somewhat more consistent, while the surface was smoother than the remaining growth region; hillocks were also obtained. However these effects were significantly less evident than for the case of pulsed laser action. Their existence can be attributed to radiation modulation from nonstationary processes although here, of course, further research is necessary. In order to test and ascertain the role of these two factors it would be best to directly measure pressure using the appropriate sensors and to obtain detailed dependencies of the observed effects on the lamp and laser intensities as well as analyze the possible role of nonstationarity in the phase transition, sample inhomogeneities and processes in the gas phase.

Overall the experimentally observed differentiating features of pulsed UV laser action include high surface quality, increased lattice constant (characterizing the reduction in the number of vacancies and other defects), a different defect concentration in the vicinity of the laser spot and outside this spot; different doping under laser action and in the absence of laser action and the appearance of non-planar formations: hillocks.

Crystal defects and stresses. A stimulus delivered to the crystal bulk can alter its structure. Laser radiation and acoustic waves are of special interest in this respect. The degree and nature of such action depend not only on the radiation parameters but also on the crystal state determined by the defect structure and interaction (including dislocations, voids, microcracks, intergrain boundaries, interstitial atom vacancies and substitutional and interstitial atoms) as well as stresses. The mechanisms here are related to two factors: the existence of a nonuniform structure and nonuniform energy release in the crystal.

Significant stresses arise in materials fabricated in severely nonequilibrium conditions; their structure is unstable and has its own characteristic relaxation times. In study [21] which identified changes over time in a number of characteristics of films obtained in nonequilibrium conditions (resistance, X-ray spectra, surface structure) the relaxation times τ ran from seconds to months. The quantities τ were correlated with the local stresses σ [22, 23]:

$$\tau = \tau_0 \exp [(U_0 - \omega\sigma)/kT].$$

Here τ_0 is the characteristic time ($\tau_0 \sim 10^{-13}$ s), U_0 is the interatomic bonding energy, ω is the activation volume ($\omega \sim (10\text{-}10^2)d^3$, d is the lattice parameter). Estimates of the local stresses obtained from temporal relations revealed that they can reach the elastic limit. With significant stresses even comparatively weak stimulation may prove to be very efficient, causing structural reconfiguration by exposing its instabilities.

The other mechanism producing changes in the structure is the nonuniform energy delivery from the propagation of acoustic waves in the crystal or from the pulsed laser-induced shock wave due to the thermooptic effect. Preferential energy release occurs at the lattice tensile stress sites where the material is less dense [24]. This is characteristic of the lattice structure in the vicinity of many defects: voids, microcracks and dislocations. Lattice melting occurs at these sites earlier than in the rest of the crystal volume. Due to this preferential energy release at high irradiation intensities the defects may break down and the total defect density of the crystal can drop. At certain intensities a redistribution effect is observed with respect to the point defects that cluster around the tensile stress points serving as drains for the impurities. A detailed survey of studies on laser stimulation of material structure (laser annealing) can be found in studies [25, 26].

2. Modeling of Binary Crystal Growth Kinetics

In order to develop methods of controlling the composition, structure, and properties of materials and for programmed film growing it is necessary to obtain quantitative results on the relation between the parameters of deposition conditions and the properties of synthesized films. Clustering and self-organization (ordering) models underlie the quantitative analysis of crystal growth kinetics (accounting for their defect structure); in the present study these models are examined based on void formation and growth from a supersaturated solution [27] and void lattice formation [28]. The equations for crystal growth are a modification of the cluster growth equations and can be written accounting for the factors influencing the structure, composition, and properties of the films: substrate temperature, degree of supersaturation, rate of growth of the particle flux density and the gas phase composition.

We will examine binary crystal growth for the case of UV irradiation of the gas phase where free atoms are formed. This growth process is barrier-free, and the free atoms attach and separate on the crystal surface, migrate along the surface and then are embedded in a jog or fissure. The growth kinetics are described by the equations

$$\frac{dN_\alpha}{dt} = K_\alpha^+ N_\alpha - K_\alpha^- N_\alpha, \tag{2.1}$$

where N_α (α = A, B) are the concentrations of the A, B components in the solid phase.

The first term in the right half of equation (2.1) corresponds to crystal growing from the attachment of components A and B, while the second term corresponds to their separation from the crystal surface. The kinetic coefficients $K_\alpha^\pm$ α depend on the interaction potentials U_{AA}, U_{BB}, U_{AB} and the concentrations C_A and C_B of components A and B in the gas. According to study [26] $K_\alpha^\pm$ can be represented as:

$$K_\alpha^+ = \frac{J_\alpha \omega}{\alpha}, \qquad K_\alpha^- = \frac{J_\alpha^{\text{th}} \omega}{d} \exp\left[\frac{2U_{AB} - E(l)}{kT}\right]. \tag{2.2}$$

The particle fluxes J_α and J_α^{th} of α-type particles to and from the surface (per unit of area and unit of time) are equal to

$$J_\alpha = \frac{\nu_\alpha C_\alpha d}{4\omega}, \qquad J_\alpha^{\text{th}} = \frac{\nu_\alpha C_\alpha^{\text{th}} d}{4\omega} \exp(\Delta\mu_\alpha / kT).$$

Here ω is the volume of the elementary cell, d is the lattice period, ω is the frequency of the α-type particle, $\Delta\mu_\alpha$ is the change in chemical potential of the α component. The quantity $E(l)$ in (2.2) characterizes the energy of the α-type particle on the surface of the solid phase, and in the nearest neighbor approximation is equal to

$$E_\alpha = \sum_i l_i U_{\alpha_i},$$

where l_i is the number of nearest neighbors. The equations for the concentrations C_A and C_B in a quasi-stationary approximation take the form

$$K = \mu_R (D_A + D_B) C_A C_B + D_A \frac{d^2 C_A}{dx^2},$$

$$K = \mu_R (D_A + D_B) C_A B_B + D_B \frac{d^2 C_B}{dx^2}.$$

Here K is the rate of formation of the free atoms; μ_R is the recombination coefficient; $D_\alpha = d^2\nu_\alpha \exp(-Q_\alpha/kT)$ is the coefficient of diffusion of the α component; Q_α is the activation energy of diffusion; ν_α is the jump frequency. These relations can be used to model the film growth of a binary crystal in irradiation conditions. However with no jogs or fissures on the surface it is necessary to analyze the generation and growth of planar (two-dimensional) nuclei (clusters) on the crystal surface.

It is also necessary to analyze binary cluster growth in order to examine film condensation from a gas phase of significantly non-stoichiometry composition. As revealed by experiments [29-31] in this case the appropriate precipitates will form as a result of precipitation of the excess component. Moreover, a significant change in the condensation rate and film morphology have been observed. The kinetics of single-component cluster formation and growth have been examined in detail in study [27]. Below we will consider binary clusters consisting of components A and B. The state of the new phase nuclei formed by time t will be characterized by the distribution function $f(n_A, n_B, t)$, where n_A and n_B is the number of A and B atoms in the cluster. We can write the following kinetic equation for this function:

$$\begin{aligned} \partial f(n_A, n_B, t)/\partial t &= f(n_A + 1, n_B) K_A^- (n_A + 1, n_B) + \\ &+ f(n_A, n_B + 1) K_B^- (n_A, n_B + 1) - \\ &- f(n_A, n_B) [K_A^+ (n_A, n_B) + K_B^+ (n_A, n_B) + K_B^- (n_A, n_B) + K_A^- (n_A, n_B)] + \\ &+ f(n_A - 1, n_B) K_A^+ (n_A - 1, n_B) + f(n_A, n_B - 1) K_B^+ (n_A, n_B - 1). \end{aligned} \tag{2.3}$$

Here $K_\alpha^\pm$ ($\alpha = A, B$) are the attachment and separation probabilities of the α components.

If we consider precipitate formation, i.e., if we model the nuclei as spherical formations the kinetic coefficients take the form

$$\begin{aligned} &K_\alpha^+ = \Delta_\alpha z_\alpha n^{2/3}, \quad K_\alpha^- = n_\alpha z_\alpha n^{-1/3} \exp[\varepsilon/(n+1)^{1/3}], \\ &z_\alpha = p_\alpha^{\text{th}} (2\pi m_\alpha kT)^{-1/2} (4\pi)^{1/3} (3\omega)^{2/3}, \\ &\varepsilon = {}^2/_3 (4\pi)^{1/3} (3\omega)^{2/3} \gamma/kT, \quad n = n_A + n_B, \end{aligned} \tag{2.4}$$

where $\Delta_\alpha = p_\alpha / p_\alpha^{\text{th}}$ is supersaturation of the α components in the gas; p_α and p_α^{th} is the real pressure and equilibrium pressure of the α components; m_α is the mass of an α-type particle; γ is the surface tension coefficient; kT is Boltzmann's factor.

The kinetic coefficients can be written in the following form for the formation of two-dimensional nuclei on the growing crystal face

$$K_\alpha^+ = \Delta_\alpha z'_\alpha n^{1/2}, \qquad K_\alpha^- = n_\alpha z'_\alpha n^{-1/2} \exp[\varepsilon'/(n+1)^{1/2}],$$
$$z'_\alpha = p_\alpha^{\text{th}} (2\pi m_\alpha kT)^{-1/2} 2\pi (\omega/4\pi)^{1/2}, \tag{2.5}$$
$$\varepsilon' = \pi^{1/2} (\omega/d)^{1/2} \gamma'/kT.$$

The following relations are valid for the numbers n_A, n_B and the total number $n = n_A + n_B$

$$\frac{dn_\alpha}{dt} = K_\alpha^+(n) - K_\alpha^-(n), \tag{2.6}$$
$$\frac{dn}{dt} = K_A^+(n) - K_A^-(n) + K_B^+(n) - K_B^-(n). \tag{2.7}$$

It is necessary to add a rate equation for the concentrations of the A and B components in the gas phase to the derived system of equations (C_A, C_B):

$$\frac{dC_\alpha}{dt} + \frac{4\pi\omega}{3} \int_0^\infty f(n,t) J_{\alpha n}\, dn = 0,$$
$$J_{\alpha n} = \omega^{-1} [K_\alpha^+(n) - K_\alpha^-(n)], \tag{2.8}$$

where $J_{\alpha n}$ is the flux density of α components to the substrate. It follows from formulae (2.4) and (2.5) that the $K_\alpha^+(n_A, nB)$ relations are complex

$$\sim n^{2/3},\ n^{1/2},\ K_\alpha^- \sim (n_\alpha/\sqrt[3]{n}) \exp[\varepsilon/(n+1)^{1/3}],\ \sim (n_\alpha/\sqrt{n}) \exp[\varepsilon'/\sqrt{n+1}].$$

Consequently equations (2.3)-(2.8) are nonlinear. An analogous system of nonlinear equations was used in study [28] to investigate the ordering of vacancy voids in the irradiation process. The instability of the homogeneous solution was demonstrated from a linear stability analysis of the quasi-stationary solution of the system and the conditions for periodic solutions were found. These results are directly applicable to analyzing ordering in the growing of binary crystals.

In order to account for the influence of nonstoichiometry of the gas phase on the condensation kinetics of binary crystals it is advantageous to introduce into system (2.3)-(2.8) the parameter γ_α characterizing the excess of a given component compared to the stoichiometric composition:

$$\gamma_\alpha = [dn_\alpha/dt + dn_\alpha^s/dt]/(dn_\alpha/dt), \tag{2.9}$$

where dn_α/dt and dn_α^s/dt are the growth rates for gas phase condensation for the nonstoichiometric and stoichiometric cases, respectively. Accounting for the parameter γ_α the equation system can be written as

$$\frac{dn_A}{dt} = K_A^+(n) - K_A^-(n),$$
$$\frac{dn_B}{dt} = (\gamma_B - 1)^{-1} [K_B^+(n) - K_B^-(n)], \tag{2.10}$$

$$\frac{dn}{dt} = \frac{dn_A}{dt} + \frac{dn_B}{dt}.$$

It follows from system (2.10) that the rate of condensation is at a maximum for $\gamma_B \sim 2$ (i.e., with a stoichiometric composition of the gas phase) and drops with growth of γ_B. This conclusion is consistent with the experimental data from film growing. During the cooling process the excess amount of the component B is precipitated out and these precipitates have been observed in PbTe and HgTe epitaxial film growing experiments. An estimate of the critical dimensions of these precipitates by (2.10) yields

$$n_{*B}^{1/3} = \varepsilon/\ln(\Delta_B/\Delta_A). \tag{2.11}$$

Precipitates with dimensions $n_B < n_{*B}$ have a tendency towards evaporation while precipitates with $n_B > n_{*B}$ have a tendency towards growth. According to formula (2.11) the critical dimensions of the precipitates are dependent on the temperature, the surface tension coefficient, supersaturation, and the equilibrium pressure of the A and B components. Increasing the C_B concentration of the B component and the equilibrium pressure p_B^{th} will reduce the critical dimensions n_{*B}. The equilibrium number of the critical precipitates can be determined by the formula

$$N_{n_{*B}} \approx C_B \exp\left[-\Delta G\,(n_{*B})/kT\right], \tag{2.12}$$

where C_B is the number of molecules of the B component per unit of volume; $\Delta G(n_{*B})$ is the height of the energy barrier to the formation of critical (stable) precipitates. The quantity $\Delta G(n_{*B})$ is dependent on supersaturations and the equilibrium pressure of the A and B components, as well as the surface tension coefficient and the temperature:

$$\Delta G\,(n_{*B})/kT = \varepsilon^3/2\ln^2(\Delta_B/\Delta_A). \tag{2.13}$$

With an excess concentration of one of the components (such as A) vacant sites or vacancies will appear in the other sublattice (for example, B). If N_B^{th} is the equilibrium vacancy concentration in sublattice B, expression $N_A/N_B N_B^{th}$ will determine the vacancy supersaturation level. With $N_A/N_B N_B^{th} \gg 1$ voids can form in sublattice B with the critical dimensions of the voids equal to

$$n_*^{1/3} = \varepsilon/\ln\,(N_A/N_B N_B^{th}). \tag{2.14}$$

The apparatus examined above makes it possible in principle to calculate the primary quantities characterizing epitaxial crystal growth (including the formation of plane nuclei) as well as precipitate and void formation . However in order to implement quantitative calculations it is necessary to carry out a specific program for finding and refining the kinetic coefficients and the characteristics of their variation under external action.

3. Laser Applications

Plane film structures. The application of laser technology in microelectronics has rapidly developed in recent years. If the first stage involved primary utilization of the thermal action of laser radiation (such as cutting, drilling, welding, and scrubbing) the primary field of application today covers the problems of stimulating material structure and crystal growth processes. Such operations include sputtering, annealing, film deposition and etching, impurity doping, gettering, and epitaxy. Analysis of solids by means of laser mass-spectrometry is beginning to play an important role. Laser technology has created new, previously inaccessible techniques for controlling the structure of materials as well as their defects and has made it possible to grow structures with given properties. Laser technology can be used in all fundamental operations of microelectronics. A number of mechanisms discussed above have found reflection in laser technology. Free atom creation by laser irradiation of the gas phase is used to fabricate silicon films [32], for SiO_2 film deposition [33] and metallization. ArF (λ = 193 nm) and KrF (λ = 248 nm) exciplex lasers were used in study [32] for photodissociation of SiH_4 and $SiCl_4$ compounds. Organometallic molecules have undergone photodissociation for metallic film deposition [34].

The high degree of locality of spatial resolution in laser irradiation of surfaces is already finding application in integrated circuit fabrication [35]. Local substrate heating in laser-enhanced chemical film deposition has led to the pyrolysis of compounds contained in the gas phase. A channel of gaseous chemical components are then formed under the substrate; these do not react with the substrate at room temperature, although upon heating they decompose into components that are then deposited onto the substrate. As a result the film is deposited in a local region such as in the form of strips or filaments with characteristic dimensions determined by the laser beam focusing.

The spatial resolution is at an optimum when using UV laser radiation and can reach fractions of a micron. This makes it possible in principle to eliminate difficult photolithography processes.

Undesirable impurities can be eliminated by impurity diffusion into a distended crystal. In studies [36] an argon laser was used to generate lattice dislocations by inducing damage to the reverse side of the substrate and then heating this substrate. As a result the impurities were localized in the vicinity of dislocations, where the stresses produce lattice distension. The advantages of laser technology also include purity of the fabricated films and the capability to control their growing processes.

Nonplanar film structures. Here, unlike planar structures, initial research is devoted to both the capabilities of nonplanar structures themselves and the techniques for fabricating such structures. Nonplanar formations (hillocks) have been obtained on a growing crystal surface exposed to both pulsed [17] and CW laser action [20]. Hillocks 3.5 mm in diameter were observed after exposing the samples to CW helium-cadmium laser radiation [20]. In study [37] where CW Ar^+ and Kr^+ lasers were used, nonplanar cavities were obtained with a minimum width of 1.3 μm together with hillocks of specific profile. In this case the laser radiation was focused on a spot 2.5 μm in diameter with a radiation power density of 10-400 kw/cm^2 and a growth rate from 0.1 to 100 $\mu m/sec$. Here free atom formation plays a significant and possibly determinant role in hillock growth. Pulsed lasers [17] (unlike CW lasers) make it possible to fabricate nonplanar structures with different local structural and electronic characteristics. We will briefly enumerate the primary properties of these structures. When a relief forms on a film surface the size of its components is determined by laser focusing capabilities. In principle it can be reduced to fractions of a micron. Hillock height is determined by laser irradiation time.

At the pulsed laser irradiation site good surface quality is observed together with an increase in the lattice constant which is related to the reduction in vacancy number.

A variable defect concentration and a varying degree of lattice structural perfection and the possible formation of space charge are all characteristic of both irradiated and nonirradiated samples. Doping also varies, i.e., it is possible to form *p-n*-junctions. Dimensional effects associated with the change in material properties as a function of thickness can also arise depending on the nature of the nonplanar formations.

Overall nonplanar structures have created new possibilities and effects compared to planar structures. Methods of fabricating such structures can be based on the simultaneous application of a UV xenon lamp and a UV pulsed laser with a pulse duration $\sim 10^{-8}$ sec as examined in study [17]. The simplicity of laser beam manipulation coupled with the automation capability makes it possible, in principle, to assure good reproducibility of nonplanar structures, including rather complex structures. Reproducible reliefs can be used for information coding. This method makes it possible to improve film fabrication for integrated optics, particularly for optimizing beam injection and extraction from films. Lattice structures and given film imperfections can be created rather easily.

Since the nature of crystal growing and doping are highly dependent on laser irradiation intensity certain possibilities exist for holographic recording. The use of interfering beams for irradiation will cause the growing process to occur at variable rates at sites with variable radiation intensity; the impurity concentration will also vary. Here we have an analogy to study [38].

We should emphasize that if the mechanical masking and photography technique has been developed comprehensively in microelectronics and integrated optics for planar structures, the most promising technique for nonplanar structures involves the use of pulsed lasers.

Controlling deviation from stoichiometry. The electromagnetic irradiation of crystals during the growing process may influence the degree of deviation from stoichiometry. For binary crystals the composition stability limits in the p-T-x-diagrams of a binary crystal (p is pressure, T is temperature, x is the relative concentration of the components) are determined by the maximum possible differences δ in the number of vacancies of both components for given p and T. The figure shows a typical T-x diagram of PbTe [29, 30]. Maximum deviation from stoichiometry lies in the tellurium-rich region (and consequently the region rich in lead vacancies). We see that the excess number of vacancies $\delta = n_{Pb} - n_{Te}$ can be as high as $\sim 10^{19}$ cm^{-3}, i.e., it can account for 10^3 of the total number of lattice sites. It follows from the relation [39]

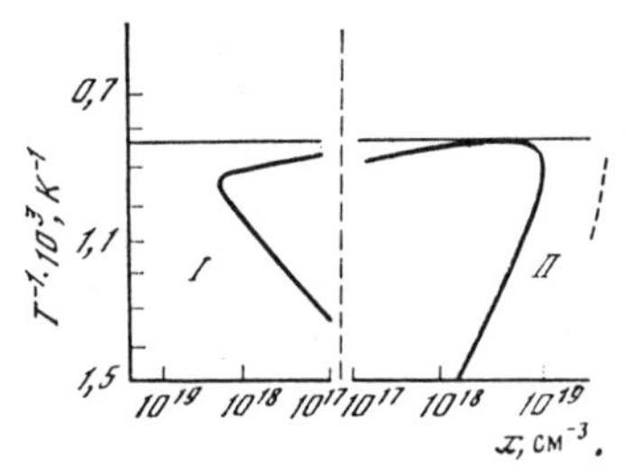

Composition stability range for a PbTe crystal I - n-type; II - p-type; dashed line - boundary of deviation from stoichiometry of a defect-free crystal

$$n_{Pb} n_{Te} = N_0 \exp\left[-(U_{Te} - U_{Pb})/kT\right] = \mu$$

that the number of vacancies is

$$n_{Pb} = \sqrt{(\delta/2)^2 + \mu} + \mu/2, \qquad n_{Te} = \sqrt{(\delta/2)^2 + \mu} - \mu/2.$$

Here N_0 is the number of lattice sites; $U_{Pb} + U_{Te}$ is the formation energy of a vacancy pair in different sublattices.

When the composition stability ranges overlap such as from crystal cooling the excess component will be precipitated out as precipitates, separated by several hundreds of angstroms [30].

The formation conditions for clusters of critical dimension for the excess component were examined for the case of an idealized lattice in Section 3. However the primary role in this case is played by the reduction in lattice strength resulting from the appearance of residual vacancies due to increasing deviation from stoichiometry. The vacancies will result in the breaking of a specific number of bonds $\sim(N_{Pb} + N_{Te})$. Investigations of the strength characteristics of crystals under irradiation have demonstrated [23] that the quantity U/kT characterizing crystal strength is inversely proportional to the irradiation intensity, i.e., it is roughly inversely proportional to the number of broken bonds. The drop in strength is caused by the effective change U determined by the relation $U = U_0 - \alpha kT \ln(n/N)$, where α is the coefficient and n is the total number of vacancies. The crystal strength at the boundary of the composition stability range drops by two orders of magnitude due to vacancies for the PbTe crystal.

The reduction in strength from increasing deviation from stoichiometry will cause the homogeneous structure to become unstable due to stresses caused by the defects in actual crystals. Such defects include voids and microcracks. Intense void formation occurs at temperatures near the melting point. Strong stresses arise around the voids due to surface tension and these stresses relax at distances of the order of the radius R from the void boundary. Pressures in a distended crystal near a void reach values of $p = -2\gamma/R$. For a specific surface energy $\gamma \sim 10^3$ erg/cm^2 and a radius $R \sim 10^{-6}$ cm the pressure $p \sim 10^9$ dyne/cm^2. As the lattice strength is reduced by more than two orders of magnitude, local lattice breakdown can occur. Such a picture is valid for a uniform vacancy distribution.

However the actual void formation mechanism of binary crystals is different from the regular mechanism in the case of significant deviation from stoichiometry. Since there can be a significant variation in the number of vacancies of the various atoms the accumulation of vacancies at the lattice tensile stress sites will result in excess atoms of one particular type at the sites (such as Te for a PbTe crystal). Small local regions will develop with an unstable lattice. A void with spherical properties will result. Its internal surface will contain primarily Te atoms. Such a situation also occurs in void formation. The excess number of lead vacancies will collect here. Voids whose walls consist of Te then arise and grow. This process can result in the formation of nuclei of a new phase.

If the upper composition stability limit in real crystals can be attributed to the intense void formation processes at high temperatures, the lower limit evidently is determined by the stresses that result from the structural inhomogeneity and grow with a drop in temperature. Microcracks can play a critical role in the breakdown of the homogeneous structure. The formation of microcracks amounting to several interatomic distances is normally the result of high stresses caused by dislocation accumulations. Microcrack growth causes a concentration of stresses at the edges; the stresses σ_{cr} at the microcrack apex are given by the relation

$$\sigma_{cr} = (1 + 2\sqrt{l/\rho})\sigma,$$

where ρ is the radius of curvature at the crack apex; l is the crack length and σ is the mean effective stress in the crack vicinity. For cracks less than 1 μm in length the stresses at the crack apex can be 10^8-10^9 dyne/cm^2. Due to the reduced strength of the binary crystal such stresses can cause local breakdown in the crystal structure and result in precipitation.

Therefore the composition stability limits in the *T*-*x*-diagram are determined by a combination of such factors as the reduced strength of the binary crystal due to deviation from stoichiometry and significant stresses resulting from structural defects (voids, microcracks, and dislocations).

Laser irradiation of the crystal growing process significantly changes the situation. Pulsed laser irradiation causes a significant change in the number of dislocations as well as the number of vacancies and suppresses void and microcrack formation compared to unradiated crystals by virtue of pressures generated by the thermooptic effect. This entire process reduces internal stresses and the crystal structure becomes more stable and in principle can end up outside the regular composition stability limits. The dashed line in the diagram represents the maximum deviation from stoichiometry for an idealized defect-free (except, of course, vacancies) crystal. The maximum deviation from stoichiometry will not exceed 0.1 at.%. In this range even weak stresses will make the structure unstable.

References

1. Kalyuzhnaya, G.A., Kiseleva, K.V. "The problem of stoichiometry in type A^2B^6 and A^4B^6 semiconductors of variable composition"

2. Medvedev, S.A., Ivannikova, G.E. "Electronic engineering" POLUPROVODNIKOVYE PRIBORY, 1970, no. 4, pp. 29-34.

3. Chistyakov, Yu.D., Kaminskiy, V.V., Grishin, V.M., Shvedova, V.V. "The application of electromagnetic radiation in gas-phase fabrication techniques for semiconductors" In: Obzovry po elektronnoy tekhnike" [Surveys on electronic engineering] Moscow: TsNII elektroniki, 1976 (Series 2, Semiconductor devices; vol. 5)

4. Kumagama, M., Suvami, H., Terasaki, T. "Epitaxial growth with light irradiation" JAP. J. APPL. PHYS., 1968, vol. 7, p. 1331.

5. Pat. 2 335 043 (France). "Method of growing semiconductor epitaxial films and a device for its implementation" Vul, B.M., Gorina, Yu.I., Ivannikova, G.E., et al.

6. Kurbatov, L.N., Britov, A.D., Karavaev, S.M., et al. "PbSnTe laser diodes for the 5-15 μm range" KVANTOVAYA ELEKTRON., 1976, vol. 3, no. 10, pp. 2238-2241.

7. Guro, G.M., Kalyuzhnaya, G.A., Mamedov, G.S., Shelepin, L.A. "Controlling crystal growth by means of electromagnetic radiation" ZhETF, 1979, vol. 77, pp. 2366-2373.

8. Guro, G.M., Kalyuzhnaya, G.A., Mamedov, T.S., Shelepin, L.A. "The influence of radiation on crystal growing kinetics" In: "Statisticheskie i kogerentnye metody issledovaniya fizicheskikh sistem"

[Statistical and coherent methods of investigating physical systems] Moscow: Nauka, 1980, pp. 1127-140 (Tr. FIAN; vol. 124).

9. Varma, A. "Rost kristallov i dislokatsii" [Crystal and dislocation growth] Moscow: IL, 1958, 216 pp.

10. Timofeeva, V.A. "Rost kristallov iz rastvorov - rasplavov" [Solution-melt growing of crystals] Moscow: Nauka, 1978, 267 pp.

11. Chernov, A.A., Givargizov, E.I., Bagdasarov, Kh.S., et al. "Crystal formation" In: Sovremennaya kristallografiya [Modern crystallography] Moscow: Nauka, 1978, vol. 3, p. 407.

12. Barton, V.K., Cabrera, N., Franck, F.T . "Elementarnye protsessy rosta kristallov" [Elementary crystal growing processes] Moscow: Izd-vo inostr. lit., 1959.

13. Baru, V.G., Vol'kenshteyn, F.F. "Vliyanie oblucheniya na poverkhnostnye svoystva poluprovodnikov" [Influence of radiation on the surface properties of semiconductors] Moscow: Nauka, 1978, 288 pp.

14. Morrison, S. "Khimicheskaya fizika poverkhnosti tverdogo tela" [The chemical physics of solid surfaces] Moscow: Mir, 1980, 488 pp.

15. Kiselev, V.F., Krylov, O.V. "Elektronnye yavleniya v adsorbtsii i katalize na poluprovodnikakh i dielektrikakh" [Electronic phenomena in adsorption and catalysis on semiconductors and dielectrics] Moscow: Nauka, 1979, 234 pp.

16. Bol'shova, K.V., Levchenko, I.S., Shubin, A.A. "Deposition of reflecting In, Zn, and Cd layers" PTE, 1968, vol. 2, pp. 228-232.

17. Aleksandrov, O.V., Zaytsev, V.V., Zelikman, I.I., et al. "The influence of pulsed laser radiation on the epitaxy of lead and tin chalcogenides" KRATKIE SOOBSHCH PO FIZIKE FIAN, 1984, no. 2, pp. 45-48.

18. "Issledovaniya po gidrofizike" [Hydrophysical investigations] Moscow: Nauka, 1984, 187 pp. (Tr. FIAN, vol. 156).

19. Korotchenko, A.N., Samokhin, A.A. "Phase transformations and the thermooptic effect from intense irradiation of absorbing media" Preprint FIAN, no. 23, Moscow, 1981, 18 pp.

20. Zaytsev, V.V., Kalyuzhnaya, G.A., Mamedov, T.S., Shelepin, L.A. "Epitaxy with laser irradiation and its capabilities" ZhETF, 1985, vol. 55, no. 5, pp. 955-956.

21. Merkulova, S.P., Shelepin, L.A., Shubin, A.A. "The influence of

stresses on the properties of superconducting films" KRATKIE SOOBSHCH PO FIZIKE FIAN, 1985, no. 4, pp. 52-55.

22. Zhurkov, S.N. "The strength of solids" VESTN. AN SSSR, 1957, no. 11, pp. 78-82.

23. Regel' V.R., Slutsker, A.I., Tomashevskiy, E.E. "Kineticheskaya priroda prochnosti tverdykh tel." [The kinetic nature of the strength of solids] Moscow: Nauka, 1974, 560 pp.

24. Kusov, A.A., Vettegren', V. I. "Calculation of the longevity of a stressed atomic chain in an anharmonic approximation" FTT, 1980, vol. 22, no. 11, pp. 3350-3357.

25. Dvurechenskiy, A.V., Kachurin, G.A., Kudaev, E.V., Smirnov, L.S. "Impul'snyy otzhig poluprovodnikovykh materialov" [Pulsed annealing of semiconductor materials] Moscow: Nauka, 1982, 208 pp.

26. Aleksandrov, L.N. "Kinetika kristallizatsii i perekristallizatsii poluprovodnikovykh plenok" [The kinetics of the crystallization and recrystallization of semiconductor films] Novosibirsk: Nauka, 1985, 225 pp.

27. Mirzoev, F.Kh., Fetisov, E.P., Shelepin, L.A. "Kinetika obrazovaniya i rosta por v kristallakh" [The kinetics of void nucleation and growth in crystals]: This volume.

28. Mirzoev, F.Kh., Fetisov, E.P., Shelepin, L.A. "O formirovanii sverkhreshetok por v tverdom tele" [The formation of void superlattices in solids]: This volume.

29. Kalyuzhnaya, G.A., Mamedov, T.S., Ragimova, G.Sh., Sentyurina, N.N. "The influence of indium on the electrical and structural properties of $Pb_{1-x}Sn_xTe$" KRATKIE SOOBSHCH PO FIZIKE FIAN, 1981, no. 6, pp. 21-26.

30. Gorina, Yu.I., Kalyuzhnaya, G.A., Kiseleva, K.V., et al. "Investigation of PbTe(Bi)-PbSnTe-PtTe laser epitaxial structures" FTP, 1979, vol. 13, pp. 305-310.

31. Gorina, Yu.I., Kalyuzhnaya, G.A., Kiseleva, K.V., et al. "The influence of a silver impurity on the electrical properties of a lead and tin telluride solid solution" KRATKIE SOOBSHCH PO FIZIKE FIAN, 1981, no. 7, pp. 42-47.

32. Boyd, I.W. "A review of laser beam applications for processing silicon" CONTEMP. PHYS., 1983, vol. 24, no. 5, pp. 461-490.

33. Boyd, L.W. "Laser-enhanced oxidation of Si" APPL. PHYS. LETT., 1983, vol. 42, no. 8, pp. 728-730.

34. Blum, S.E., Brown, K.H., Srinivasan, R. "Photo-oxidation of silicon monoxide to silicon dioxide with pulsed far ultraviolet (193 nm) laser radiation" APPL. PHYS. LETT., 1983, vol. 43, no. 11, pp. 1026-1027.

35. Ehrlich, D.I. "Long-term prospects for lasers in microfabrication" LASER FOCUS (ELECTRO-OPT. TECHNOL.), 1984, vol. 20, no. 10, p. 108.

36. Eggermont, G.E.J., Falster, R.J., Hahn, S.K. "Laser induced back-side" SOLID STATE TECHNOL., 1983, vol. 26, no. 11, pp. 171-178.

37. Bäuerke, D. "Laser-induced chemical vapor deposition" EUROPHYS. NEWS, 1983, vol. 14, no. 10, pp. 9-12.

38. Shtyrkov, E.I., Khaybulin, I.B., Galyautdinov, M.F., Zaripov, M.M. "An ion-doped layer: a new material for holograms" OPTIKA I SPEKTROSKOPIYA, 1975, vol. 38, pp. 1031-1036.

39. Geguzin, Ya.E. "Diffuzionnaya zona" [Diffusion zone] Moscow: Nauka, 1979, 343 pp.

THE KINETICS OF VOID NUCLEATION AND GROWTH IN CRYSTALS

F.Kh. Mirzoev, E.P. Fetisov, L.A. Shelepin

ABSTRACT

The kinetics of the formation and development of a new phase in supersaturated systems are investigated theoretically. Quasi-equilibrium functions are used to find general expressions for the nonstationary distribution function and the flux of new phase nuclei in dimensional space. The application of this method to investigating the nucleation of voids in crystals supersaturated by vacancies has made it possible to present a uniform kinetic picture of the process and to determine its principal characteristics (the critical void nucleation time; the duration of the initial stage; the characteristic and total condensation time of the vacancies and maximum void size). The role of interstitial atoms in void nucleation kinetics in irradiated metals is investigated. The influence of an external mechanical stress on void nucleation kinetics is considered. It is demonstrated that the rate of void nucleation grows linearly with growth of stress at a low stress level and diminishes exponentially at high stress levels.

Introduction

Interest in investigating the kinetics of phase transformations that occur in a variety of supersaturated systems resulting in the formation and growth of nuclei has increased in recent years. This is due primarily to the fact that in a number of cases (crystallization of solids, nucleation of voids from a supersaturated solution of point defects; gas condensation in supersonic wind tunnels, etc.) it is important to know the kinetics of phase transitions. This knowledge makes it possible to predict the nonequilibrium states of the system that occur after different thermal or mechanical treatments and also to describe their behavior as a function of external conditions and their evolution in time. Investigating the defect structure of crystals and its relation to the physical and mechanical properties is an important task in solid state physics and has been the result of the development of new fields of technology, primarily atomic power engineering, space science and solid state electronics. The kinetics of void nucleation and growth has been critical in developing new methods of fabricating monocrystals and also holds significant promise in the search of efficient techniques for deliberate manipulation of the properties of crystal materials.

The quasi-steady-state theory of nucleation has been rather well developed to date [1-6]; this theory makes it possible to obtain a stationary distribution function of the nuclei together with their rate of nucleation as a function of supersaturation, temperature, pressure, radiation dosage and other factors. The results from this theory are applicable only in limiting cases where a quasi-steady-state has already been reached in the system. In many experiments the test system rapidly goes from a stable state to a metastable state. Here the nucleation rate will be time-dependent until a quasi-stationary distribution is achieved.

If the observation time of the process is longer than the time to steady state in the system, the influence of the nonstationarity period is not significant. However in many processes the transient period will be quite extensive and is observed experimentally. Such processes will primarily include the phase transformations that occur in condensed media (the formation of glasses, crystal growth in a supercooled melt and vacancy cluster formation from rapid cooling or heating of solids, etc.). The nonstationarity effect plays an important role in investigating gas condensation processes in the supersonic nozzles of jet engines and wind tunnels where supersaturation occurs rather rapidly. The investigation of nonstationary void nucleation and growth is therefore a very important and timely problem. However this problem has not been investigated in sufficient detail from the theoretical viewpoint. The analysis has largely been done using numerical techniques [7-9]. Existing analytical analyses [10-14] have been carried out in rather narrow time and size intervals. Moreover, it is difficult to apply these results to investigating the phase transformations that occur in multicomponent systems.

The present study proposes a new approach to analyzing the kinetics of the nonstationary nucleation and development of the new phase that is valid over long time intervals and size ranges and with rather general assumptions. This method is based on the quasi-stationary distribution function technique for solving Fokker-Planck-type equations. Such equations describe phase transitions.

According to this method the solution of the kinetic equations is given as a time-asymptotic series. In physical terms this is an expansion in powers of the slowness of the process and the degree of proximity to an equilibrium state. This technique is often used to derive results that cannot be achieved by using other methods; in other cases analytic solutions are obtained in place of numerical solutions. A wide range of problems have been solved in plasma physics, relaxational processes, statistical radiophysics and the theory of chemical reactions [15-18]. The advantage of this technique is that it produces an analytic solution over a broad time interval and provides a uniform and formal approach.

This technique also makes it possible to analyze the second order phase transition described by the Fokker-Planck equation. (This process is examined as Brownian motion for thermodynamic parameters [19].) It is assumed in Landau theory [20] that the times of variation in the system parameters significantly exceed the characteristic times τ to equilibrium distributions. The case of large times τ is examined in study [21] using the quasi-stationary distribution function technique. In this case long-lived stationary distributions arise in the system whose lifetimes may exceed the experimental time. Such metastable states are dependent on the initial conditions.

In the present study the quasi-stationary distribution function technique is applied to first order phase equations. Section 1 gives a mathematic condensation (cluster formation and growth). This model is independent of the nature of the condensing particles and can be used for describing a wide range of systems undergoing a first order phase transition. Here it is applied to the problem of void nucleation and growth. Section 2 considers void condensation from a supersaturated vacancy solution. Such a solution results from rather high-energy particle irradiation of solids (electron, neutron, heavy ion irradiation) and from a variety of mechanical and thermal material treatment processes (heavy quenching, thermal annealing, deformation, etc.).

The void nucleation mechanism is analogous to the vapor condensation mechanism. It involves the formation of vacancy or multivacancy clusters at high temperatures due to their high mobility as indicated from estimates of the diffusion coefficient.

As we know there are three stages in the void nucleation process, as in any condensation process [12]. In the first stage fluctuations result in supercritical void nucleation. These voids are capable of further growth. In this stage the number of vacancies in the voids is significantly less than the total number of "free" vacancies in the crystal. In the second stage the number of vacancies in the voids becomes comparable to the total number of "free" vacancies in the crystal. In this case the majority of voids are of sufficiently large size and their number is virtually constant.

In the third stage large void growth resulting from the dissolution of small voids (coalescence) predominates. In this case fluctuation nucleation is not likely. The analysis carried out in Section 2 describes in a uniform manner all stages of void nucleation and growth. Many studies have been devoted to void nucleation in conditions of elementary particle irradiation. The results from these studies are covered in studies [5, 6, 23-25]. The vacancies formed in irradiated materials (at temperatures pT ~ (0.2-0.6) T_M, T_M is the melting point) are vacancy clusters tens and hundreds of angstroms in size.

The void phenomenon has a significant influence on the various physical properties of solids. Specifically it causes radiation swelling of solids used in modern equipment systems. Extensive efforts have been made in recent years to develop methods of eliminating swelling of different parts and components operating in nuclear reactor cores.

In the general case in analyzing the kinetics of void nucleation and growth in irradiation conditions we cannot limit our examination to vacancy condensation. In addition to vacancies irradiation also creates interstitial atoms having significant mobility. Under pulsed irradiation the interstitial atoms, due to their high mobility, will rapidly exit onto the surface or will be annihilated with the vacancies in the crystal bulk. In thermal annealing when vacancy mobility is still higher than the thermal equilibrium mobility, void growth can be examined as vacancy condensation. However in conditions of continuous irradiation it is necessary to analyze a two-component system consisting of vacancies and interstitial atoms. Such an analysis is carried out in Section 3.

In Section 4 we investigate the influence of tensile stresses on the kinetics of void nucleation in irradiated crystals. The derived results are compared to experiment. A satisfactory agreement is obtained. The mathematical model formulated below represents the starting point for analyzing the applied problems of void nucleation examined in Sections 2-4.

1. Mathematical Model. Solution of Kinetic Equations

We will consider a system in which a phase transformation occurs. At initial time $t = 0$ all particles (such as atoms, molecules, vacancies) are in the metastable phase. We assume that there are no nuclei of the new phase in this system. Nucleation will occur following either the homogeneous-fluctuation mechanism or heterogeneously: at centers of different types with inhomogeneities functioning as such centers, including impurity atoms, grain boundaries, dislocations, etc. The present study will consider a homogeneous model for the appearance of the new phase. Fluctuation nucleation largely occurs with weak supersaturation where the size of the critical nucleus is significantly greater than the lattice constant. According to Volmer's theory [2] the homogeneous nucleation process can be described by the following bimolecular reactions:

$$E_{n-1} + E_1 \underset{K_n^-}{\overset{K_{n-1}^+}{\rightleftarrows}} E_n, \quad E_n + E_1 \underset{K_{n+1}^-}{\overset{K_n^+}{\rightleftarrows}} E_{n+1},$$

where the symbol E_n represents a nucleus consisting of n particles, while E_1 is a single nucleus. These reactions infer that the nucleus grows or shrinks from the attachment or separation of separate particles at the interface. The rate constants of these reactions are given by the symbols K_n^+ and K_n^-, respectively. We ignore the change in the quantity and dimensions of nuclei from coagulation.

We will describe the behavior of the ensemble of nuclei of size n in time by means of the size distribution function $Z_n(t)$ at time t. The radius, volume, or surface area of the nucleus or the number of particles making up the nucleus can be used as the dimensions. For definiteness we shall henceforth use the number of particles or the radius for nucleus dimensions.

If we plot the number of particles n in the nuclei on a certain axis, a shift on this axis of ±1 will correspond to each particle attachment or separation (due to the weak supersaturation we neglect the mobility of nuclei with $n \geqslant 2$). Then it is possible to describe the dynamics of the variation of the function $Z_n(t)$ in time by the following finite-difference equations:

$$dZ_n(t)/dt = I_{n-1}(t) - I_n(t). \tag{1}$$

The flow of nuclei $I_n(t)$ in dimensional space is determined by the expression

$$I_n(t) = Z_n(t)K_n^+ - Z_{n+1}(t)K_{n+1}^-. \tag{2}$$

The chain of equations (1), (2) is supplemented by the following initial and boundary conditions:

$$Z_1(0) = N_1, \ Z_n(0) = 0 \ (n \geqslant 2), \tag{3}$$

$$Z_1(t) = N_1. \tag{4}$$

Equation (4) infers that we are ignoring the change in concentration of particles of the metastable phase. This is valid in the initial stage of nucleation when the number of nuclei is still small and their intense growth has not yet begun. The law of conservation of metastable phase particles takes the form:

$$N_1 + \sum_{n \geqslant 2} nZ_n(t) = N_0, \tag{5}$$

where N_0 is the total particle concentration of the system for $t = 0$.

Equations (1), (2), and (5) with conditions (3) and (4) yield a closed description of the problem of nonstationary nucleation and development of the new phase nuclei.

In the general case it is not possible to analyze equations (1)-(5) due to the dependence of the rate constants on particle number n and time t. Therefore numerical methods are largely used to carry out an analysis of the nonstationary formation and growth process of the new phase based on (1) and (2). An approximate analytic solution of the problem will be derived below after transformation from the system of finite-difference equations to a differential equation. The functions $Z_n(t)$, K_n^+, K_n^- figuring into (1) and (2) are given only for discrete values of n. However if we take $n \gg 1$, i.e., if we assume that the nuclei have a macroscopic size, the dependence of these functions on n can be taken as continuous. Then, expanding $Z(n \pm 1, t)$, $K^+(n + 1)$, and $K^-(n - 1)$ into a Taylor series we have from (1) and (2)

$$\frac{\partial Z(n,t)}{\partial t} = \frac{\partial}{\partial n}\{Z(n,t)[K^-(n) - K^+(n)]\} + \frac{1}{2}\frac{\partial^2}{\partial n^2}\{Z(n,t)[K^+(n) + K^-(n)]\}. \quad (6)$$

The following initial and boundary conditions must be satisfied in solving equation (6):

$$Z(n, 0) = 0, \ (n > 0), \quad Z(0, t) = N_1, \quad (7)$$

that replace more exact conditions (3), (4). In the literature equation (6) has been solved by incorporating in the right half either the term with the first derivative for the large times [13, 23] or with the second derivative with small t [3, 7, 10]. A comparatively complete analysis is given in study [14]. However the results given in study [14] are applicable only for nuclei in the vicinity of their critical dimensions and for sufficiently large t. At the same time the literature does not contain a consistent description of the nucleation and growth kinetics of new phase nuclei that would provide a unified picture of the process in time. The present study proposes a unified analysis of the kinetics of phase transitions in condensed media.

Equation (6) is the Fokker-Planck equation normally employed in Brownian motion theory. It can be represented as

$$\frac{\partial Z(n,t)}{\partial t} = -\frac{\partial I(n,t)}{\partial n},$$

$$I(n,t) = F(n)Z(n,t) - \frac{\partial}{\partial n}[D(n)Z(n,t)], \quad (8)$$

where $F(n) = K^+(n) - K^-(n)$ and $D(n) = [K^+(n) + K^-(n)]/2$ represent the growth rate and the coefficient of "diffusion" of the nuclei in their dimensional space, respectively.

If the total number of stable nuclei remains fixed during their entire growth process, $I(0, t) = 0$. Indeed, integrating equation (8) within the limits from zero to infinity and accounting for the fact that $I(\infty, t) = 0$ we find:

$$dN/dt = I(0, t)$$

It is obvious that when $N = \text{const}$, $I(0, t) = 0$. However when nuclei are continuously formed in the system, $dN/dt = I(0, t) = I_0 > 0$. It is this case that is examined in the present study.

We will obtain a solution of equation (8) by first analyzing the limiting cases. Let $Z(n, t)$ be small compared to the particle number in the metastable phase. Then the distribution function is explicitly time-independent $Z(n, t) = Z_s(n)$ and from (8) we have

$$-\frac{d}{dn}[D(n)Z_s(n)]+F(n)Z_s(n)=I_s=\text{const}.$$

From here we find

$$Z_s(n)=\exp[-G(n)/kT]\left\{I_s\int_n^\infty\frac{\exp[G(n_1)/kT]\,dn_1}{D(n_1)}+C\right\},$$

$$G(n)=-kT\left[\int_0^n\frac{F(n_1)}{D(n_1)}dn_1-\ln\frac{D(n)}{D(0)}\right]. \quad (9)$$

The integration constant $C = 0$ is determined from the condition $Z_s(\infty) = 0$. According to boundary condition (7) I_s takes the form

$$I_s=N_1\left[\int_0^\infty\frac{\exp(G/kT)\,dn}{D(n)}\right]^{-1}. \quad (10)$$

We introduce the distribution function $f(n)$ for which the flux $I(n, t)$ is equal to zero for all n:

$$\frac{d}{dn}[D(n)f(n)]=F(n)f(n).$$

From here

$$f(n) = N_1\exp[-G(n)/kT]. \quad (11)$$

In formulae (9), (10), and (11) the function $G(n)$ models the potential energy for nucleation. With small dimensions it grows, when $n = n_{cr}$ it reaches a maximum and when $n > n_{cr}$ it decays.

The critical dimension n_{cr} corresponds to the maximum of the $G(n)$ function and is found from the condition $(dG/dn)|_{n=n_{cr}} = 0$.

Nuclei with dimensions $n < n_{cr}$ have a tendency to "evaporate", while when $n > n_{cr}$ they have a tendency to grow. The nonstationary solution of equation (8) will be found by the quasi-stationary distribution function technique [18]. Carrying out the substitution

$$Z(n, t) = f(n)\varphi(n, t). \quad (12)$$

Substituting (12) into (8) we have

$$f\frac{\partial\varphi}{\partial t}=-\frac{\partial I}{\partial n},\qquad I=-Df\frac{\partial\varphi}{\partial p}, \quad (13)$$

$$\varphi(0, t)=1. \quad (14)$$

Doubly integrating the right and left halves of (12) with respect to n and remembering boundary condition (14) we obtain

$$\varphi = \varphi_0+\hat{E}\varphi_s \quad (15)$$

where

$$\hat{E}=\int_0^n\frac{d\,1}{fD}\int_0^{n_1}f\,dn_2\frac{\partial}{\partial t},\qquad \varphi_0=1\quad I_0(t)\int_0^n\frac{dn_1}{fD},$$

$I_0(t)$ is the parameter of the problem coinciding with the flux of nuclei at the point $n = 0$.

From the iteration procedure the solution of equation (15) is represented as a series in powers of the operator E or as a series in the temporal derivatives of the parameter $I_0(t)$:

$$\varphi = \varphi_0 + \hat{E}\varphi_0 + \ldots + (\hat{E})^m \varphi_0 + \ldots = 1 - \sum_{m=0}^{\infty} \beta_m \frac{d^m I_0}{dt^m}, \tag{16}$$

where the coefficients of the expansion β_m are determined by the following n_1 recurrent relations

$$\beta_m(n) = \int_0^n \frac{dn_1}{fD} \int_0^{n_1} f\beta_{m-1}(n_2) dn_2, \quad \beta_0 \equiv \int_0^n \frac{dn_1}{fD}.$$

The distribution function obtained by truncation of infinite series (16) in the mth iteration step is called the mth order quasi-stationary distribution function. This truncation is based on the Markovian nature of the process: in time the system quickly loses information on its initial state. In terms of its physical content solution (16) is a power expansion of the proximity to an equilibrium state. The higher the order of the quasi-stationary distribution function, the larger the dimension and time intervals required for it to describe the system's process in achieving a quasi-steady-state. We will consider the kinetics of the formation and development of the new phase based on a first order quasi-stationary distribution function:

$$\varphi^{(1)} = 1 - I_0 \int_0^n \frac{dn_1}{fD} - \frac{dI_0}{dt} \int_0^n \frac{dn_1}{fD} \int_0^{n_1} f\, dn_2 \int_0^{n_2} \frac{dn_3}{fD}. \tag{17}$$

The corresponding expression for the flux takes the form:

$$I = I_0 + \frac{dI_0}{dt} \int_0^n f\, dn_1 \int_0^{n_1} \frac{dn_2}{fD}. \tag{18}$$

Numerical calculations revealed that with fixed and rather large times t the distribution function drops and at certain values $n = n_0(t)$ corresponding to supercritical nuclei, the distribution function vanishes. Therefore the following boundary condition holds:

$$\varphi^{(1)}(n, t)\,|_{n=n_0(t)} = 0. \tag{19}$$

The limit of $n_0(t)$ is a monotonically-increasing function of time and will be determined below together with the parameter $I_0(t)$.

From the absence of sufficiently large new formations with a finite analysis time of the process we have

$$fD \left. \frac{\partial \varphi^{(1)}}{\partial n} \right|_{n=n_0(t)} = 0. \tag{20}$$

We will obtain the time dependencies of the functions $I_0(t)$ and $n_0(t)$ by substitution of (17) into (19) and (20). As a result we obtain a system of two equations for $I_0(t)$ and $n_0(t)$

$$I_0 \int_0^{n_0} \frac{dn_1}{fD} + \frac{dI_0}{dt} \int_0^{n_0} \frac{dn_1}{fD} \int_0^{n_1} f\, dn_2 \int_0^{n_2} \frac{dn_3}{fD} = 1, \tag{21}$$

$$I_0 + \frac{dI_0}{dt} \int_0^{n_0} f\, dn_1 \int_0^{n_1} \frac{dn_2}{fD} = 0, \tag{22}$$

$$I_0(0) = 0. \tag{23}$$

Solving equations (21) and (22) jointly we find

$$I_0 = \int_0^{n_0} f\, dn_1 \int_0^{n_1} \frac{dn_2}{fD} \Big/ W(n_0), \quad W(n_0) = \int_0^{n_0} f\, dn_1 \left(\int_0^{n_1} \frac{dn_2}{fD} \right)^2, \tag{24}$$

$$dI_0/dt = -W^{-1}(n_0). \tag{25}$$

The integrals in (24) have a variable upper integration limit. Carrying out time differentiation of (24) and setting the derived equation equal to derivative (25) we obtain an equation describing the variation of $n_0(t)$ in time:

$$\frac{dn_0}{dt} \left[I_0(t) \int_0^{n_0} \frac{dn}{fD} - 1 \right] = \left[f(n_0) \int_0^{n_0} \frac{dn}{fD} \right]^{-1}. \tag{26}$$

Solving this equation with the initial condition $n_0(0) = 0$ we find the variation of $n_0(t)$ in time which together with (5), (17), and (24) makes it possible to comprehensively describe the formation and development of the new phase.

We will now consider the qualitative dependence of the distribution function $\varphi(n, t)$, the flux $I(n, t)$, the parameter $I_0(t)$, and the number of nuclei $N(t)$ on time and dimensions n without fixing the form of the potential barrier $g(n)$ for nucleation. We will be interested in the behavior of these quantities in a relatively neglected time interval, specifically for $t > \tau_0$ (τ_0 is the critical nucleation time). Substituting (24) and (25) into (18) we have

$$I(n, n_0) = \left[\int_n^{n_0} f\, dn_1 \int_0^{n_1} dn_2/fD \right] \Big/ W(n_0). \tag{27}$$

From (27) we obtain

$$\partial I/\partial n = -\left[f(n) \int_0^{n} dn_2/fD \right] \Big/ W(n_0) < 0.$$

We can derive the sign of the derivative $\partial\varphi/\partial n$ from (13): $\partial\varphi/\partial n = -I/Df < 0$, since $I > 0$.

We now evaluate

$$\frac{\partial I}{\partial t} = \frac{dn_0}{dt} f(n_0) \int_0^{n_0} \frac{dn_1}{fD} [1 - I(n, n_0) \int_0^{n_0} \frac{dn_1}{fD} W^{-1}(n_0).$$

Substituting (27) into here and changing the integration order in the numerator we have

$$\frac{\partial I}{\partial t} = \frac{dn_0}{dt} f(n_0) \int_0^{n_0} \frac{dn_1}{fD} \int_0^{n_0} \frac{dn_2}{fD} \int_{n_2}^{n} f\, dn_3 \int_0^{n_3} \frac{dn_4}{fD} W^{-1}(n_0). \tag{28}$$

According to (26) $dn_0/dt > 0$. We can easily see from (28) that for the subcritical nuclei $\partial I/\partial t < 0$, while for the supercritical nuclei $\partial I/dt > 0$. We determine the sign of $\partial\varphi/\partial t$ from (13):

$$\frac{\partial \varphi}{\partial t} = -f^{-1}(n) \frac{\partial I}{\partial n} = W^{-1}(n_0) \int_0^{n} \frac{dn_1}{fD} > 0.$$

We then have

$$\frac{dI^0}{dt} = -\frac{I_0}{W(n_0)} < 0, \qquad \frac{dN}{dt} = \frac{d}{dt}\left(\int_0^{n_0} f\varphi\, dn\right) = f(n_0)\, \varphi(n_0) \frac{dn_0}{dt} > 0.$$

Thus with growth of t the distribution function and nucleus number grow. The flux of subcritical nuclei drops while the flux of supercritical nuclei increases. With fixed t the distribution function and flux drop with growth of nucleus dimensions, approaching a zero value. We note that these qualitative conclusions are in good agreement with results from numerical integration of equation (1) (see, for example, Fig. 53 in study [14]). The derived general expressions for the temporal evolution of the distribution function and the flux of new phase nuclei make it possible to provide a unified kinetic picture of the formation and growth processes of the new phase.

In order to establish the applicability criterion of the derived solution we will substitute (24) into expansion (16). This is an alternating series. Therefore the error from the substitution of the infinite sum with its partial sum has a smaller magnitude than the next dropped term. Accounting for this as well as equations (24) and (26) the applicability condition of solution (17) can be written as:

$$\beta_1(n)/\beta_2(n) \gg \beta_0(n_0)/\beta_1(n_0).$$

2. Kinetics of Void Nucleation and Growth in a Vacancy-Supersaturated Crystal

In applying the derived general expressions of the distribution function of the nuclei an their flux in supersaturated systems as well as the temporal characteristics of the new phase formation and growth process to specific systems it is necessary to know the probabilities of the elementary processes. We will consider a crystal exposed to preliminary high-energy particle (electrons, neutrons, ions, etc.) irradiation at low temperatures. In such a system irradiation produces a supersaturated solution of vacancies in which nucleation occurs at high temperatures (post-radiation annealing). The probabilities of the elementary processes are given here by the rate of attachment of the vacancy to the vacancy cluster $K^+(n)$ and the rate of separation $K^-(n)$, respectively.

The problem of calculating these rates is quite complex and is not completely solved at present. The primary difficulty is the need to describe nuclei containing a few vacancies (for high degrees of supersaturation). Macroscopic quantities are also used, such as the density, surface tension, etc. Normally $K^+(n)$ and $K^-(n)$ are found from an analysis of diffusion processes occurring near and on the surface of the nuclei by solving a macroscopic equation for the rate of growth of the supercritical nuclei ignoring fluctuations. Normally the derived results are used for a qualitative description of nuclei up through unitary dimensions ($n = 1$).

Overall, however, the application of macroparameters to microsystems is far from always successful. This problem remains one of the most significant difficulties in the modern theory of homogeneous nucleation. Attempts have been made to express the desired probabilities through the parameters of the intermolecular interaction potential of the molecules comprising the cluster (see, for example, [22]). A great deal of confusion still exists here. Below we will proceed from study [3] bearing in mind that the results are easily modified for other dependencies on n. According to study [3] the functions $K^+(n)$ and $K^-(n)$ are determined by the following expressions:

$$K^+(n) = J_0 A(n), \qquad K^-(n) = J_{0e} A(n-1), \tag{29}$$

where J_0 and J_{0e} are the vacancy fluxes per unit of void area; $A(n)$ is the effective void surface, $\Phi(n)$ is the Gibbs free energy of void formation from n vacancies; kT is the Boltzmann factor. In the case of a spherical void the functions $A(n)$ and $\Phi(n)$ take the form [3]:

$$A(n) = (36\pi\Omega^2)^{1/3} n^{2/3} = A(1)\, n^{2/3}, \quad \Phi(n) = -nkT \ln(\Delta + 1) + \gamma A(1)\, n^{2/3}, \tag{30}$$

where $\Delta_v = (N_v - N_{ve})/N_{ve}$ is the degree of supersaturation of the crystal, N_v and N_{ve} are the real and equilibrium vacancy concentrations; γ is the surface tension; Ω is the vacancy volume; $A(1)$ is the surface of a single void. The vacancy fluxes J_0, J_{0e} are equal to

$$J_0 = D_v N_v / 4d\Omega, \quad J_{0e} = (D_v N_{ve} / 4d\Omega) \exp\left(\frac{1}{kT}\frac{\partial \Phi}{\partial n}\right),$$

where D_v is the coefficient of diffusion of the vacancies; d is the lattice period. Generally speaking the functions $K^+(n)$ and $K^-(n)$ are functions of time since they are dependent on the concentration of vacancies $N_v(t)$. However the time to quasi-equilibrium vacancy fluxes in the space between the voids is much less than the characteristic time of variation in void dimensions and therefore in these conditions we can take $K^+(n, t)$ and $K^-(n, t)$ to be only functions of size, and the dependence on the vacancy concentration in this case is a parametric dependence.

Analysis reveals that equilibrium distribution $f(n)$ with kinetic coefficients (29) is very close to an equilibrium solution of equation system (1) and in the case of weak supersaturation coincides with this solution. The equilibrium solution of system (1) takes the form

$$f = N_v \exp[-\Phi(n)/kT]. \tag{31}$$

The critical void dimensions correspond to the maximum of the function $\Phi(n)$ and are found from a solution of the equation $(d\Phi/dn)_{n=n_{cr}} = 0$:

$$n_{\text{кр}}^{1/3} = \sigma/\ln(\Delta_v + 1), \quad \sigma = {}^2/_3\, A\,(1)\,\gamma/kT. \tag{32}$$

If we know $f(n)$ we can determine the parameter $I_0(t)$. Substituting $f(n)$ into equation (21) and integrating it accounting for initial condition (23) we find that with small times I_0 is determined by the following formula:

$$I_0(t) \approx 2I_s[1 - \exp(-t/\tau_0)], \tag{33}$$

$$\tau_0 \approx \int_1^{n_{\text{кр}}} \exp(-\Phi/kT)\,dn \int_1^{n} \frac{\exp(\Phi/kT)}{D}\,dn_1, \tag{34}$$

where the quantity I_s is the stationary void nucleation rate (i.e., the void flux through the critical size $n = n_{cr}$). According to (10) this rate is equal to

$$I_s = \frac{N_v D(n_{\text{кр}}) \exp[-\Phi(n_{\text{кр}})/kT]}{\sqrt{6\pi n_{\text{кр}}^{4/3}/\sigma}}, \quad D(n_{\text{кр}}) = J_0 A\,(1)\,n_{\text{кр}}^{2/3},$$
$$\Phi(n_{\text{кр}})/kT = \sigma n_{\text{кр}}^{2/3}/2. \tag{35}$$

(in formulae $n_{cr} = n_{\text{кр}}$)

Expression (34) determines the characteristic time to a stationary value of the parameter $I_0(t)$ in the subcritical range of dimensions $(1 < n < n_{cr})$. It corresponds in magnitude to the critical nucleus formation time. Evaluating the integrals in (34) by the Laplace method yields the following result:

$$\tau_0 = {}^1/_4 \sqrt{\frac{6\pi}{\sigma}}\, \frac{n_{\text{кр}}^{2/3} \exp[\Phi(n_{\text{кр}})/kT]}{D(n_{\text{кр}})\,\varkappa}, \quad \varkappa = \frac{1}{kT}\frac{d\Phi}{dn}\bigg|_{n=1}. \tag{36}$$

Knowing $I_0(t)$ relations (17) and (18) can be used to determine the temporal progression of the flux and the distribution function of voids of any size in the range $1 < n \leqslant n_{cr}$. As an example we will calculate the flux through the critical dimensions $n = n_{cr}$. Substituting $n = n_{cr}$ into (18) and evaluating the integrals by the Laplace method we have, subject to (33),

$$I(n_{кр}, t) = I_s [1 - \exp(- t/\tau_0)].$$

This expression determines the temporal dependence of the void nucleation rate with small times. It exactly coincides with the known expression found from a solution of the diffusion equation in the subcritical region [7].

We find from formula (17) in an analogous manner that the value of the distribution function of voids of critical dimensions with small times takes the form

$$Z(n_{кр}, t) = Z_s [1 - \exp(- t/\tau_0)], \quad Z_s \approx {}^1/_2 f(n_{кр}).$$

We will now examine the evolution of the void distribution function for times $t > \tau_0$. Substituting (24) and (25) into (17) and altering the integration order in the numerator of (17) we obtain

$$Z(n, t) = f(n) W^{-1}(n_0(t)) \int_n^{n_0(t)} f\, dn_1 \int_n^{n_1} \frac{dn_2}{fD} \int_1^{n_2} (fD)^{-1} dn_3. \tag{37}$$

Subject to (27) expression (37) will be represented in the following convenient form for calculation:

$$Z(n, t) = f(n) \int_n^{n_0(t)} \frac{I(n_1, n_0(t))\, dn_1}{f(n_1) D(n_1)}. \tag{38}$$

The integrand in (38) in the range $n_{cr} < n < n_0(t)$ drops off rapidly with growth of n which is related to the growth of the functions $f(n)$ and $D(n)$ and the reduction in the flux $I(n, n_0(t))$ consistent with (27). Hence in the subcritical range the vicinity of the point n makes the primary contribution to the integral. Accounting for this we obtain the following estimate from (38):

$$Z(n, t) \approx I(n, n_0(t))/D(n). \tag{39}$$

We will determine the void flux in the supercritical range. Evaluating the integrals in (27) by the Laplace method we have

$$I(n, t) = I_s \{1 - \exp[\ln(\Delta_v + 1)(n - n_0(t))]\}. \tag{40}$$

Substituting (40) into (39) we obtain the following expression for the distribution function of supercritical voids:

$$Z(n, t) = Z_s(n) \{1 - \exp[\ln(\Delta_v + 1)(n - n_0(t))]\}. \tag{41}$$

We can easily see that with a fixed end the distribution function of supercritical voids is an increasing function of time and as $\tau \to \infty$ takes the form

$$Z_s(n) = I_s / K^+(n). \tag{42}$$

With a given t it diminishes with growth of n, approaching zero. The variation of the maximal boundary number $n_0(t)$ in time is determined from equation (26). Substituting (24) into (26) and altering the integration sequence in the numerator we obtain the equation

$$\frac{dn_0}{dt} \int_0^{n_0} \frac{dn}{fD} \int_0^{n} f\, dn_1 \int_0^{n_1} \frac{dn_2}{fD} \left[\int_0^{n_0} f\, dn \left(\int_0^{n} \frac{dn_1}{fD} \right)^2 \right]^{-1} = \left[f(n_0) \int_0^{n_0} \frac{dn_1}{fD} \right]^{-1}. \tag{43}$$

This equation is simplified when $n_0(t) \gg n_{cr}$ and takes the form

$$dn_0/dt = -D(n_0)\Phi'(n_0)/kT. \tag{44}$$

So far we have considered the temporal evolution of the void distribution in the case of constant supersaturation. This condition holds, as discussed in the preceding section, only as long as the void number is not too large and intensive void growth does not begin. Over time supersaturation diminishes while the critical dimensions become a monotonically increasing function of time. Therefore for a complete description of the process it is necessary to add an expression to determine the variation in supersaturation in time to solutions (40) and (41).

We will first find the temporal dependence of supersaturation with small times. For this we will substitute distribution (41) into rate equation (5). The problem is, however, significantly simplified if we account for the fact that the duration of the initial condensation stage is significantly greater than the time to a quasi-stationary distribution in the subcritical and near-critical regions. Therefore we can substitute the quasi-stationary distribution of supercritical voids into equation (5).

We can easily see from formula (41) that in the supercritical region the evolution of the ensemble of voids is determined by the difference $n_0(t) - n$. Using this fact we introduce a new variable g:

$$g = n_0(t) - n, \quad 0 < g < n_0(t), \quad n_0(t)|_{t=0} = 0.$$

We will expand the function in the exponent of distribution (42) into a Taylor series in the vicinity of the initial supersaturation $\Delta(g)|_{g=0} = \Delta_0$

$$\Phi(\Delta) = \Phi(\Delta_0) + \left.\frac{d\Phi}{d\Delta}\right|_{\Delta=\Delta_0} (\Delta - \Delta_0) + \dots$$

Substituting the derived result into relation (42) we have:

$$Z_s(g) = \frac{I_s(\Delta_0)\, kT \exp[-B\Pi(g)/kT]}{\Phi'(n_0(t) - g)\, D(n_0 - g)}, \tag{45}$$

where $B=-\frac{\Delta_0}{kT}\frac{d\Phi}{d\Delta}\Big|_{\Delta=\Delta_0}$.

The function $\Pi(g) = 1 - \Delta(g)/\Delta_0$ characterizes the relative drop in supersaturation.

Accounting for the determination of Δ and making the transformation from summation over n to integration, we represent equation (5) as

$$\Delta_0=\Delta(t)+N_{ve}^{-1}\int_0^{n_0(t)} nZ(n,t)\,dn,$$

$$\Pi(n_0)=-(N_{ve}\Delta_0)^{-1}\int_0^{n_0}(n_0-g)\,Z(g)\,dg. \tag{46}$$

We will decompose the integration range into two parts: $(0, n_{cr})$ and $(n_{cr}, n_0(t))$ and we assume that the majority of the vacancies are in supercritical voids. Then substituting (45) into (46) and accounting for the fact that with small times $\Pi(g) \ll 1$, we obtain

$$\Pi(n_0)=\frac{3I_s(\Delta_0)\,n_0^{4/3}}{4N_{ve}\Delta_0^2J_0A(1)} \tag{47}$$

or

$$\Delta(t)=\Delta_0\left[1-\frac{3I_s(\Delta_0)\,n_0^{4/3}(t)}{4N_{ve}\Delta_0^2J_0A(1)}\right]. \tag{48}$$

In order to find the variational law of the function $n_0(t)$ we will use equation (44) that, accounting for the determination of $\Pi(n_0)$ takes the form

$$dn_0/dt=J_0A(1)\,[\Delta_0(1-\Pi(n_0))]/\,n_0^{2/3}. \tag{49}$$

From here with small times

$$dn_0/dt=J_0A(1)\,\Delta_0 n_0^{2/3}.$$

Solving this equation we obtain

$$n_0^{1/3}(t)=J_0\,A(1)\,\Delta_0 t/3. \tag{50}$$

Finally, substituting $n_0^{1/3}(t)$ into (48) we find the following expression for the temporal variation of supersaturation:

$$\Delta(t)=\Delta_0\left[1-\left(\frac{t}{\tau_1}\right)^4\right],\qquad \tau_1=\frac{3}{J_0A(1)\,\Delta_0}\left(\frac{4N_{ve}\Delta_0^2J_0A(1)}{3I_s(\Delta_0)}\right)^{1/4}. \tag{51}$$

The time τ_1 determines the duration of the initial vacancy condensation stage. The dimensions of the void nucleus corresponding to time τ_1 is, according to (50), equal to

$$n_1^{1/3} = J_0 A\ (1)\ \Delta_0\ \tau_1/3. \tag{52}$$

In order to determine the total number of voids formed in a crystal at the end of the first vacancy condensation stage, we substitute (51) into expression (45) and integrate it from zero to τ_1. We then find

$$N = I_s(\Delta_0) \int_0^{\tau_1} \exp\left[-B\,(t/\tau_1)^4\right] dt \approx {}^1/_4 I_s(\Delta_0)\,\tau_1 \Gamma\,({}^1/_4)/B^{1/4},$$

where $\Gamma(1/4)$ is the gamma-function.

It follows from equation (49) that with large times a rapid growth of the function $\Pi(n_0)$ causes deceleration of the front as well as the entire dimensional spectrum of the subcritical voids. We then arrive at the second vacancy condensation stage. Characteristically this stage does not result in the formation of new voids, but rather the existing voids grow from the absorption of vacancies. By the end of the second stage supersaturation drops to zero, corresponding to a maximum value of the function $\Pi(r_{02}) = 1$, where r_{02} is the coordinate of the void distribution front at the end of the second stage, which will be determined below.

It is clear that equations (47) and (49) hold not only in the initial stage but also in the second stage. Therefore using these equations we find the following expression for the temporal dependence of supersaturation with large times:

$$\Delta\,(t) = \Delta_0 \exp\,(-\,t/\tau_2),\ \tau_2 = ({}^3/_4)\ \tau_1. \tag{53}$$

We note that in the second condensation stage, unlike the initial stage, supersaturation is a rapid function of time and decays exponentially with relaxation time τ_2. Consistent with (49) the supersaturation rate of the distribution front drops and when pt > $\tau_2 r_0$(pt) it approaches its own time-independent maximum value r_{02}.

It is possible to determine the maximum void radius from the condition $\Pi(r_{02}) = 1$. Substituting (47) into this expression we have

$$r_{02} = \left(\frac{3\Omega}{4\pi}\right)^{1/3} n_{02}^{1/3} = \left(\frac{3\Omega}{4\pi}\right)^{1/3} \left(\frac{4N_{ve}\Delta_0^2 J_0 A\,(1)}{3I_s\,(\Delta_0)}\right)^{1/4}.$$

We will estimate the characteristic time of the process. The characteristic time t_c normally refers to either the time in which it is possible to ignore the change in void dimensions or the time over which saturation drops by a factor of 2. Using the second definition we obtain according to (53) the estimate

$$t_c \approx \tau_2. \tag{53a}$$

In conclusion we address the applicability condition of equation (44). In deriving this equation we assumed that $n_0(t) > n_{cr}$ or $r_0(t) > r_{cr}$. However as indicated by formula (32) with sufficiently large times the drop in supersaturation will produce an increase in the critical dimensions. Therefore the critical dimensions approach in value the coordinate $r_0(t) - r_1$ (r_1 is the spectral width). Accounting for this fact the applicability condition of (44) can be written as:

$$r_{cr}(\Delta(t)) \lesssim r_0(t) - r_1. \tag{54}$$

We will determine the time τ_p in which this inequality is violated. Substituting into (4) the sign of the approximate equality and using equations (32) and (50) we find

$$t_p \approx t_2 \ln \tau_2. \tag{55}$$

At times $t \gtrsim t_p$ a certain number of supercritical voids become subcritical, which in turn causes a growth of the large voids from the dissolution of the small voids. As a result the number of supercritical voids begins to change in the crystal. We then arrive at the final, third vacancy condensation stage: coalescence. The ratio $n_0(t)/n_{cr}(t)$ adopts a fixed value and will then change only slightly. Taking advantage of this fact as well as equation (49) we find that in this condensation stage

$$\Delta t \sim i^{-1/3}, \quad n_{cr}^{1/3}(t) \sim t.$$

The coalescence process has been investigated in detail in article [26]; the quasi-stationary distribution function method yields analogous results.

Simple analytic void nucleus size distributions are found for all temporal ranges of the vacancy condensation process. Experimentally-verified characteristics of the process are derived together with the degree of supersaturation in the void nucleation process in crystals as a function of time, the critical void nucleation time, the duration of the initial stage and the characteristic and total condensation time. The theory developed above makes it possible to explain a number of experimental factors. In study [27] devoted to an investigation of void dynamics in 304 stainless steel samples of this material were initially irradiated at low temperatures and were then exposed to isothermic annealing at a variety of temperatures. The following picture was observed: in the $T < 1000^\circ$C range voids were still discovered, while at $T \gtrsim 1000^\circ$C no voids were observed. According to the present theory this is due to the fact that the characteristic time of the annealing process is significantly dependent on the annealing temperature. We will provide estimates of this time for 304 stainless steel at $T = 800^\circ$C and $T = 1000^\circ$C using the following stainless steel parameters [28]: $n_1^{1/3} \sim 900$, $D_\upsilon N_{\upsilon e} = \exp(-\varepsilon/kT)$, where $\varepsilon = 3$ eV, $\Delta_\upsilon \sim 10^4$, $d \sim 3.57$ Å. As a result we obtain for $T = 800^\circ$C, $\tau \sim 10^2$ s, and for $T = 1000^\circ$C, $\tau \sim 10^{-3}$ s. Therefore the voids have a rather long life at $T = 800^\circ$C, which has been observed experimentally. When $T = 1000^\circ$C the voids vaporize virtually instantaneously which complicates their observation at this temperature.

We note in concluding this section that although formulae (51), (52), (53), and (55) are similar in external form to the results from study [12] they yield estimates that are significantly (2-3 orders of magnitude) different from study [12].

3. Void Nucleation Kinetics Under Continuous Irradiation

A supersaturated solution of vacancies and interstitial atoms is formed from the high-energy particle irradiation of crystals. This solution is thermodynamically unstable and over time decays, forming new phase nuclei and a variety of voids and loops. Under continuous irradiation both types of point defects form in equal quantities, although there is more intense dislocation absorption of interstitials than vacancies. This produces an additional flow of interstitial atoms to the dislocations. This produces excess vacancies that scatter to the voids, causing their growth. Therefore unlike the condensation of a purely vacancy "gas" the rate of growth of a separate void nucleus will be determined by the excess vacancy flux to the void compared to the interstitial atom flux. Further we note that the high concentration of interstitial atoms increases the potential barrier for void nucleation as well as their critical dimensions. Thus the kinetics of void nucleation and growth in the case of a two-component solution is significantly different from a one-component solution. Below we will derive simple analytic expressions for the nonstationary void nucleation rate with both small and large times and for the critical void nucleation time.

As noted above formulae (17) and (18) are applicable for both a one-component and a two-component solution. In order to determine the explicit form of the distribution function $\varphi(n, t)$ and the flux $I(n, t)$ it is necessary to know the potential barrier $G(n)$ for void nucleation. According to [5]

$$G(n) = kT \sum_{m=n_m}^{n} \ln\left[\frac{\beta_i}{\beta_v} + \exp\left(\frac{1}{kT}\frac{\partial\Phi}{\partial m}\right)\right], \tag{56}$$

where $\Phi(m)$ is the free void nucleation energy from m vacancies in the absence of interstitial atoms $\beta 2i/\beta 2v = N_i D_i/N_v D_v$, N_i, D_i are the concentration and diffusion coefficients of the interstitial actions, respectively; n_m is the minimum nucleus size.

Accounting for (30) expression (56) will be represented in the following form:

$$G(n) = kT \sum_{m=n_m}^{n} \ln\left[\frac{\beta_i}{\beta_v} + \frac{1}{\Delta}\exp\left(\sigma m^{-1/3}\right)\right]. \tag{57}$$

The critical dimensions n'_{cr} in this case correspond to the maximum $G(n:)$

$$\left.\frac{dG}{dn}\right|_{n=n'_{\text{кр}}} = 0,$$

$$n'_{\text{кр}} = \sigma^3/\ln^3\left[\Delta\left(1 - \beta_i/\beta_v\right)\right].$$

Void growth is a slow process compared to the characteristic time to the quasi-stationary point defect distribution. The vacancy and interstitial atom concentrations therefore satisfy stationary kinetic equations describing the compensation of the rate of creation K of new point defects by corresponding losses due to recombination or diffusion to the various free surfaces and drains within the crystal. These equations can be written as:

$$K = \mu (D_i + D_v) N_v N_i + \sum (K^+_{nv} - K^-_{nv}) Z(n, t) + D_v (N_v - N_{ve}) B_v \rho_d,$$

$$K = \mu (D_i + D_v) N_v N_i + \sum K^+_{ni} Z(n, t) + D_i N_i B_i \rho_d,$$

where μ is the recombination coefficient of the vacancy-interstitial pair; ρ_d is the total density of linear dislocations; B_v and B_i are the coefficients accounting for the interaction of vacancies and interstitial atoms with the dislocation: $B_v = 1.00$, $B_i = 1.01$, K^+_{ni} is the absorption rate of the interstitial atoms by the void. According to study [3] $K^+_{ni} = D_i N_i A(1)\, n^{2/3}/4d$. We note that we have neglected point defect losses at the grain boundaries as well as the vacancy- and interstitial-type dislocation loops. A homogeneous spatial distribution of the point defect was assumed on the microscopic level. Knowing $G(n)$ we can use equation (21) to determine the temporal behavior of the parameter $I'_0(t)$ at $t < \tau'_0$. From (21) we have

$$I'_0(t) \approx 2I'_s\,[1 - \exp(-t/\tau'_0)], \tag{58}$$

$$\tau'_0 = \int_{n_m}^{n'_{\text{кр}}} \exp[-G(n)/kT]\, dn \int_{n_m}^{n} \frac{\exp[G(n_1)/kT]\, dn_1}{D(n_1)},$$

$$I'_s = N_v \left[\int_{n_m}^{\infty} \exp[G(n)/kT] \frac{dn}{D(n)} \right]^{-1}, \tag{59}$$

where the quantity I'_s is stationary void nucleation rate in the vacancy and interstitial atom solution; τ'_0 is the critical void nucleation time.

Substituting (57) into (18) and accounting for (58) we find that the temporal progression of the rate of void nucleation with small times is determined by the following formula:

$$I'(n'_{\text{кр}} t) = I'_s[1 - \exp(-t/\tau'_0)].$$

We find analogously that the distribution function of the critical voids is equal to

$$Z'(n'_{\text{кр}}, t) = Z_s(n'_{\text{кр}})\,[1 - \exp(-t/\tau'_0)],\ Z_s(n'_{\text{кр}}) \approx {}^1/_2 f(n'_{\text{кр}}).$$

We will now evaluate the integrals in formulae (59) and (60). For this we will first represent the function $G(n)/kT$ by means of the Euler-Maclaurin summation formula as:

$$G(n)/kT = \int_{n_m}^{n} \Psi(z)\,dz - {}^1/_2[\Psi(n) + \Psi(n_m)] + {}^1/_{12}[\Psi'(n) - \Psi'(n_m)] - \dots,$$

$$\Psi(n) = \ln\left[\frac{\beta_i}{\beta_v} + \frac{1}{\Delta}\exp(\sigma n^{-1/3})\right]. \tag{60}$$

Substituting (61) into (60) and accounting for the fact that the function $G(n)/kT$ has its maximum at $n = n'_{cr}$, by the Laplace method we find

$$I'_s = N_v D(n'_{\text{кр}}) \exp[-G(n'_{\text{кр}})/kT] \times$$
$$\times \left[\left(\frac{6\pi}{\sigma}\frac{\beta_v}{\beta_v - \beta_i}\right)^{1/2} n'^{2/3}_{\text{кр}}\left(1 + \frac{1}{20a} + \frac{1}{21a^2} + \dots\right)\right]^{-1}, \tag{61}$$

$$a = \frac{2n'^{4/3}_{\text{кр}}}{\sigma}(1 - \beta_i/\beta_v),$$

$$G(n'_{\text{кр}})/kT = \int_{n_m}^{n'_{\text{кр}}} \Psi(z)\,dz - \Psi(n_m) - {}^1/_2\Psi'(n_m) + O(1/n'_{\text{кр}}).$$

Analogously we obtain the following expression for τ'_0:

$$\tau'_0 = \frac{1}{4}\sqrt{\frac{6\pi}{\sigma}\frac{\beta_v}{\beta_v - \beta_i}}\;\frac{n'^{2/3}_{\text{кр}}\exp[G(n'_{\text{кр}})/kT]}{D(n'_{\text{кр}})\,\varkappa} \times$$
$$\times\left(1 + \frac{1}{20a} + \frac{1}{21a^2} + \dots\right), \tag{62}$$

$$\varkappa = \frac{1}{kT}\frac{dG}{dn}\bigg|_{n=n_m}.$$

When the condition $a \gg 1$ is satisfied we can limit the analysis to the first term in expansions (61) and (62). Based on (62) we will analyze τ'_0 as a function of the properties of a two-component supersaturated solid solution: the concentrations of interstitial atoms N_i and vacancies N_v, the diffusion coefficients D_i and D_v, and the surface energy γ and temperature T.

The need to consider this problem is related to the specific features of a number of construction materials used in modern equipment. Study [29] carried out an investigation of the influence of interstitial atoms on the void nucleation rate. However as indicated by a numerical analysis [20] the influence of interstitial atoms on τ'_0 is significantly stronger than on the void nucleation rate I'_s. Following study [29] we will obtain simple analytic expressions for τ'_0 for the cases of greatest practical interest. It is clear from formula (62) that the time τ'_0 is determined primarily by the height of the potential barrier $G(n'_{cr})$ for void formation. Therefore it is necessary to carry out a detailed analysis of this function. Applying integration by parts we represent (61) as:

$$G(n'_{\text{кр}})/kT = \sigma^3\int_{\eta_k}^{\eta_m}\frac{e^x}{e^{\eta_v} + e^x}\frac{dx}{x^3} - G(n_m)/kT,$$

$$G(n_m)/kT = (n_m + {}^1/_2)\Psi(n_m) + {}^1/_{12}\Psi'(n_m) + \dots,$$

$$\eta_k = \sigma n_{\text{кр}}^{\prime -1/3}, \qquad \eta_m = \sigma n_m^{-1/3}, \ \eta_v = \sigma n_v^{-1/3} \equiv \ln \frac{\beta_i \Delta}{\beta_v}.$$

Beginning with the case where the primary dissolution mechanism in the entire subcritical range ($n_m < n < n'_{cr}$) is the absorption of interstitial atoms by the void, i.e., $\beta_i/\beta_v > \Delta_v^{-1} \exp \sigma n_m^{-1-3} > 1-\beta_i/\beta_v$. Here $G(n'_{cr})/kT$ can be represented as:

$$G(n'_{\text{кр}})/kT = \sigma^3 \sum_{l=1}^{\infty} (-1)^{l-1} \exp(-l\eta_v) \int_{l\eta_k}^{l\eta_m} e^{\theta}\, d\theta/\theta^3 - G(n_m)/kT. \tag{63}$$

Expanding the integral in (63) into a Taylor series in $1/\theta$ limited to terms of the order of $1/\theta^3$ when $\sigma/4$ o $n_{cr}^{1}/3$, we obtain

$$G(n'_{\text{кр}})/kT \approx \frac{\pi^2}{3\sigma} n_0^{4/3} \exp(\eta_m - \eta_v) - \left(\frac{3n_{\text{кр}}^{\prime 4/3}}{\sigma} + n_{\text{кр}} - n_m\right) \exp(\eta_k - \eta_v).$$

We note that $G(n'_{cr})$ and, consequently, τ'_0 in the case of a two-component solution, unlike a one-component solution, is determined not only by the critical dimensions n'_{cr} but also by the parameter $n_v (\eta_v = \sigma n_v^{-1/3})$ [29]. The parameter n_v for $\beta_i\Delta/\beta_v > 1$ has the following physical meaning: for a void consisting of $n = n_v$ vacancies the interstitial atom absorption and vacancy emission processes occur at an identical rate. A void with dimensions $n < n_v$ releases vacancies more rapidly while a void with $n > n_v$ absorbs interstitial atoms more extensively [29].

We will now consider $G(n_{cr})/kT$ when

$$\Delta_v^{-1} \exp \sigma n_m^{-1/3} > \beta_i/\beta_v > 1 - \beta_i/\beta_v.$$

Different dissolution mechanisms operate depending on the relations between the transition rates in the different integration ranges. Therefore in the range $n_m < n < n_v$ the dominant dissolution mechanism is intensive emission of vacancies at the same time that in the region $n_v < n < n'_{cr}$ intense absorption of the interstitial atoms by the void occurs. Consistent with this approach we will decompose the integration range (η_k, η_m) into parts (η_k, η_v), (η_v, η_m) and will carry out the calculations separately in each of the intervals and analogous to the preceding case we obtain

$$G(n'_{\text{кр}})/kT \approx \frac{\pi^2 n_v^{4/3}}{3\sigma} + \frac{1}{3}\sigma n_v^{2/3} - \left(n'_{\text{кр}} + \frac{3n_{\text{кр}}^{\prime 4/3}}{\sigma}\right) \exp(-\eta_v + \eta_k) -$$
$$- \frac{3n_m^{4/3}}{\sigma} \exp(\eta_v - \eta_m).$$

We note that in this case $G(n'_{cr})$, and consequently, the average critical void nucleation time, is largely determined by the parameter n'_v and not by the critical dimensions n'_{cr}. This means that the interstitial atoms have a significant influence on the stable void nucleation time.

We will now obtain an expression for the flux of critical voids with large times ($t > \tau_0'$). Substituting (57) into formula (27) and evaluating the integrals by the Laplace method we find

$$I_{n'_{\kappa p}}(t) = I_s\{1 - \varkappa_1 \exp[G(n'_0(t))/kT]\},$$

$$\varkappa_1 = -\frac{G'(n'_0(t))}{G'(n_m)} > 0, \quad G(n'_0(t))/kT = \int_{n_m}^{n'_0(t)} \Psi(s)\,ds - \tag{64}$$

$$- {}^1/_2[\Psi'(n'_0(t)) + \Psi(n_m)] + \ldots .$$

We obtain the temporal relation of the change in the boundary number $n_0'(t)$ (for $t > \tau_0'$) from the solution of equation (44). This solution is

$$n_0'^{1/3}(t) = [(J_{0v} + J_{0i})\ A\ (1) \ln(\beta_i/\beta_v + \Delta_v^{-1})]\ t/3. \tag{65}$$

It follows from formulae (64) and (65) that the rate of void nucleation with large times ($t \to \infty$) grows, approaching its own steady-state value equal to I_s'.

In order to determine the supercritical void flux ($n_{cr}' < n < n_0'(t)$) we obtain from (27) the following relation:

$$I'(n, t) = I_s'\{1 - \exp[(\ln(\beta_i/\beta_v) + \Delta_v^{-1})(n_0'(t) - n)]\}. \tag{66}$$

The value of the flux in (66) with a given n grows with time and as $t \to \infty$ ($n_0'(t) \to \infty$) it adopts its maximum value I_s' while with a fixed t it diminishes with growth of the dimensions n, approaching zero. In conclusion we note that the function $G(n)$ for a single-component solution has a sharp maximum, which makes possible easy evaluation of the integrals in formulae (59), (60). In the case of a two-component solution this maximum is continuous, which results in the conditions $a \geq 1$ and $\sigma/4 \geq n_{cr}'^{1/3}$ which is easily implemented in the most interesting range of values of the critical void dimensions.

Therefore it is also possible to obtain an analytic expression for the temporal evolution of the void nucleation rate in the case of a two-component solution of vacancies and interstitial atoms (assuming quasi-stationarity of the distribution) as well as to estimate the average stable void formation time in conditions of continuous irradiation accounting for the influence of interstitial atoms.

In conclusion we will compare τ_0' to experimental data. We will use the following parameters of neutron-irradiated stainless steel to estimate τ_0':

$D_{v0} \simeq 0.6$ cm²/s, $N_{ve} = \exp(-\varepsilon^v_f/kT)$, $D_v = D_{v0}\exp(-\varepsilon^v_m/kT)$, $\varepsilon^v_m \simeq 1.6$ eV, $\varepsilon^v_f \simeq 1.4$ eV, $\Delta \simeq 10^5$, $n_v \simeq 2$, $n_{cr}' \simeq 11$, $\beta_i/\beta_v \simeq 0.99$, $\gamma \simeq 10^3$ erg/cm². At $T = 500°$C the estimate by formula (62) yields: $\tau_0' \approx 2 \cdot 10^4$ s. This value of τ_0' is in good agreement with computer-generated calculations [30].

4. Influence of Stresses on Homogeneous Void Nucleation Kinetics in Irradiated Crystals

Recently extensive efforts have been made to identify the formation and growth mechanisms of the vacancy voids responsible for the swelling of many structural materials used in modern nuclear reactors. The nature of the void nucleation and growth processes depends on temperature, the radiation dosage, the properties of the material and a number of other factors, including the influence of tensile stress σ_H. An investigation of the influence of this factor on the void nucleation process can yield exhaustive information in the comprehensive search for the means to improve the strength of materials exposed to a variety of complex mechanical stresses in reactor cores. Studies [31-35] have investigated the influence of stresses on the radiation swelling rate both experimentally and theoretically. It has been established that at elevated temperatures tensile stress causes a significant change in void concentration. The influence of external stress on the void nucleation process has been established experimentally in nickel [36]. This study has shown that at low stresses the nucleation rate I_s' grows linearly, and at high stresses it drops, with the maximum of the $I_s'(\sigma_H)$ relation approximately corresponding to the yield strength of the material. It has also been established that the voltage has a stronger influence on I_s' with growth of the irradiation temperature: a stress of σ_H = 30 MPa results in an increase in the void nucleation rate (void concentration) by a factor of 1.0, 1.1, 1.7, and 2.0 at irradiation temperatures of 450, 550, 625, and 650°C, respectively. The influence of stresses on the void nucleation process has been examined numerically in study [37]. Below we derive analytic expressions for the dependencies of the nucleation rate and critical void nucleus formation time on stresses; these expressions make it possible to theoretically predict the kinetics of the void nucleation process in stressed, irradiated metals.

We will consider a crystal containing dislocations of density ρ_d at temperature T. In this system continuous irradiation produces voids of density ρ_s and radius $\bar{r}$. We will take the dislocations to be the dominant drain for defects ($B_v\rho_d \gg 4\pi\rho 2s\bar{r}$, where B_v is the vacancy trapping efficiency by the dislocations) and we assume that the volume recombination of point defects is small compared to the their drain losses.

It has been established to date that an external stress can influence the formation and growth of vacancy voids in metals in many different ways. First the tensile stress influences the nature of the interaction of point defects with the dislocations, thereby changing the efficiency of vacancy and interstitial trapping by the dislocations. It has been demonstrated in study [35] that in a stressed crystal these efficiencies obey a linear law with growth of stress:

$$B_{i,v}(\sigma_H) = B_{i,v}(0)(1 + w_{i,v}\sigma_H/E), \tag{67}$$

where $B_{i,v}(0)$ are the trapping efficiencies in an unstressed crystal; σ_H is uniaxial tensile stress; E is Young's modulus; $w_{i,v}$ is a constant of the order of 10^{-1}. Second, the stress reduces the equilibrium vacancy concentration by the factor $\exp[\Omega\sigma_H Å^3 kT]$ and also reduces the potential barrier $G(r_{cr})$ of void formation compared to an unstressed sample by $4\pi\ r^3_{cr}(0)\ \sigma_H/9$, where $r_{cr}(0)$ are the critical dimensions of the void nucleus in an unstressed sample.

Finally, the third mechanism for external stress action on the void nucleation process is the change in dislocation density at high stresses.

According to classical nucleation theory [5, 6] and formula (62) of the present study the rate of void nucleation and the critical void nucleus formation time are determined by the height of the activation barrier $G(r_{cr})$

$$I_s \sim \exp[-G(r_{кр})/kT], \qquad \tau_0 \sim \exp[G(r_{кр})/kT], \tag{68}$$

whose value is dependent on such factors as the vacancy supersaturation Δ_v, the surface void tension energy γ, the efficiencies of point defect capture by the dislocations B_i and B_v and the dislocation density ρ_d. A change in the activation barrier height is the channel for the external stress altering the nucleation rate and critical void formation time. We will estimate $G(\sigma_H)$ in a stressed crystal. Consistent with these mechanisms the actions σ_H on the nucleation process $G(\sigma_H)$ can be represented as a sum of four components:

$$G(\sigma_H) = G_0 + G_1 + G_2 + G_3, \tag{69}$$

where G_0 is the height of the energy barrier in an unstressed crystal; G_1, G_2, G_3 characterized the contribution to $G(\sigma_H)$ from the change in the efficiency of dislocation trapping of point defects, the equilibrium vacancy concentration and the dislocation density.

We should note that the contribution to $G(\sigma_H)$ from changes in B_i, B_v is small and, as indicated by calculations, will not result in experimentally observed changes in the quantities I_s and τ_0 and also explains the increase (decrease) in the nucleation rate (critical void formation time) with growth of the radiation temperature. Henceforth we will ignore the change in $G(\sigma_H)$ from $B_{i,v}(\sigma_H)$.

Study [33] obtained the contribution of the change in equilibrium vacancy concentration to $G(\sigma_H)$:

$$G_2 = 4\pi r^3_{кр}(0)\ \sigma_H/9.$$

In order to estimate the contribution to $G(\sigma_H)$ from changes in the dislocation density we will use the following expression [38] for the void nucleus growth rate in a stressed, irradiated crystal

$$\frac{dr}{dt}=\frac{(B_i-B_v)K}{rB_iB_v\rho_d}+\frac{D_vN_{ve}\Omega}{rkT}\left(\frac{\sigma_H}{3}-\frac{2\gamma}{r}\right).$$

We will determine the critical nucleus dimensions in the stressed sample. Setting the derivative dr/dt equal to zero we find

$$r_{кр}(\sigma_H)=2\gamma\left[\frac{(B_i-B_v)KkT}{B_iB_vD_vN_{ve}\rho_d\Omega}+\frac{\sigma_H}{3}\right]^{-1}.$$

The estimate shows that at high temperatures $(B_i - B_v)KkT/B_iB_vD_vN_{ve}\rho_d \gg \sigma_H/3$. Hence

$$r_{кр}(\sigma_H)=\frac{2\gamma\Omega B_iB_vD_vN_{ve}\rho_d(\sigma_H)}{kTK(B_i-B_v)}\left[1-\frac{\Omega B_iB_vD_vN_{ve}\sigma_H\sigma_d(\sigma_H)}{3(B_i-B_v)kTK}\right]. \tag{70}$$

Under high stresses the total dislocation density in the irradiated crystal obeys the following law with growth of stress [31]:

$$\rho_d(\sigma_H)=\rho_d(0)(1+\alpha\sigma_H^2), \tag{71}$$

where $\rho_d(0)$ is the total dislocation density in the irradiated unstressed crystal. The quantity α depends on the radiation temperature. When T = 450°C α = 3.6·10^{-4} MPa^{-2}, while when T = 650°C α = 3·10^{-4} MPa^{-2} [36]. Substituting (71) into (70) we find that at high stresses the critical void nucleus dimensions obeys the following law with growth of stress:

$$r_{кр}(\sigma_H)=r_{кр}(0)(1+\alpha\sigma_H^2)\left[1-\frac{r_{кр}(0)}{6\gamma}\sigma_H(1+\alpha\sigma_H^2)\right].$$

Therefore we can represent the contribution from the change in the total dislocation density to $G(\sigma_H)$ as:

$$G_3=\frac{4\pi r_{кр}^3(0)\sigma_H}{9}\left[\frac{3\gamma\alpha}{r_{кр}(0)}(2\sigma_H+\alpha\sigma_H^3)-(1+\alpha\sigma_H^2)^3\right].$$

We then have the following expression for $G(\sigma_H)$:

$$G(\sigma_H)=G_0-\frac{4\pi r_{кр}^3(0)\sigma_H}{9}+\frac{4\pi r_{кр}^3(0)\sigma_H}{9}\left[\frac{3\gamma\alpha}{r_{кр}(0)}(2\sigma_H+\alpha\sigma_H^3)-(1+\alpha\sigma_H^2)^3\right]. \tag{72}$$

Substituting (72) into formulae (68) we find that the void nucleation rate and critical void formation time in irradiated samples obey the following laws with growth of stress:

$$I_s^{\sigma_H}=I_s^0\exp\{A\sigma_H[1-B(2\sigma_H+\alpha\sigma_H^3)+(1+\alpha\sigma_H^2)^3]\}, \tag{73}$$

$$\tau^{\sigma_H}=\tau^0\exp\{A\sigma_H[B(2\sigma_H+\alpha\sigma_H^3)-(1+\alpha\sigma_H^2)^3-1]\}, \tag{74}$$

where I_s^0 and τ^0 are the rate of void nucleation and the critical void formation time in an unstressed crystal as determined by formulae (35) and (36). The constants A and B depend on the irradiation conditions and are equal to

$$A = 4\pi r_{\text{кр}}^3(0)/9kT, \quad B = 3\gamma\alpha/r_{\text{кр}}(0).$$

According to expression (73) the void nucleation rate at low stresses increases linearly $I_s^{\sigma_H} \sim I_s^0(1 + A\sigma_H)$, and under high stresses diminishes exponentially $I_s^{\sigma_H} \sim \exp\{-A[B(2\sigma_H^2 + \alpha\sigma_H^4) + (1 + \alpha\sigma_H^2)^3]\}$. We will now compare results obtained on the basis of this model to experiment. The most detailed experimental data from an investigation of the influence of external stress on the void nucleation process in irradiated crystals were obtained in studies [31, 36]. Study [36] experimentally determined the nonmonoticity of the dependence of the void nucleation rate (void concentration) on the applied stress and the following empirical formula was proposed for its description:

$$I_s^{\sigma_H} = I_s^0 \exp[A\sigma_H(1 - P\sigma_H^2)], \tag{75}$$

where the constants A and B are dependent on the irradiation temperature. At 650°C $A = 3.1 \cdot 10^{-2}$ MPa^{-1} and $P = 3 \cdot 10^{-3}$ Pa^{-2}. Fig. 1 shows a theoretical plot of the void nucleation rate plotted as a function of the applied stress in preannealed nickel irradiated by 1 meV nitrogen ions together with the relation calculated by formula (75). Here the following radiation parameters were used: $K = 7 \cdot 10^{-3}$ SNA/s, $T = 650$°C, radiation dosage, 20 SNA. The following parameters were selected for the nickel: $D_\upsilon = D_{\upsilon 0}\exp(-\varepsilon/kT)$, where $\varepsilon = 1.3$ Ev, $D_{\upsilon 0} = 0.9$ cm^2/s, $N_{\upsilon e} = \exp(-\varepsilon/kT)$, where $\varepsilon = 1.2$ eV, $\Omega = 1.1 \cdot 10^{-23}$, $\gamma = 400$ erg/cm^2, $\rho_d^0 = 2 \cdot 20^{10}$ cm^{-2}, $(B_i - B_\upsilon)/B_iB_\upsilon = 2 \cdot 10^{-1}$. It is clear from Fig. 1 that the void nucleation rate at stresses less than 30 MPa grows and is identical to the experimentally-observed relation. The growth of the void nucleation rate in this range of stresses is related to the reduction of the energy barrier of critical void formation in the stressed sample. When $\sigma_H > 30$ MPa $I_s^{\sigma_H}$ drops, due to the increase in dislocation density resulting from stress action. The

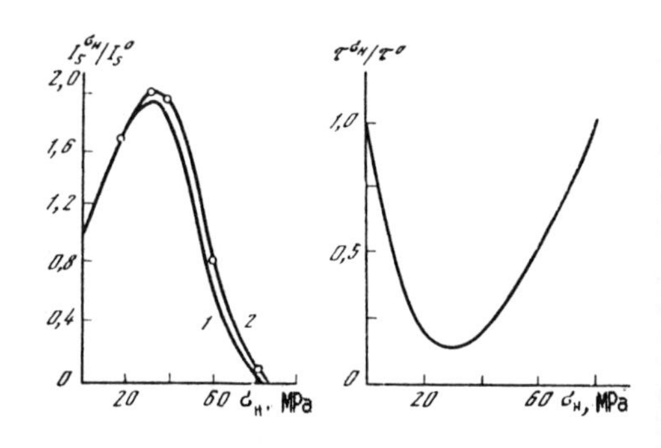

Fig. 1. Relative change in the void nucleation rate plotted as a function of applied stress in nickel
1 - calculated curve; 2 - curve from empirical formula

Fig. 2. Relative change in the critical void formation time plotted as a function of nickel stress (radiation dosage 20 SNA, $T = 650$°C)

small discrepancy between calculated values and experiment at high stresses is evidently related to ignoring the change in the trapping efficiency of point defects by dislocations with growth of the dislocation density.

It follows from Fig. 2 that the critical void nucleus formation time at low stresses drops, and when $\sigma_H = \sigma_H^{min}$ it adopts a minimum value, growing when $\sigma_H > \sigma_H^{min}$. The minimum value σ_H^{min} corresponds to the minimum of the functions $G(\sigma_H)$ and is found from a solution of the equation

$$(1+\alpha\sigma_H^2)^3 + 6\alpha\sigma_H^2(1+\alpha\sigma_H^2)^2 - 4\sigma_H B(1+\alpha\sigma_H^2) + 1 = 0. \tag{76}$$

Solving equation (76) numerically we find that when $T = 650°C$ $\sigma_H^{min} \sim 30$ MPa. This value approximately corresponds to the yield strength of nickel. Overall the calculated data are in agreement with experimentally observed relations. This section has derived analytic expressions for the $I_s^{\sigma_H}$, $\tau_0^{\sigma_H}$ relations that make it possible to describe void nucleation kinetics in an irradiated crystal under an external stress. A subject of equal interest is investigating the influence of a stress $\sigma_H(t)$ variable over time on void kinetics. A temporal dependence of $\sigma_H(t)$ may cause a significant change in the nonstationary void nucleation rate and the radiation swelling rate of the material, influence the formation dynamics and stability of void superlattices and create certain possibilities for controlling these processes.

References

1. Becker, R., Doring, W. "Kinetische Behandlung der Keimbildung in übersättigten Dämpfen" ANN. PHYS., 1935, vol. 24, pp. 719-730.

2. Volmer, M., Weber, A. "Keimbildung in übersättigten Gebilden" ZTSCHR. PHYS. CHEM., 1926, vol. 119, pp. 277-300.

3. Zel'dovich, Ya.B. "Theory of new phase formation: cavitation" ZhETF, 1942, vol. 12, no. 11/12, pp. 525-533.

4. Frenkel' Ya.I. "Kineticheskaya teoriya zhidkostey" [Kinetic theory of fluids] Leningrad: Nauka, 1975, 592 pp.

5. Russel, K.C. "Nucleation of voids in irradiated metals" ACTA MET., 1971, vol. 19, no. 8, pp. 753-758.

6. Katz, J.L., Widersich, H. "Nucleation of voids in material supersaturated with vacancies and interstitials" J. CHEM. PHYS., 1971, vol. 55, no. 3, pp. 1414-1425.

7. Abragam, F.F. "Non-steady-state nucleation" J. CHEM. PHYS., 1969, vol. 51, pp. 1632-1636.

8. Bondarenko, A.I., Konobeev, Yu.V. "Void growth kinetics irradiated metals" PHYS. STATUS SOLIDI (a), 1976, vol. 34, pp. 195-205.

9. Ryazanov, A.I., Sidorenko, A.D. "Kinetics of nucleation of clusters of defects in materials under continuous and cyclic irradiation conditions" RADIAT. EFF., 1984, vol. 81, pp. 89-100.

10. Kantrovits, A. "Nucleation in very rapid vapor expansions" J. CHEM. PHYS., 1951, vol. 19, no. 9, pp. 1097-1102.

11. Binder, K. "Theory for the dynamics of 'clusters'" PHYS. REV. B, SOLID STATE, 1977, vol. 15, no. 9, pp. 4425-4447.

12. Mikhaylova, Yu.V., Maksimov, L.A. "The kinetics of void nucleation from a supersaturated vacancy solution" ZhETF, 1970, vol. 59, no. 10, pp. 1368-1377.

13. Suhl, H., Turner, P.A. "Nucleation of voids and their growth during electromigration" J. APPL. PHYS., 1973, vol. 44, no. 11, pp. 4891-4895.

14. Lyubov, B.Ya. "Teoriya kristallizatsii v bol'shikh ob"emakh" [Large volume crystallization theory] Moscow: Nauka, 1975, 256 pp.

15. Reshetnyak, S.A. "Problems in kinetics in plasma and rotational transition lasers" In: Teoreticheskie problemy spektroskopii i gazodinamicheskikh lazerov [Theoretical problems of spectroscopy and gasdynamic lasers] Moscow: Nauka, 1975, pp. 146-215 (Tr. FIAN; vol. 83).

16. Reshetnyak, S.A., Shelepin, L.A. "The quasi-equilibrium function technique in physical kinetics" In: Konstruktivnye i kineticheskie problemy statisticheskoy fiziki [Structural and kinetic problems of statistical physics] Moscow: Nauka, 1979, pp. 90-118 (Tr. FIAN; vol. 106).

17. Shelepin, L.A. "Plasma as an active medium of lasers" In: Plazma v lazerakh [Plasma in lasers] Moscow: Energoizdat" 1982, pp. 8-70.

18. Reshetnyak, S.A., Kharchev, S.M., Shelepin, L.A. "Asymptotic techniques in kinetic equation theory" In: Gruppovye razlozheniya i kineticheskie metody v teorii gazov [Group expansions and kinetic methods in gas theory] Moscow: Nauka, 1984, pp. 3-36 (Tr. FIAN; vol. 144).

19. Klimontovich, Yu.L. "Statisticheskaya fizika" [Statistical physics] Moscow: Nauka, 1982, 608 pp.

20. Landau, L.D., Lifshits, E.M. "Statisticheskaya fizika" [Statistical physics] Moscow: Nauka, 1976, Ch. 1.

21. Reshetnyak, S.A., Kharchev, S.M., Shelepin, L.A. "Asymptotic solutions in Landau theory for second order phase transitions" In: Kvantovaya mekhanika i statisticheskie metody [Quantum mechanics and statistical methods] Moscow: Nauka, 1986 (Tr. FIAN; vol. 173).

22. Lushnikov, A.N., Sutugin, A.G. "Current state of homogeneous nucleation theory" USPEKHI KHIMII, 1976, vol. 40, no. 3, pp. 385-415.

23. Sears, V.F. "Nucleation of voids in irradiated metals" J. NUCL. MATER., 1971, vol. 39, pp. 18-22.

24. Mayer, R.M., Brown, L.M. "Nucleation and growth of voids by radiation" J. NUCL. MATER., 1980, vol. 95, pp. 46-82.

25. Akhiezer, I.A., Davydov, L.N. "The theory of vacancy swelling of metals and alloys, Part II" Ch II, METALLOFIZIKA, 1981, vol. 3, no. 1, pp. 3-14.

26. Lifshits, I.M., Slezov, V.V. "The kinetics of the diffuse decay of supersaturated solid solutions" ZhETF, 1958, vol. 35, no. 2(8), pp. 479-492.

27. Ehrlich, K., Packan, N.H. "Void resulting from fast-neutron irradiation of stainless steel" J. NUCL. MATER., 1973, vol. 46, pp. 77-85.

28. Harkness, S.D., Li Ch.Y. "A study of void formation in fast neutron-irradiated metals" MET. TRANS., 1971, vol. 2, pp. 1457-1470.

29. Subbotin, A.V. "Void nucleation" ATOM. ENERGIYA, 1978, vol. 45, no. 4, pp. 276-280.

30. Powell, R.W., Russel, K.C. "Computer evaluation of nucleation of voids in irradiated metals" RADIAT. EFF., 1972, vol. 12, pp. 127-131.

31. Bates, J.F., Gilbert, E.R. "Experimental evidence for stress enhanced swelling" J. NUCL. MATER., 1976, vol. 59, pp. 59-102.

32. Boltax, A., Foster, J.P., Weiner, R.A. "Void swelling and irradiation creep relationships" J. NUCL. MATER., 1977, vol. 65, no. 1/3, pp. 174-181.

33. Brager, H.R., Garner, F.A., Guthrie, G.L. "The effect of stress on the microstructure of neutron irradiated type 316 stainless steel" J. NUCL. MATER., 1977, vol. 66, pp. 301-321.

34. Lebedev, S.Ya., Bogdanov, V.G., Surkov, S.Ya. "The influence of a uniaxial tensile stress on radiation swelling of nickel" FMM, 1982, vol. 54, no. 3, pp. 595-596.

35. Heald, P.T., Speight, M.V. "Point defect behavior in irradiated materials" ACTA MET., 1975, vol. 23, pp. 1389-1399.

36. Bagdinov, V.G., Lebedev, S.Ya. "The influence of tensile stress on radiation void nucleation in metals" FMM, 1984, vol. 58, no. 3, pp. 518-521.

37. Fisher, S.B., White, R.J., Harbottle, J.E. "Growth of voids under stress" RADIAT. EFFICIENT., 1979, vol. 41, pp. 65-70.

38. Fisher, S.B., White, R.J., Harbottle, J.E. "The effect of stress on the incubation and growth of voids" RADIAT. EFFICIENT., 1979, vol. 40, pp. 87-90.

FORMATION OF VOID SUPERLATTICES IN SOLIDS

F.Kh. Mirzoev, E.P. Fetisov, L.A. Shelepin

ABSTRACT

A vacancy void lattice is examined as a dissipative structure appearing in open, nonequilibrium systems as a result of cooperative phenomena (self-organization). It is proven that the formation process of void superlattices in irradiated metals has a threshold nature based on a linear stability analysis of a homogeneous quasi-stationary solution of a system of nonlinear equations for the point defect concentration and the void distribution function, normally employed in radiation void theory for the case of irradiated metals. A phase transition occurs in the system at specific defect generation rates (and low dislocation densities): a homogeneous void distribution becomes inhomogeneous and a three-dimensional void lattice arises. The period and characteristic lifetime of the resulting structure are determined. The influence of an external stress (distension) on these characteristics is briefly examined. The derived estimates are in agreement with the experimental data.

1. Defect Superlattices in Solids. Void Superlattices

Interest has increased in recent years in three-dimensional superlattice structures formed by defects in solids. These include interstitial atom, vacancy, dislocation, dislocation loop, precipitate and void superlattices. Superlattice structures significantly change the properties of matter and are of interest for a number of applications.

Vacancy and N, O, C impurity atom superlattices have been observed in the vanadium group transition metal crystals (V, Nb, Ta) [1-3]. Investigations have been carried out in an impurity range from 0.1 at.% to 10 at.%. The period is three times greater than the matrix period in a number of superlattices such as those of the Ta-N system. Very large quantities of N, O, and C dissolve in the vanadium group metals at high temperatures. These solubilities drop significantly with reduction in temperature. During the cooling process the solid solution becomes supersaturated which often causes precipitation to a second phase. Study [2] has observed a high density of ordered precipitates ~2 nm in diameter.

A superlattice significantly changes the physical properties of materials, particularly the mechanical, electrical, and electronic properties. Therefore metals containing a superlattice of impurity atoms have optimum microhardness and residual resistance. Study [4] has discussed the influence of dislocation systems on superconducting properties. Superconducting states are formed in such structures at temperatures and fields below critical values. If the dislocations form a periodic structure, the superconducting states are delocalized, forming a zone. Point defect, dislocation and precipitate superlattices arise from the interaction of defects. These interactions are determined by the their crystal lattice distortions and stresses and can be estimated approximately.

Another situation has evolved in the analysis of the spatial ordering of vacancy voids. There is no unified approach at present here. Voids in solids are formed from the confluence of mobile vacancies formed from high-energy particle irradiation of the specimen [5] or from a variety of mechanical and thermal treatments. One chapter of the present volume [6] investigates the kinetics of void nucleation and growth in detail. As a rule the spatial distribution of vacancy voids in irradiated material is arbitrary and near-uniform. However in certain experimental conditions the voids can form a three-dimensional ordered structure with a high degree of spatial symmetry. Evans discovered a vacancy void lattice for the first time in pure molybdenum irradiated at T = 870°C with 2 MeV nitrogen ions. This phenomenon was then observed in other metals with a face- and body-centered cubic structure: W, Ta, Nb, Ni, Al, etc. According to results from research carried out using the X-ray small-angle scattering technique and by electron microscopy [8] spatial void ordering is observed even in certain hardened nonirradiated metals such as hardened polycrystalline molybdenum foil. For the majority of metals the average void size in the lattices is $\bar{x}$ ~ 2-4 nm while the void lattice period pd_0 ~ 20-60 nm. Extensive experimental and theoretical studies have been devoted to an investigation of this effect [7-21]. It was established that a variety of external actions influence the void lattice period. Such factors as the radiation temperature, the defect generation rate (or radiation dosage), pressure, etc. have a significant influence. It is important to obtain the corresponding dependencies for practical application. Knowledge of the temperature dependence of d_0 makes it possible to determine the radiation temperature at points in fast neutron reactors where thermocouples or other instruments cannot be used.

The voids alter the mechanical [25], optical [24], and superconducting [26] properties of the metals. It was demonstrated in a detailed investigation [26] of superconductors with a void superlattice that in certain conditions Josephson effects will be observed; these are associated with the tunneling of Cooper pairs between the voids.

In the present study we will consider the void superlattice formation process. Section 2 contains a survey of experimental data and discusses the mechanisms of void interaction. In Section 3 the conditions for spatial ordering are derived based on an analysis of a superlattice as a dissipative system and the primary relations are estimated. Section 4 compares the derived results to experiments.

Experimental and Theoretical Data on Void Superlattices

We will consider the primary experimental results [7-15, 27] from investigations of void lattices in irradiated metals. Electron microscopy has revealed that radiation voids will form in certain conditions a one-dimensional chain in equidistant planes [11] or will form in a three-dimensional void lattice whose symmetry and crystallographic axes are identical to those of the lattice of the host material. The degree of perfection of the void superlattices and their period are highly dependent on the properties of the irradiated material and the irradiation conditions. (The irradiation conditions normally include the following factors: type of bombarding particles; their energy; the flux density; temperature of the irradiated specimen; radiation duration (fluence); and radiation mode.) For example, for molybdenum irradiated by fast neutrons with a fluence up to $\sim 3 \cdot 10^{26}$ neutrons/m^2 at $T \approx 650^\circ$C, $d_0 \sim 30$ nm for $T \approx 585^\circ$C; $d_0 \sim 26$ nm for $T = 470^\circ$C, $d_0 \sim 23$ nm [9].

Growth of d_0 with growth of sample temperature was observed even in refractory alloys [14]. It was established experimentally that the period of the void superlattices in neutron-irradiated body-centered cubic metals have a complex dependence on $T_{\text{пл}}/T$ [9, 13]. The period d_0 in niobium alloys and the mean void radius in the lattice increased with growth of the irradiation temperature up to 800°C. A drop in the void lattice period is observed with an increase in the point defect creation rate [9, 13].

The degree of perfection and the very existence of void superlattices are largely determined by the rate of defect formation or the radiation dosage as well as the impurity atom concentration, the dislocation density and the number of dislocation loops. The degree of perfection of the void superlattices can be enhanced by increasing the defect creation rate; growth of the impurity concentration aids in void disordering. An excess number of dislocations at the separate sites of a metallic matrix can also cause local void disordering. Void superlattice formation is directly related to the generation of vacancies and interstitial atoms from irradiation. At sample temperatures above (0.2-0.3) $T_{\text{пл}}$ ($T_{\text{пл}}$ is the melting point of the metal) these defects are mobile and effectively scatter throughout the crystal bulk. After diffusion they either recombine or are absorbed by a variety of inhomogeneities (grain boundaries, dislocations, atom impurities, etc.) or form clusters (interstitial- and vacancy-type voids or dislocation loops). A quasi-steady-state distribution of vacancies and interstitial atoms is established at some time after the beginning of irradiation. Cluster growth has been investigated in detail in one chapter [6] of the present volume although in this case ignoring their spatial distribution. This is sufficient with random (homogeneous) spatial void distribution, which is normally the case. However in order to analyze void lattice formation it is necessary to account for their interaction.

Many studies [16-19] have proposed void lattice stability mechanisms. Study [17] attributed superlattice stability to elastic void interaction determined by the stresses arising in the vicinity of the void [17]. The initial expression for the interaction energy of the two voids is:

$$E(r) = \frac{8\gamma^2 (g^2 - 1)}{3\pi C_{44}} \left\{ \left[(x/r)^3 - \frac{28 C_{44}}{9(1-g)\delta} (x/r)^5 \right] [3 - 5S(4)] \right\}, \tag{1}$$

where γ is the surface energy; $S(4)$ is the geometric factor; δ and g are expressed through the elastic constants C_{11}, C_{12}, and C_{44} and are equal to

$$g = (C_{12} + C_{44})/(C_{12} + 2C_{44}),\ \delta = C_{11} - C_{12} - 2C_{44} = (C_{11} - C_{12})\ (1 - A),$$

$A = 2C_{44}/(C_{11} - C_{12})$ — is the anisotropy factor.

It follows from (1) that at large distances the voids attract as $\sim(x/L)^3$ (L is the average distance between voids) while at short distances they repel as $\sim(x/L)^5$. According to (1) the void lattice stability is related to the anisotropy of the elastic lattice constants and is determined by the condition $\delta > 0$, i.e., $A < 1$. However the elastic interaction mechanism cannot explain the entire set of experimental data. According to this mechanism a void lattice will not be formed in such metals as tungsten (which is virtually isotropic), tantalum (has a negative anisotropy parameter), and nickel (for which $A \sim 1.2$), although a void lattice has been obtained experimentally in these very metals.

In study [9] it was assumed that superlattices form as a result of the influence of stress fields created by the voids on the emission of vacancies from the void surfaces. According to Tewary's model [16] in which the anisotropy role is no longer critical, the crystal-matrix is divided into blocks of equal size to the voids and analogous to Wigner-Seitz cells for regular lattices. In this case the interaction energy between blocks, analogous to the Born-von Kármán model is calculated by the Green's function technique. A value of $d_0/\bar{x} \sim 10$ in agreement with experimental data [7] was obtained for molybdenum at the minimum of the binding energy of two voids. The block model makes it possible to explain the stability of a void lattice compared to thermal fluctuations. However this model does not account for the dynamics of the change in size occurring during the formation of the superlattices and, consequently, cannot explain the spread of the ratio of the void lattice period to the average void radius in different experiments.

Maksimov and Ryazanov [18, 19] formulated a theory of void interaction in a point defect supersaturated crystal for the case of a constant vacancy generation rate. Underlying this theory is an analysis of the diffusion interaction of two voids by the volumetric diffusion of point defects in the intervoid space. A void whose radius is greater than the critical radius ($x > x_c$) causes a reduction in the vacancy density in its vicinity and thereby slows down the growth of the neighboring void. On the other hand a void whose radius is subcritical ($x < x_c$) will, in evaporating, create excess vacancies in its vicinity, which accelerates growth of the neighboring void. Within the framework of this model the voids "do not interact" when their dimensions are critical. The superlattices are formed by the suppression of growth of neighboring voids by the large voids. Under continuous irradiation the vacancy concentration as well as the probability of homogeneous formation of new voids are at a maximum at the center between neighboring large voids. In the initial stage of the void formation process (until the large voids reach a level characteristic of the coalescence stage) the large voids have a tendency to arise at a maximum distance from one another which, in turn, produces mutual correlation of the voids [18]. However this theory cannot answer the question of why the symmetry and crystallographic axes of the superlattices coincide with the symmetry and crystallographic axes of the host lattice with isotropic diffusion coefficients of the cubic crystals.

Study [8] has attempted to explain the formation of void lattices within the framework of the long-range void diffusion interaction mechanism (ignoring elastic interaction) as the anisotropy of diffuse fluxes caused by the internal reflections of the interstitial atoms off the close-packed faces.

In order to establish the void superlattice formation mechanisms it is important to identify the dependencies of the void lattice parameters on the irradiation conditions. From this viewpoint study [12] is of significant interest; this study proposed a void ordering mechanism for irradiated materials based on minimizing the configurational energy of the system in the formation of the lattice which includes both voids and point defects. The following expression was derived for the period d_0

$$d_0 = [144\gamma (2x)^3 D_V / K\varepsilon_f^v]^{1/4}, \tag{2}$$

where $D_v = D_{v0}\exp(-\varepsilon_m^v/kT)$ is the diffusion coefficient of the vacancies; ε_m^v and ε_f^v is the migration energy and the vacancy formation energy. Relation (2) gives the dependence of the void lattice parameter on the rate K of generation of the point defects. The very derivation of this experimentally-observed relation is a certain achievement itself.

However a comparison of $d_0(K)$ given by formula (2) to the experimental data indicates a discrepancy. Thus, for example, for molybdenum and niobium irradiated by accelerator ions ($K = 2 \cdot 10^{-2}$–$8 \cdot 10^{-3}$ SNA/s) and neutrons ($K = 7 \cdot 10^{-7}$–10^{-6} SNA/s) the $d_0(K)$ relation is not in agreement with experimental data, since in fact $d_0(K) \sim K^{-1/16}$. Moreover, expression (2) does not explain the complex dependence of the void lattice period on the irradiation temperature. As indicated by study [9] ε_m^v must be 3-4 times smaller than the value from study [9] in order to achieve satisfactory agreement between experimental and calculated values of d_0. Other attempts to describe the temperature dependence of the void lattice parameter have been unsuccessful. Thus, for example, studies [10, 15] have derived the following expressions for d_0:

$$d_0 = \alpha T, \ \alpha = 0.032 \text{ nm/K}, \tag{3}$$

$$d_0 = 5.457 \exp(T/530.79), \text{ nm}. \tag{4}$$

However at high temperatures the calculated values of d_0 obtained from (3), (4) differ from the experimental value by a quantity significantly greater than the experimental error. The experimental dependencies of the void superlattice parameter on the temperature and rate of generation of point defects remain unclear, and the problem of developing a consistent theory of void superlattice formation has not yet been solved.

A significant step forward was made in recent years when void superlattice formation came to be examined as a self-organization process in disordered open systems in nonequilibrium conditions. Void self-ordering is a "stage-by-stage" process. Initially many small, chaotically distributed voids are formed; some of these grow from accumulation of the smaller voids and then locally ordered regions arise that spread to the neighboring regions of the crystal.

Study [20] considered void formation during irradiation to result from the bifurcation of the point defect and cluster spatial distribution, which is initially homogeneous. The primary factor behind the rise of instability is the high density of the vacancy dislocation loops. However approach [20] does not give an explanation of void lattice formation in crystals with a low dislocation vacancy loop density and ignores the dependence of the void superlattice formation process on the defect generation rate. The experimental fact of superlattice formation when $K > K_{min}$ and their vanishing when $K \geqslant K_{max}$ therefore remains unexplained. Moreover, diffusion void mobility in the coordinate space is not accounted for.

In Sugakov's theory of point defect density ordering in irradiated materials a higher point defect density is achieved at the nodes of a specific lattice. Conditions arise here for the formation of dislocation loop voids. The corresponding superlattices are then formed. Such an approach made it possible to explain the interrelationship between the symmetry of the voids and the crystal. Although this theory made an indisputable and significant contribution to the understanding of the general regularities of void lattice formation it does not account for their characteristics (for example, the average void dimensions in the lattice). The derived formula for the period has a significant deviation from experiment. The point defects and voids are examined below within the framework of a uniform self-organization process.

3. Void Superlattice as a Dissipative Structure

A vacancy void superlattice is an example of open nonequilibrium dissipative structures appearing in open systems as a result of cooperative phenomena (self-organization). The formation of such spatial structures in physics, chemistry, biology, ecology and other fields of science has been widely investigated to date [22, 23].

We will examine the specific conditions behind the spatial ordering of vacancy voids in irradiated metals. The kinetics of voids and the kinetics of point defects formed by irradiation are intimately interrelated. For simplicity we will first limit our examination to a model of chaotically distributed vacancy clusters (voids) in a vacancy "gas." Such a model can be described by a system of nonlinear equations for vacancy concentrations $N_{\upsilon}(r, t)$ and void distribution functions $f(x, r, t)$ in terms of their radii x:

$$\frac{\partial N_v}{\partial t} = K - 4\pi \int_0^{\infty} x^2 V(x) f(x, r, t)\, dx - D_v \rho_d (N_v - N_{\infty}) + D_v \nabla^2 N_v, \tag{5}$$

$$\frac{\partial f(x, r, t)}{\partial} = -\frac{\partial}{\partial x}[V(x) f(x, r, t)] + D_0(x) \nabla^2 f(x, r, t). \tag{6}$$

Here r is the spatial coordinate, $V(x) = \theta/x$, $\theta = D\upsilon(N_{\upsilon} - N_S(x))$ is the rate of void growth; $N_S(x)$ is the equilibrium vacancy concentration near a void of radius x, $N_S(x) = N_{\infty} \exp(2\gamma\Omega/xkT)$, N_{∞} is the thermodynamically equilibrium vacancy concentration. The first term in the right half of equation (5) corresponds to vacancy generation from source K; the second and third terms represent vacancy losses to the internal drains: the voids and dislocations, respectively (the quantity ρ_d is the total dislocation length per unit volume), the fourth term characterizes spatial diffusion.

Equation (6) describes the dynamics of the variation in the void radius distribution function. The first term in the right half corresponds to the change in $f(x, r, t)$ due to void growth at rate $V(x)$ determined by the excess vacancy flux, while the second term corresponds to spatial void diffusion, $D_0(x) = \beta/x^2$, $\beta = \Omega N_S(x) D_{\upsilon} + D_i C_i$. Fluctuation void formation is ignored in equation (6) and it is assumed that void density N_0 = const. Henceforth we take the factor $\exp(2\gamma\Omega/xkT) \simeq 1$.

Equation system (5) and (6) has a quasi-stationary spatially-homogeneous solution ($f^S(x)$, N_{υ}^S) determined by the following relations:

$$f^s(x) = N_0\delta(x - \bar{x}), \tag{7}$$

$$K = 4\pi\theta \int_0^\infty x f^s(x)\,dx + D_v\rho_d(N_v^s - N_\infty). \tag{8}$$

We will determine the stability of these stationary solutions. We will represent the quantities f and N_v as:

$$f = f^s + \delta f(x, r, t), \quad N_v = N_v^s + \delta N_v(r, t),$$

$$|\delta f/f^s| \ll 1, \; |\delta N_v/N_v^s| \ll 1,$$

where f^S and N_v^S are the roots of equations (7) and (8); $\delta f(x, r, t)$ and $\delta N_v(r, t)$ are variations of the quantities f and N_v.

Substituting these expressions into equations (5) and (6) and retaining only those terms linear in δf and δN_v we obtain

$$\frac{\partial(\delta N_v)}{\partial t} + 4\pi \int_0^\infty \left(V\delta f + \frac{D_v}{x} f_v^s \delta N_v\right) x^2\,dx + D_v\rho_d\delta N_v = D_v\nabla^2(\delta N_v), \tag{9}$$

$$\frac{\partial(\delta f)}{\partial t} + \frac{\partial}{\partial x}\left(V\delta f + \frac{D_v}{x} f^s\delta N_v\right) = D_0\nabla^2\delta f. \tag{10}$$

Accounting for the fact that the characteristic variation time of the void radius greatly exceeds the time to a quasi-stationary distribution of the vacancy concentration we can take $\delta\dot{N}_v = 0$. We will represent the fluctuations δN_v and δf as:

$$\delta N_v, \; \delta f \sim \exp(\lambda t + i\mathbf{q}\mathbf{r}). \tag{11}$$

Substituting (11) into equation system (9) and (10) and accounting for the fact that $\delta\dot{N}_v = 0$, we find

$$D_v\delta N_v = -4\pi\theta(\mathbf{q}^2 + \rho_{tot})^{-1} \int_0^\infty x\delta f(x, \mathbf{q}, \lambda)\,dx,$$

$$[\lambda + D_0\mathbf{q}^2]\,\delta f + \theta\frac{d}{dx}\left(\frac{\delta f}{x}\right) = 4\pi\theta(\rho_{tot} + \mathbf{q}^2)^{-1}\frac{d}{dx}\left(\frac{f^s}{x}\right)\int_0^\infty x\delta f(x, \mathbf{q}, \lambda)\,dx. \tag{12}$$

Here $\rho_{tot} = \rho_d + 4\pi N_0\bar{x}$, $\bar{x}$ is the mean void radius. In order to obtain the dispersion relation for equation system (5) and (6) we first determine from (12) the function $\delta f(x, \mathbf{q}, \lambda)/x$ and then multiply it by x^2 and carry out integration within the limits from zero to infinity. As a result we obtain

$$\rho_d + \mathbf{q}^2 = -(4\pi/\theta)\int_0^\infty \frac{x^2\,dx}{p(x)} \int_{x_c}^x \left(\lambda + \frac{\beta\mathbf{q}^2}{y^2}\right) p(y) f^s(y)\,dy, \tag{13}$$

where $p(x) = (x/x_c)^{\frac{\beta\mathbf{q}^2}{\theta}} \exp\left[\frac{\lambda}{2\theta}(x^2 - x_c^2)\right]$ is the integrating multiplier. For characteristic values of the quantities $(\lambda/2\theta)(x^2 - x_v^2) \sim 10^{-1} \ll 1$; hence $\exp[(\lambda/2\theta)(x^2 - x_v^2)] \simeq 1$. Substituting $p(x) \simeq (x/x_c)\beta\mathbf{q}^2/\theta$, and $f_s(x) = N_0\delta(x - \bar{x})$ into formula (13) we obtain

$$\lambda_{q^2} = [(3\theta - \rho_{tot}\beta)\, q^2 - \beta q^4 + 3\theta\rho_d]/4\pi N_0 \bar{x}^3. \quad (14)$$

Analysis reveals that when $3\theta > 4\pi N_0\bar{x}\beta$, i.e., $K > K_{min} = (4\pi N\bar{x})^2\beta/3$ and $\rho_d \ll 4\pi N_0\bar{x}$, the homogeneous quasi-stationary solution is unstable: there exists at least one positive eigenvalue of q^2 such that $\lambda_{q^2} > 0$ or

$$\beta q^4 - (3\theta - 4\pi N_0 x\beta)\, q^2 - 3\theta\rho_d < 0.$$

It follows from the latter inequality that the unstable eigenvalues of q^2 will satisfy the conditions

$$q^2_{c(+)} < q^2 < q^2_{c(-)}, \quad (15)$$

where $q^2_{c(\pm)}$ are solutions of the biquadratic equation

$$\beta q^4 - (3\theta - 4\pi N_0 \bar{x}\beta)\, q^2 - 3\theta\rho_d = 0$$

and are equal to

$$2\beta q^2_{c(\pm)} = (3\theta - 4\pi N_0\bar{x}\beta) \pm \sqrt{3\theta - 4\pi N_0\bar{x}\beta)^2 + 12\theta\beta\rho_d}.$$

$\lambda_{q^2} > 0$ for each q^2 satisfying condition (15). Therefore the spatially homogeneous solution $(f_s(x),\ N_v^s)$ is in certain conditions unstable with respect to fluctuations with the wave numbers from the interval $q^2_{c(+)} < q^2 < q^2_{c(-)}$. We can easily see that in this range, δq_2 grows and $d\lambda/dq^2 > 0$, and therefore the $\lambda(q^2)$ relation will have a maximum. We will determine the value of q^2 where the function $\lambda(q^2)$ adopts a maximum value. Following the regular routine we find

$$q^2_{max} = (3\theta - 4\pi N_0\bar{x}\beta)\ /\ 2\beta.$$

The maximum value of the function $\lambda(q^2)$ is equal to

$$\lambda_{max} = \lambda(q^2_{max}) = (3\theta - 4\pi N_0\bar{x}\beta)^2\ /\ 8\pi N_0\bar{x}^3\beta. \quad (16)$$

Therefore when certain conditions in a system of randomly distributed vacancy voids are satisfied a transition to an inhomogeneous steady-state is possible. The characteristic lifetime of such an unstable state $\tau_0 \sim \lambda^{-1}_{max}$ is several orders of magnitude greater than the lifetime of the stable homogeneous steady-state (102-10^3 s). The reason for the appearance of the ordered state is that the sharp maximum of $\lambda(q^2)$ causes a rapid growth of fluctuations with wave vectors close to $|\mathbf{q}|_{max}$. As a result we have a periodic structure with a characteristic wave number $k \sim 2\pi/d_0 \sim |\mathbf{q}|_{max}$. We can write the following formula for the period of the structure:

$$d_0 = 2\pi \sqrt{2\beta}/\sqrt{3\theta - 4\pi N_0\bar{x}\beta}. \quad (17)$$

The characteristic lifetime of the void superlattices consistent with (16) is

$$\tau_0 = \frac{8\pi N_0 \bar{x}^3 \beta}{(3\theta - 4\pi N_0 \bar{x}\beta)^2}\,. \tag{18}$$

The stability of the homogeneous distribution of vacancy voids in the supersaturated vacancy "gas" was examined above. However, as we know, during the irradiation process both vacancies and interstitial atoms are generated; the latter can influence d_0 and τ_0 as well as the conditions for the generation of instabilities in the vacancy void distribution.

Initial equation system (5) and (6) accounting for the interstitial atoms takes the following form:

$$\begin{aligned}
\frac{\partial N_v}{\partial t} &= K - 4\pi \int_0^\infty x^2 V_1(x) f(x, \mathbf{r}, t)\, dx - D_v\rho (N_v - N_\infty) + D_v \nabla^2 N_v, \\
\frac{\partial N_i}{\partial t} &= K - 4\pi \int_0^\infty x^2 V_2(x) f(x, \mathbf{r}, t)\, dx - \eta\rho D_i N_i + D_i \nabla^2 N_i, \\
\frac{\partial f}{\partial t} &= -\frac{\partial}{\partial x}[V_3(x) f] + D_0(x)\nabla^2 f(x, \mathbf{r}, t).
\end{aligned} \tag{19}$$

Here N_i and D_i are the concentration and diffusion coefficients of the interstitial atoms, respectively; K is the rate of generation of the point defects due to irradiation; η is the parameter characterizing the asymmetry of interstitial atom trapping by the dislocations compared to the vacancies ($\eta > 1$, $\eta - 1 \ll 1$), $\rho = \rho_d + 4\pi x_\upsilon \rho_\upsilon + 4\pi x_s \rho_s$, ρ_d is the total length of dislocations per unit of volume, x_υ, x_s, ρ_υ, ρ_s are the mean radius and the density of the vacancy- and interstitial-type dislocation loops, respectively.

The quantities $V_1(x)$ and $V_2(x)$ in equations (19)-(21) characterize the rates of change in void radii due to the vacancy and interstitial atom fluxes, respectively, and are equal to $V_1(x) = \theta/x$, $\theta = D_\upsilon(N_\upsilon - N_\infty)$, $V_2(x) = \nu/x$, $\nu = D_i N_i$. The quantity $V_3(x)$ determines the void growth rate from the excess vacancy flux compared to the interstitial atoms and is equal to: $V_3(x) = \varepsilon/x$, $\varepsilon = \theta - \nu$. For simplicity we have ignored the change in vacancy and interstitial atom concentrations from their recombination as well as the nucleation of new voids during the irradiation process (N_0 = const) in (19)-(21). The quasi-stationary homogeneous solution (f^S, N_υ^S, N_i^S) in this case is determined from the solution of the following system of equations:

$$\begin{aligned}
K &= 4\pi\theta \int_0^\infty x f^s(x)\, dx + D_v (N_v^s - N_\infty)\rho, \\
K &= 4\pi\nu \int_0^\infty x f^s(x)\, dx + D_i N_i \rho\eta, \\
f^s(x) &= N_0 \delta(x - \bar{x}).
\end{aligned} \tag{20}$$

A linear analysis of the stability of homogeneous solution (f_S, N_υ^S, N_i^S) based on equation system (19) assuming quasi-stationarity of the N_υ and N_i concentrations for the fluctuation Fourier components $\delta N_\upsilon(\mathbf{q})$, $\delta N_i(\mathbf{q})$, $\delta f(\mathbf{q}x)$ yields the following system of equations:

$$D_v \delta N_v(\mathbf{q}) = -4\pi\theta(\mathbf{q}^2 + \rho_{\text{tot}})^{-1} \int_0^\infty x \delta f(x, \mathbf{q}, \lambda)\, dx, \tag{21}$$

$$D_i \delta N_i(\mathbf{q}) = -4\pi\nu[\mathbf{q}^2 + \rho_{\text{tot}} + (\eta - 1)\rho]^{-1} \int_0^\infty x \delta f(x, \mathbf{q}, \lambda)\, dx, \tag{22}$$

$$[\lambda + D_0(x)\mathbf{q}^2]\delta f(x, \mathbf{q}, \lambda) + \varepsilon \frac{d}{dx}\left[\frac{\delta f(x, \mathbf{q}, \lambda)}{x}\right] =$$

$$= -\frac{d}{dx}\left[\frac{f^s(x)}{x}\right](D_v \delta N_v(\mathbf{q}) - D_i \delta N_i(\mathbf{q})). \tag{23}$$

When $\eta - 1 \ll 1$ equation (22) can be transformed in the following manner:

$$D_i \delta N_i(\mathbf{q}) = -4\pi\nu(\mathbf{q}^2 + \rho_{\text{tot}})^{-1}\left[1 - \frac{(\eta - 1)\rho}{\mathbf{q}^2 + \rho_{\text{tot}}}\right] \int_0^\infty x \delta f(x, \mathbf{q}, \lambda)\, dx. \tag{24}$$

Substituting (21) and (24) into equation (23) we have

$$[\lambda + D_0(x)\mathbf{q}^2]\delta f(x, \mathbf{q}, \lambda) + \varepsilon \frac{d}{dx}\left[\frac{\delta f(x, \mathbf{q}, \lambda)}{x}\right] =$$

$$= -4\pi(\mathbf{q}^2 + \rho_{\text{tot}})^{-1}\left[\theta - \nu\left(1 - \frac{(\eta - 1)\rho}{\mathbf{q}^2 + \rho_{\text{tot}}}\right)\right]\frac{d}{dx}\left[\frac{f^s(x)}{x}\right] \int_0^\infty x \delta f(x, \mathbf{q}, \lambda)\, dx. \tag{25}$$

Proceeding analogously to the case above we find from (25) the following dispersion equation:

$$(\mathbf{q}^2 + \rho_{\text{tot}})\left[1 - \frac{\nu}{\varepsilon}\frac{(\eta - 1)\rho}{\mathbf{q}^2 + \rho_{\text{tot}}}\right] =$$

$$= 4\pi\left[N_0 \bar{x} - \frac{1}{\varepsilon}\int_0^\infty \frac{x^2\, dx}{p(x)} \int_{x_c}^x \left(\lambda + \frac{\beta \mathbf{q}^2}{y^2}\right) f^s p\, dy\right]. \tag{26}$$

The integrating multiplier $p(x)$ in this case is equal to $p(x) = (x/x_c)^{\frac{q^2\beta}{\varepsilon}} \exp\left[\frac{\lambda}{2\varepsilon}(x^2 - x_c^2)\right] \simeq (x/x_c)^{\frac{q^2\beta}{\varepsilon}}$. Substituting $p(x)$ and $f^S(x) = N_0 \delta(x - \bar{x})$ into (26) and evaluating the derived integrals we obtain the following dispersion equation:

$$\lambda_{\mathbf{q}^2} = \left[3\varepsilon\rho + \left(3\varepsilon - \rho_{\text{tot}}\beta + \frac{\nu\beta(\eta - 1)\rho}{\varepsilon}\right)\mathbf{q}^2 - \beta \mathbf{q}^4\right] \Big/ 4\pi_0\ \bar{x}^3. \tag{27}$$

Analysis of dispersion equation (27) reveals than when $3\varepsilon > 4\pi N_0 \bar{x}$, i.e., $K > K_{\min} = (4\pi N_0 \bar{x})^3 \beta / 3\rho(\eta - 1)$ and $\rho \ll 4\pi N_0 x$ quasi-stationary homogeneous solution (f^S, N_v^S, N_i^S) is unstable: there exists a range of values of the wave vector $\mathbf{q}^2$ such that $\lambda_{\mathbf{q}^2} > 0$ or

$$\beta q^4 - [3\varepsilon - \beta 4\pi N_0 \bar{x} + \nu\beta(\eta - 1)\rho/\varepsilon]\mathbf{q}^2 - 3\varepsilon\rho < 0. \tag{28}$$

It follows from inequality (28) that the unstable solutions will satisfy the conditions

$$q_{c(-)}^2 < q^2 < q_{c(+)}^2,$$

where $q^2_{c(\pm)}$ are determined from the solution of the equation

$$\beta q^4 - [3\varepsilon - 4\pi N_0 \bar{x}\beta + \nu\beta(\eta - 1)\rho/\varepsilon] q^2 - 3\varepsilon\rho = 0$$

and are equal to

$$2\beta q^2_{c(\pm)} = \left(3\varepsilon - 4\pi N_0 \bar{x}\beta + \frac{\nu\beta\rho(\eta - 1)}{\varepsilon}\right) \pm$$
$$\pm \sqrt{\left(3\varepsilon - 4\pi N_0 \bar{x}\beta + \frac{\nu\beta\rho(\eta - 1)}{\varepsilon}\right)^2 + 12\beta\varepsilon\rho}\,.$$

We will determine the maximum of the function $\lambda(\mathbf{q}^2)$. Setting the derivative of $\lambda(\mathbf{q}^2)$ equal to zero, we find

$$\mathbf{q}^2_{\max} = [3\varepsilon + \nu\beta\rho(\eta - 1)/\varepsilon - 4\pi N_0 \bar{x}\beta]/2\beta, \tag{29}$$

$$\lambda_{\max} = [3\varepsilon + \nu\beta\rho(\eta - 1)/\varepsilon - 4\pi N_0 \bar{x}\beta]^2/8\pi N_0 \bar{x}^3\beta,$$
$$d_0 = 2\pi/|\mathbf{q}|_{\max} = 2\pi\sqrt{2\beta}/\sqrt{3\varepsilon + \nu\beta\rho(\eta - 1)/\varepsilon - 4\pi N_0 \bar{x}\beta},$$
$$\tau_0 = 8\pi N_0 \bar{x}^3\beta/[3\varepsilon + \nu\beta\rho(\eta - 1)/\varepsilon - 4\pi N_0 \bar{x}\beta]^2. \tag{30}$$

Therefore this analysis has revealed that the void superlattice formation process in irradiated metals is a stage-by-stage process. In the initial void nucleation stage the voids are distributed chaotically and there is no void lattice. Over a long time period the voids reach dimensions such that the conditions $4\pi N_0\bar{x} \gg \rho$, $K \gg K_{min}$ are satisfied. A unique phase transition then occurs in the system: the homogeneous void distribution becomes inhomogeneous and a periodic structure arises with a characteristic scale determined by expression (29). It follows from formula (29) that the period of the resulting structure, aside from the crystal parameters, depends on the irradiation conditions: the rates of generation of the point defects and the irradiation temperature T through θ, ε, and β. At a given temperature T with growth of K the period d_0 drops, while with fixed K the period grows with growth of T. These relations are in good qualitative agreement with experimentally-measured regularities.

Formulae (15), (16), (29), and (30) make it possible to theoretically predict the possible influence of an external stress on the period and lifetime of void superlattices. Such influence can also be manifest through the equilibrium vacancy concentration near a void of radius $x(N_s(x))$, i.e., through the parameters ρ and η on which these quantities depend. As noted in study [6] a tensile stress will cause a reduction in $N_s(x)$ and growth of ρ and η. According to formulae (15), (29) the superlattice period will grow with growth of an external mechanical stress. Here consistent with formulae (16) and (30) their characteristic lifetime will drop.

4. Comparison to Experiment

As indicated by this analysis a theoretical model of void lattice formation (lattice self-organization) must include the fundamental processes occurring in the metal under irradiation, i.e., the interrelated kinetics of the voids, vacancies, and interstitial atoms. The physical causes for lattice self-organization are as follows. Due to the different interstitial atom and vacancy fluxes excess vacancies arise in the dislocation. As the excess vacancies multiply a supersaturated solution of vacancies is formed and condensation - void nucleation - begins. With a further growth in vacancy number, their excess number will be compensated by an increase in the flux to the voids. This vacancy flux is determined by the void surface area and also by void configuration; it is at a maximum with void ordering. Here there is a direct analogy to Bénard cell formation in the fluid layer where the thermal flux passing through the layer will grow with growth of the temperature gradient; this produces ordering of the thermal fluxes that helps to increase the capacity of the layer. Specific conditions were identified above where a periodic structure will appear in the system; its period is determined and its characteristic lifetime is estimated. Accounting for the expression for K_{min} we can represent formula (29) as:

$$d_0 = d_v \sqrt{K_{min}/(K - K_{min})}, \qquad d_v = \sqrt{8\pi^2/N_0 x}. \tag{31}$$

Formula (31) reveals the existence of a special phase transition when $K = K_{min}$ related to the development of a vacancy void superlattice. The quantity K_{min} is dependent on the irradiation temperature and determines the lower limit of the point defect generation rate in the formation of the void superlattices. Numerical estimates for molybdenum revealed that at a dislocation density $\rho_d \sim 10^9$ cm^{-2}, a void density $N_0 \sim 10^{16}$ cm^{-3} and a void radius $x \sim 50$ Å the minimum defect generation rate lies in the 10^{-4}-10^{-7} SNA/s range at temperatures $T \sim (0.44$-$0.52)\ T_{пл}$.

In order to test the adequacy of the approximations used in Section 3 it is necessary to compare the derived results to existing experimental data, particularly with respect to the radiation temperature dependencies of the lattice period d_0. The figure gives the theoretical and experimental $d_0(T)$ dependencies in 3.1 MeV $^{51}V^+$ ion-irradiated molybdenum. The following molybdenum parameters were used: $\varepsilon^{v}_{m} = 2.25$ eV, $\varepsilon^{v}_{f} = 3.3$ eV, $D_{v0} = 10^{-4}$ cm^2/s, $D_{v0}N_{v0} = \exp[-\varepsilon^{v}_{m} + \varepsilon^{-v}_{f})/kT]$, $\eta - 1 = 8 \cdot 10^{-2}$ [9]. The rate of creation of point defects K was $5 \cdot 10^{-3}$ SNA/s. It is clear from the figure that the theoretical relation $d_0(T)$ (curve 1) calculated by (31) is in good agreement with the experimental relation (curve 2) obtained in study [27]. The small discrepancies at low temperatures can be attributed to the simplifying assumption.

The derived agreement between the $d_0(T)$ theoretical relation and experiments can be considered satisfactory if we account for the insufficient reliability of the energy parameters of the test sample used in estimating $d_0(T)$. We emphasize that the estimate of $d_0(T)$ by formula (2) produces significant discrepancies from experiments. In order to achieve a satisfactory agreement between the calculated values of $d_0(T)$ and the experimental values it is necessary to use a vacancy migration energy level 3-4 times smaller than the value obtained from the literature [9].

The attempt to describe the temperature dependence $d_0(T)$ by formula (38) was also unsuccessful; it produced values 3-4 orders of magnitude off the experimental data. To date the results from the present study are the only results in satisfactory agreement with experiments.

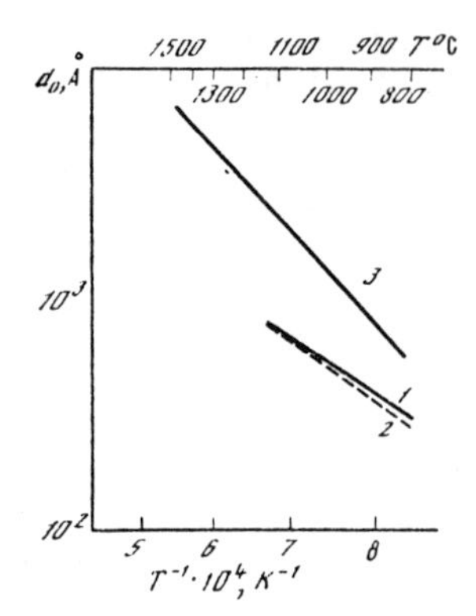

Theoretical and experimental radiation temperature dependencies of the void superlattice period

We will now obtain a numerical estimate of the characteristic lifetime of void superlattices τ_0. This quantity is one of the primary parameters of superlattice nucleation theory and it characterizes the lifetime of the resulting structure. It has a significant dependence on the radiation temperature T and the rate of point defect formation. According to formula (30) with fixed G and growth of radiation temperature, τ_0 diminishes; with a given T and growth of G, d_0 drops. These qualitative conclusions are in good agreement with experimentally-observed regularities. Taking T = 900°C, $\beta \sim 10^{-29}$ cm²/s, $G \sim 5\cdot10^{-3}$ st. u/s, $\rho_d \sim 3\cdot10^9$ cm^{-2}, $N_0 \sim 6\cdot10^{16}$ cm^{-3}, $\bar{r} \sim 35$ Å, $\eta - 1 = 8\cdot10^{-2}$ in the formula we find that τ_0 is of the order of $3\cdot10^3$ s.

In conclusion we note that the analysis of void superlattice formation carried out in the present study is rather general in nature and can be used to investigate the formation of other defect superlattices (such as dislocations, dislocation loops, and interstitial atoms).

References

1. De Diego, N., Mirn, M., Rühl, M. "Transmission electron microscopy studies of the ordering if nitrogen" ACTA MET., 1979, vol. 27, pp. 1445-1451.

2. Hörz, G., Popovic, M. "Precipitation processes in supersaturated tantalum nitrogen solid-solutions" ACTA MET., 1979, vol. 27, pp. 1453-1460.

3. Sidorenko, F.A., El'ner, V.Ya., Gel'd, P.V. "The energy aspect of stabilization of crystals with vacancies" FTT, 1980, vol. 22, no. 2, pp. 619-621.

4. Nabutovskiy, V.M., Shapiro, B.Ya. "Superconducting states in dislocation systems" FIZIKA NIZKIKH TEMPERATUR, 1979, vol. 5, no. 10, pp. 1128-1137.

5. Cawthorne, C., Fulton, E.J. "Voids in irradiated stainless steel" NATURE, 1967, vol. 216, no. 5115, pp. 575-576.

6. Mirzoev, F.Kh., Fetisov, E.P., Shelepin, L.A. "Kinetics of void nucleation and growth in crystals": This volume.

7. Evans, J.H. "Observation of regular void array in high purity molybdenum irradiated with 2 MeV nitrogen ions", NATURE, 1971, vol. 229, no. 5284, p. 403.

8. Norris, D.I.R. "Voids in irradiated metals" RADIAT. EFF., 1972, vol. 15, no. 1/2, pp. 1-22.

9. Shcherbak, V.I., Zakharova, M.I., Bykov, V.N. "Void lattice in molybdenum and tungsten" VOPR. ATOM. NAUKI I TEKHNIKI. FIZIKA RADIATS. POVREZHDENIY I RADIATS. MATERIALOVEDENIYA, 1976, vol. 1/3, pp. 61-65.

10. Sikka, V.K., Moteff, J. "Superlattice of voids in neutron-irradiated tungsten" J. APPL. PHYS., 1972, vol. 43, no. 12, pp. 4942-4944.

11. Risbet, A., Levy, V. "Ordre de cavités dans le magnetism et l'aluminium irradiés aux neutron" J. NUCL. MATER., 1974, vol. 50, no. 12, pp. 116-118.

12. Brown, L.M. "A simple explanation for the stability of voids in mat. under irrad." - SER. MAT., 1972, vol. 6, No. 5, pp. 387-394.

13. Winter, A.T. "A possible explanation of the BCC form of void lattice" SER. MET., 1973, vol. 7, no. 1, pp. 49-51.

14. Moteff, J., Sikka, V.K., Yang, H. "The influence of neutron irradiation temperature on the void characteristics of BCC metals and alloys" In: The physics of irradiation produced voids: Proc. cons. symp., Harwell, Sept. 9-11, 1974, Harwell: AERE, 1975, pp. 181-186.

15. Sikka, V.K., Moteff, J. "Damage in neutron-irradiated molybdenum (1) Characterization of as-irradiated microstructure" J. NUCL. MATER., 1974, vol. 54, no. 2, pp. 325-345.

16. Tewary, V.K., Bullough, R. "Theory of the void lattice in molybdenum" J. PHYS. F: METAL. PHYS., 1972, vol. 2, pp. L69-L72.

17. "Void lattices" COMMENTS SOLID STATE PHYS., 1976, vol. 7, no. 5, pp. 105-115.

18. Maksimov, L.A., Ryazanov, A.I. "Diffusion interaction of voids" FMM, 1976, vol. 41, pp. 884-891.

19. Maksimov, L.A., Ryazanov, A.I. "Kinetic equation for vacancy voids. Void lattice as a dissipative structure, stable in irradiation conditions" ZhETF, 1980, vol. 79, no. 6, pp. 2311-2327.

20. Krishan, K. "Void ordering in metals during irradiation" PHILOS. MAGNETIC. A, 1982, vol. 45, no. 3, pp. 401-417.

21. Suzakov, V.I. "Defect density superlattices in irradiated crystals" ITF, no. 70P, Kiev, 1984, 30 pp.

22. Khaken, G. "Sinergetika" [Synergetics], Moscow: Mir, 1980, 480 pp.

23. Nikolis, G., Prigozhin, I. "Samoorganizatsiya v neravnovesnykh sistemakh" [Self-organization in nonequilibrium systems] Moscow: Mir, 1979, 512 pp.

24. Sugakov, V.I. "Appearance of void lattice in the optical properties of metals" VOPR. ATOM. NAUKI I TEKHNIKI. FIZIKA RADIATS. POVREZHDENIY I RADIATS. MATERIALOVEDENIYA, 1981, vol. 2(16), pp. 71-72.

25. Kryuchenko, Yu.V., Sugakov, V.I. Electrodynamics of a spatially dispersive medium with spherical cavity" PHYS. STATUS. SOLIDI (b), 1982, vol. 111, no. 1, pp. 177-185.

26. Rudko, V.N., Sugakov, V.I. "Superconductivity of metals with a superlattice of voids" PHYS. STATUS SOLIDI (b), 1984, vol. 126, pp. 703-712.

27. Motelf, J. "Void swelling behavior of vanadium ion irradiated molybdenum" J. NUCL. MATER., 1981, vol. 101, no. 1, pp. 64-77.

STRUCTURAL RELAXATION IN IMPACTED SOLIDS

S.P. Merkulova, L.A. Shelepin, A.A. Shubin

ABSTRACT

Various relaxational processes of the defect structure in films under pulsed irradiation are examined. The relief features of surface periodic structures obtained under short laser pulse irradiation are investigated experimentally. The effects arising from dispersion of the film material are investigated, particularly the formation of narrow bands due to pulsed energy contribution.

Introduction

Problems associated with pulsed action on solids are of significant interest today [1, 2]. Laser, beam, current, and acoustic pulses and impact loads can significantly change material properties and, specifically, improve their characteristics. Laser annealing, for example, has found broad applications. However, unlike the comprehensively developed series of pulsed action on gaseous media, in this case the level of understanding of the phenomena is still insufficient. The relaxational processes that occur as a result of pulsed action for a gas include energy exchange between the different molecular degrees of freedom (electronic, vibrational, rotational, and translational) [3]. In the majority of cases a change in the spatial configuration of the gas molecules does not play a significant role in the relaxational processes. The properties of a solid are determined by its structure: the type of lattice, as well as the configuration and interaction of defects: voids, interstitial boundaries, dislocations, microcracks and point defects. Therefore in examining pulsed action it is no longer possible to avoid an analysis of structural relaxation. The role of stresses responsible for the degree of stability and the nature of the distribution of many defects play an important role here.

One of the fundamental principles in the kinetic theory of a nonequilibrium gas is the hierarchy of the characteristic energy transport times in the molecular systems. This has made it possible to differentiate the determinate processes from secondary or past processes and to formulate a general quasi-stationary distribution technique for molecular gas.

The situation is more complex in analyzing structural relaxation in solids, since it is necessary to fully incorporate the space-time picture of the processes. The type and configuration of defects is important as well as the nature of the pulse, and the formation of new defects resulting from the spatial inhomogeneity of the action, the degree of inhomogeneity of the defect system, etc. The pulsed impact action itself, in interacting with the solid, undergoes its own spatial and temporal structural changes. Even a single spatially-homogeneous pulse can become a complex pulse in this case.

The purpose of the present study is to carry out a brief analysis of the primary features of structural relaxation and to consider the possibilities for using the developed experimental technique and derived data for developing a theory of structural relaxation.

1. Characteristic Structural Relaxation Times

A wide range of processes including excitation and relaxation of the electronic subsystem, electron-phonon, vibrational, and structural relaxation occur in a solid under sufficiently powerful pulsed action. Structural relaxation is a slower process for a wide range of conditions and consistent with the general principle of hierarchical characteristic times it can be considered in a certain approximation as independent of other processes.

In turn in order to establish regularities of the structural relaxation in solids it is necessary to measure the observed characteristic times and relate these to specific kinetic processes. Pulsed experiments on V-Si and Nb-Al alloy films have revealed, specifically, the existence of relaxational processes that occur over time periods up to several months.

These films were obtained by $\sim 10^{-5}$ torr vacuum deposition onto a hot sapphire substrate and were then exposed to current pulses of the order of tens of microseconds in duration. Since under pulsed action the sapphire substrates were at room temperature and a mass significantly higher than that of the film (the test films had a thickness ≤ 4 μm), the cooling rates were quite high which in turn produced significant local stresses. Broadening of the peaks from X-ray structural analyses of the samples revealed their existence.

Changes in time in such film characteristics as resistance, X-ray spectra, surface structure and superconducting properties were investigated. The time intervals over which noticeable changes in these characteristics were identified extended from seconds to months, which reveals a significant nonuniformity of samples obtained from pulsed action. It turns out that unlike properties occurring in a gas medium it is much more difficult to interpret the temporal characteristics of structural relaxation. A large statistical sampling of experimental data is required. In study [4] the quantities τ were correlated with the local stresses σ in accordance with the relation [5] $\tau = \tau_0 \exp[(U_0 - \gamma\sigma)/kT]$, where τ_0 is the characteristic time ($\tau_0 \sim 10^{-13}$ s), U_0 is the interatomic bonding energy; γ is the activation volume ($\gamma \sim (10\text{-}10^{-2})\ d^3$, d is the lattice parameter). An estimate of local stresses obtained from the temporal relations reveals that they can reach the yield strength. This makes possible plastic flow and crystallization in films observed at room temperature.

Repeated experiments to measure a number of relaxation times in similar conditions, in spite of the spread, have also identified elements of a certain reproducibility of measurement results. In order to recover information on the characteristic lifetimes of specific phases and defects and their associated stresses and instabilities it is necessary to accumulate and compare data on τ. Here it is necessary to account for the spatial and temporal structure of the action and its transformation as the action propagates through the solid as well as the degree of inhomogeneity of the material itself.

2. Space-Time Structure Under Pulsed Action

The results of pulsed action on solids are highly dependent on different structural features: defects, stresses, surface morphology, etc. In recent years interest has been focused on the effects of periodic structure formation on solid surfaces resulting from irradiation by powerful laser pulses [6-9]. It has been established that this phenomenon has a general, universal nature and is activated when a specific laser irradiation intensity is achieved. Surface lattices have been observed in semiconductors (Si, Ge, GaAs), metals (Ni, Cu, Pb, for example), and dielectrics (NaCl, SiO_2). They have different formation mechanisms. They can be produced by the interference of an incident laser beam and a secondary wave field: surface electromagnetics, acoustics, capillary, and hydrodynamic waves. An integral characteristic of all mechanisms is the nonlinear nature of the process and the existence of feedback. Since diffraction occurs from laser pulse action on an irregular surface, a Stokes wave arises with the wave vector $k_s = k - q$ together with an anti-Stokes wave with the vector $k_{as} = k + q$ (k is the projection of the wave vector of the radiation onto the surface; q is the Fourier-component of the surface relief).

When the specific intensity is achieved the pump-over from the incident wave to the diffracted wave sharply increases due to non-linearity. Interference of the pump wave and the diffracted wave generates a space-time periodicity in the light intensity distribution in the surface layer. We note that the dependence of the amplitude of the diffracted waves on the wave vectors is resonant in nature; resonance corresponds to equal modulo values of the wave vectors of the free surface electromagnetic waves. We have the relation $d = \lambda/(1 \pm \sin\theta)$ from phase synchronism for wave vectors for a lattice period d formed by a wave at an angle θ to the normal.

This range of phenomena has created new capabilities in the fabrication technology of periodic structures for applications in integrated optics. Moreover, comparatively little attention has been focused on the possibility of using the information contained in the surface structural changes after pulsed laser action. In order to illustrate the nature of the change in a surface from irradiating a periodic lattice we will provide data obtained for V-Si alloy films (deposited using the technique from study [10]) for the case of single neodymium laser pulses (λ = 1.06 μm, $\tau \approx$ 20 ns).

The observed surface relief picture was found to have a direct relation to both the material and the structure of the initial sample as well as the surface quality. Specifically when using irregular films obtained by deposition onto the mat surface of sapphire substrates only a few broad, heavily blurred bands were observed in the vicinity of the laser spots (Fig. 1). In the case of relatively smooth bands deposited on polished sapphire substrates a clearly expressed periodic relief with a period of the order of 5-25 μm was obtained.

Fig. 2 shows the periodic structure on the film surface. The light bands are nonplanar formations rising above the average film level by 0.5-1 μm. The ripple is evidently formed by the thermooptic effect producing pressure in the material. The hillocks can be attributed to the extrusion of material from sites containing the radiation antinodes. The spatial inhomogeneity of this action will also form microcracks observed in the photomicrographs.

It is important to emphasize the wide variety of the relief structure itself. Fig. 3 shows a double band rising over the general film level. The width of the band is much less than the gap between them. The width of the bulge in the photomicrograph shown in Fig. 4 is comparable to the width of the depressions. There are a significant number of transverse cracks. Distortions caused by the structural inhomogeneities of the film are visible against the background of the regular picture. Fig. 5 shows a type of band structure that is qualitatively different from the preceding cases: extended formations with sharply differentiated boundaries and slack tops. Evidently these are attributable to the crystallization process.

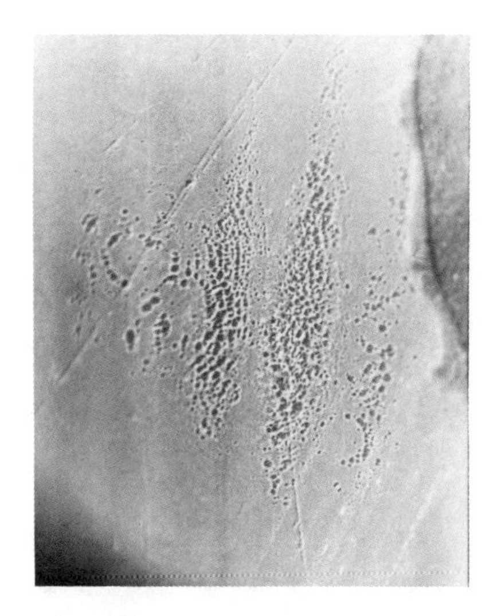

Fig. 1. Photomicrograph of the surface relief on a course-grained film deposited on the mat surface (magnification: X150)

Fig. 2. Photomicrograph of surface relief bands on V-Si alloy film deposited on a polished substrate (magnification: X3000)

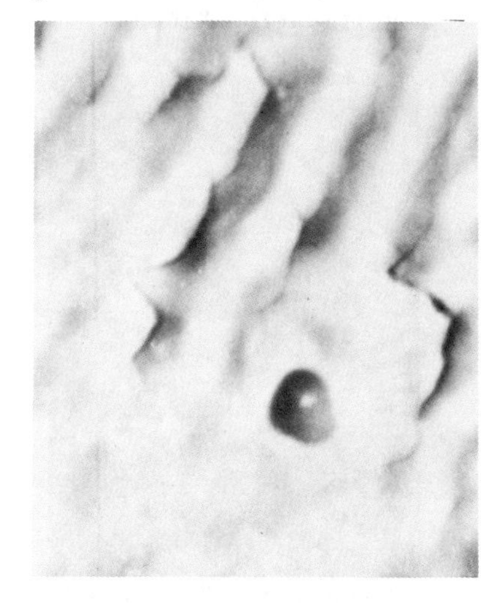

Fig. 3. Photomicrograph showing an isolated double band on a V-Si alloy film (magnification: X4500)

Fig. 4. Photomicrograph of a periodic relief containing cracks (magnification X4500)

Fig. 6 shows the results from the application of two successive pulses with a ~ 1 min. gap. The second pulse acted on a displaced sample. A two-dimensional picture in the form of a grid consisting of

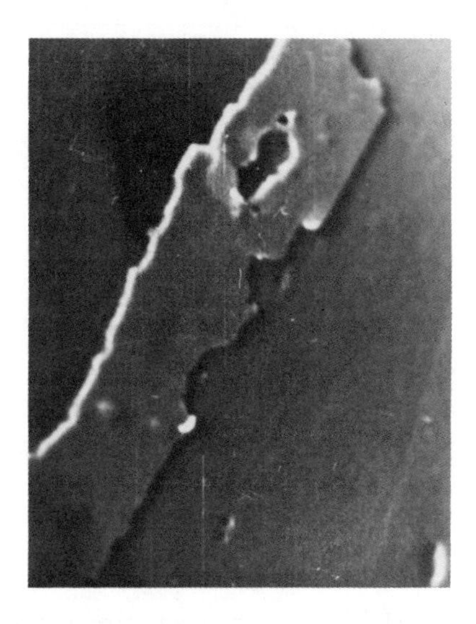

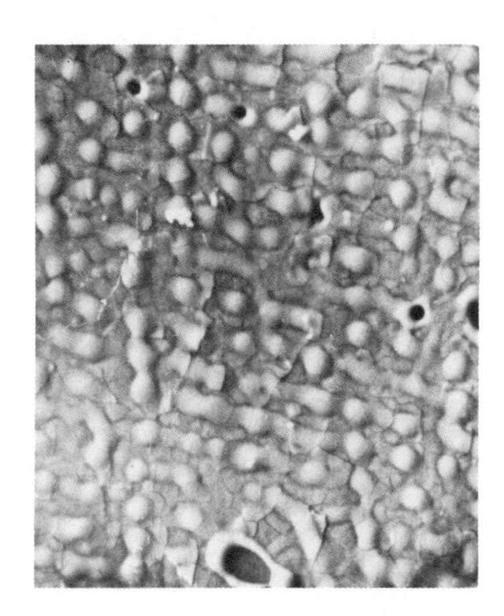

Fig. 5. Structure of an isolated band consisting of extended insular formations with sharply differentiated edges (magnification X1500)

Fig. 6. Film surface relief after irradiation by two successive laser pulses (magnification X1500)

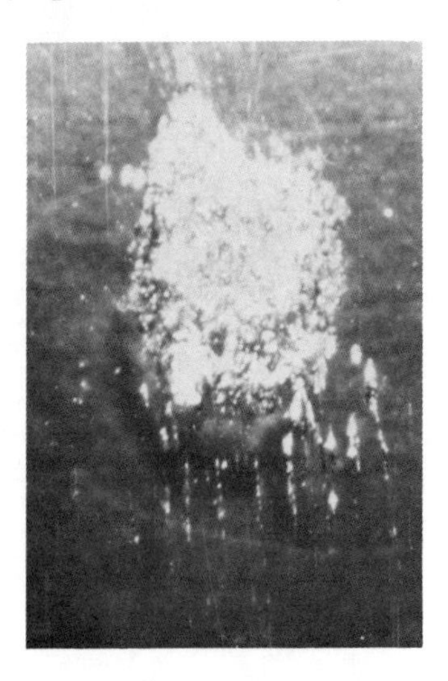

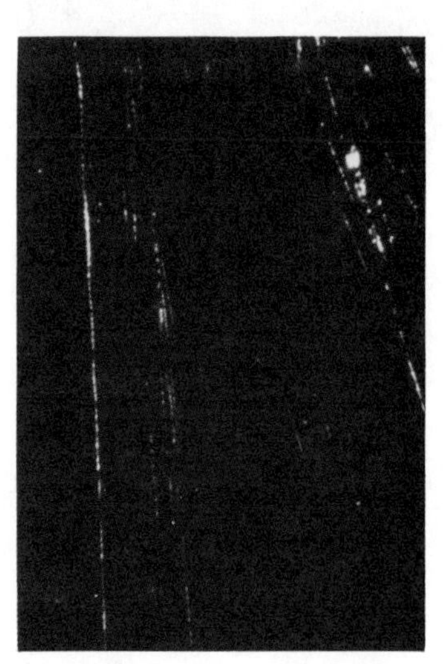

Fig. 7. Photograph of film deposited by a thin palladium foil microburst (magnification X30)

Fig. 8. Photomicrograph of isolated palladium microbands (magnification X45)

fused hillocks was formed. The inhomogeneity of the action and its temporal nature caused a developed system of cracks. Caverns formed at isolated film sites due to structural inhomogeneities and their related stresses.

As we see from these results the surface periodic structures formed after laser pulse irradiation contain a wide variety of data on the periodic structures observed on the surface as well as nonplanar formations and defects. The surface structure is highly sensitive to the specific nature and features of both the materials and the radiation pulses. It can yield extensive information on the dynamics of processes in the surface region under laser pulse action. In this connection further accumulation and systemization of experimental data is required.

3. Influence of Voids on the Nature of Pulsed Action

Under pulsed irradiation in addition to surface waves there are waves propagating through the material bulk. These include acoustic waves and, with sufficient power, soft waves. The nature of acoustic and soft wave propagation and energy release in a solid is largely determined by the structure of the actual material. Voids (their size, shape and spatial configuration) play a significant role here together with stresses and their associated condensed and sparse regions, regions containing various inclusions, etc. The spatial and temporal structure of the waves change from interaction with such structural defects and nonuniform energy delivery occurs. Moreover, by developing given defect structures it is possible in principle to control the spatial and temporal characteristics of the shock waves. The propagation of such waves in an actual crystal (accounting for all forms of defects) is a complex, nonlinear process. Its theory has not yet been developed sufficiently. In this stage it is therefore necessary to model the interaction of specific defects with the shock waves. Below we briefly discuss the overall scheme of this model and procedure based on porous materials.

We will first consider the behavior of a single, spherically-symmetric void under pulsed action. With a sufficiently rapid rise in pressure (and temperature) of the material around the void, the latter begins to compress. Here we have a direct analogy to the behavior of a bubble in a fluid. According to cavitation theory [11] spherical bubbles form in a liquid medium in the expansion phase, and their volume decreases in the compression phase. The latter process is characterized by a change in the radius R of the bubble and is described by the equation

where ρ, p, and r are the density, pressures, and radius (coordinate). This equation is obtained by substituting into the Bernoulli equation for an ideal incompressible fluid the potential φ of the velocities satisfying the Laplace equation $\nabla^2\varphi$ and taking the form $\varphi = R^2\dot{R}/r$. We have the expression $\Delta T \approx R_0\sqrt{\rho/p_0}$ from here for the collapse time of an empty cavity. The maximum pressure developed in the collapse stage is estimated by the relation $p_{max} \sim p_0(R_0/R_k)^3$ for an empty bubble, where p_0 is the pressure outside the bubble; R_0 and R_k are the initial and final radii. There are a number of factors limiting the compression of matter in actual fluids (the existence of vapor in the cavity and violation of compression sphericity).

For a solid with sufficiently intense energy contribution the material is heated and forms a liquid state, and the cavitation approach becomes applicable [11]. A differentiating characteristic of a solid from a liquid is the existence of cavities of not only spherical shape but also, in principle, any other shape; the interaction process with the shock wave is significantly dependent on this shape. We will consider as an example the compression of thin cylindrical cavities for which the results at the collapse process are significantly dependent on the angle θ between the cylinder axis and the direction of the shock wavefront. The velocity υ at which the collapse region travels along the cylindrical cavity is given by the relation $\upsilon = \upsilon_{\Phi}/\cos\theta$, where υ_{Φ} is the velocity of the shock wavefront. At values of θ close to 90° simple radial cylindrical collapse is observed: $n_k \approx n_0(R_0/R_k)^2$.

If the angle θ is sufficiently small then in addition to compaction of the products in the radial direction there is also directional motion along the axis resulting from the difference in pressures due to the temporal nonuniformity of the collapse process and cumulative streamlines are formed [11]. Such microstreamlines can also form for conical cavities with a base parallel to the front.

In model experiments to investigate cavity compression it is possible to use negative crystals that have a high degree of perfection and consistency of shape as well as fewer defects in the surrounding material compared to other types of cavities. Experiments with voids in films and in the intergrain boundary regions are of special interest. In order to approximately simulate cylindrical cavities we can apply abrasions either to a polished sapphire surface followed by deposition or directly to the film. Nb foils and Nb-Al, V-Si alloy films of various thicknesses deposited on sapphire substrates were used as the test specimens in model experiments to investigate the behavior of cavities under pulsed action.

Preliminary pulsed laser irradiation was used in addition to mechanical techniques to obtain sets of cracks, abrasions and cavities of various shape in the films. Subsequent pulsed irradiation of the samples using current and laser pulses caused significant changes in the surface structure. An investigation of the morphological features of the laser- and current-pulse-treated films and foils revealed microscopic from tenths of a micron through tens of microns in size. Traces of microstructure oval spikes were observed at certain film sites where abrasions and extended cavity formations existed prior to the pulsed irradiation.

In addition to investigating pulsed irradiation of a single cavity it is also interesting to investigate the behavior of the entire system of cavities. A differentiating feature of voids in solids is their fixed position in space, their long lifetime, the fundamental possibility of ordering and, in the case of an asymmetric shape, a specific orientation with respect to a selected direction.

A shock wave impacting the microcavity system has its own features and depends on the spatial configuration of the microcavities. After the wavefront passes, these cavities begin to collapse and, in turn, create diffusing microshock waves with a certain delay which in the case of a spherical cavity depends on its radius.

With certain relations between the dimensions of the microcavities and their position it is possible in principle to achieve a situation where the secondary waves from each of the cavities arrive at a given point at the same time. For the case of the void centers configured in a plane focusing will occur with a cavity distribution as

$$R = R_0 + (F_0/v)\sqrt{p_0/p}\,(1 - 1/\cos\theta),$$

where F_0 is the focus, R_0 is the radius of the microcavity at the center of the distribution; θ is the angle at which a microcavity of radius R is visible from the focus; v is the rate of propagation of the shock wave.

The pressure p_F at the focus is determined by the superposition of pressures from each microcavity

$$p_F \simeq {}^1/_3 p_0 R^3 F_0 / R_K^2 \cos\theta,$$

where R_K is the radius of the microcavity at the final compression stage.

In addition to the transformation of the spatial structure of the initial wave, its temporal structure is converted as well. The energy going to void collapse is taken from the leading edge of the pulse and is rereleased as secondary waves, delayed by the collapse time. The secondary pulses can be superimposed on the initial pulse or can follow the pulse with a certain time delay depending on the pulse duration and the characteristic collapse time.

As is clear from even a brief model analysis of pulsed action on porous bodies there is very complex interaction between the shock wave and the structure of the defects. Acoustic or shock waves formed from pulsed laser irradiation generate secondary waves in propagating within the solid. In addition a spatial structure with nonuniform energy release is set up. The temporal structure of the shock waves is also transformed. A comprehensive analysis of the processes in actual crystals requires modeling of the interaction (and transformation) of shock waves with different types of defects and collections of defects.

4. Breakdown Processes Under Pulsed Action

Under sufficiently strong pulse action not only local (for example, from cavity collapse) but also total material breakdown is possible. The experiment was carried out on V-Si films as well as Nb and Pd foils. The degree and nature of the breakdown of samples depended on the density of the energy contribution as well as the test material and its structural features. V-Si alloy films were fabricated by spraying pre-prepared V-Si alloy droplets onto a hot polished sapphire substrate in a vacuum of ~10^{-5} torr. The geometry of the sample consisted of two triangles connected at the apex. Samples were cut from the films in a similar manner so that in the center region their width was many times narrower compared to the peripheral sections and were then deposited on the sapphire wafers. When a current pulse of sufficient power is passed through samples with the geometry mentioned above a highly nonuniform breakdown occurs; the nature of this breakdown varies with distance from the central region. If the contributed energy is sufficient to vaporize the central region, then fused regions and regions with a developed cracking network would extend outward. Films fabricated by deposition of the products from the microburst in the central region of the sample onto the sapphire substrate have a morphology significantly different from films deposited by the regular technique. Fig. 7 shows a photomicrograph of a palladium film obtained by the technique described above. A homogeneous surface relief is characteristic here together with a rather strong bond to the substrate (evidently due to microfractures in the surface layers of the substrate).

An X-ray structural analysis was carried out on films obtained by a similar technique from a V-Si alloy; this analysis revealed significant broadening of the diffraction peaks. After annealing at 700°C the film structure was virtually identical to films obtained by regular deposition.

The melted metallic microdroplets resulting from the breakdown of the central region of the sample leave traces in the form of narrow and long bands as they slip along the sapphire substrate surface (Fig. 8). The strip width varies from fractions of a micron to tens of microns. The narrower the strip then, as a rule, the shorter the strip length and vice versa (Fig. 9). Their formation mechanism is related to the excitation of the surface by "hot" primary electrons.

Fabricating such narrow films having the properties outlined above is also of significant practical interest. Moreover, this technique is convenient for investigating relaxational processes, since the specific energy contribution to the material varies from a maximum level at the center of the triangle apexes where the dispersion occurs to very weak energy contribution along the periphery (approximately proportional to the film cross-section). This makes it possible to investigate the structural relaxation of a single crystal with different energy contributions.

As we see from a brief examination of the results of pulsed action of sufficient power in various stages the total picture of processes occurring here is rather complex. For example, under pulsed laser irradiation a variety of secondary surface waves arise (diffraction by the surface irregularities; surface acoustical and electromagnetic waves; capillary waves; and vaporization waves) together with waves propagating into the crystal bulk (acoustic and shock waves) caused by thermooptic and vaporization effects. The interaction between these waves and different structural defects (particularly the voids) can form tertiary waves of a local nature. As a result of the interaction of such waves and with structural defects (voids, extensions) local energy release occurs.

As indicated by experimental research processes from pulsed action are largely determined by the properties of the sample overall (the surface state, the spatial configuration of defects, stresses, the existence and properties of boundaries, particularly boundaries with the substrate and the degree of structural inhomogeneity) and the related nature of wave transformation and their space-time structure. The derived experimental data are largely the result of a large number of processes that vary significantly from sample to sample, from experiment to experiment.

Fig. 9. Photograph of microband group (magnification X45)

Today the problem of investigating structural relaxation largely involves calculating from a single common pulsed process the separate elementary interactions and then finding their temporal characteristics.

References

1. Dvurechenskiy, A.V., Kachurin, G.A., Nidaev, E.V., Smirnov, L.S. "Impul'snyy otzhig poluprovodnikovykh materialov" [Pulsed annealing of semiconductor materials] Moscow: Nauka, 1982, 208 pp.

2. Aleksandrov, L.N "Kinetika kristallizatsii i perekristallizatsii poluprovodnikovykh plenok" [The kinetics of crystallization and

recrystallization of semiconductor films] Novosibirsk: Nauka, 1985, 225 pp.

3. Gordiets, B.F., Osipov, A.I., Shelepin, L.A. "Kineticheskie protsessy gazakh i molekulyarnye v lazery" [Kinetic processes in gases and molecular lasers] Moscow: Nauka, 1985, 512 p

4. Merkulova, S.P., Shelepin, L.A., Shubin, A.A. "The influence of stresses on the properties of superconducting films" KRATKIE SOOBSHCH. PO FIZIKE FIAN, 1985, no. 4, pp. 52-455.

5. Rozel', V.R., Slutsner, A.I., Tomashevskiy, E.E. "Kinetichsekaya priroda prochnosti tverdykh tel. [The kinetic nature of the strength of solids] Moscow: Nauka, 1974, 560 pp.

6. Isenor, N.R. "CO_2 laser produced ripple patterns on NI_xP_{1-x} surfaces" APPL. PHYS. LETT., 1977, vol. 31, pp. 148-150.

7. Prokhorov, A.M., Sychugov, V.A., Tishchenko, A.V., Danilov, A.A. "Excitation of electromagnetic waves by high-power laser irradiation of surfaces" PIS'MA V ZHETF, 1982, vol. 8, pp. 961-966.

8. Prokhorov, A.M., Sychugov, V.A., Tishchenko, A.V., Danilov, A.A. "The kinetics of ripple formation on a germanium surface under high-power laser irradiation" PIS'MA V ZhETF, 1982, vol. 8, pp. 1409-1413.

9. Emel'yanov, V.I., Seminogov, V.N. "Stimulated scattering of electromagnetic waves on the surface of a conducting medium" ZhETF, vol. 62, no. 2, pp. 496-504.

10. Merkulova, S.P., Motulevich, G.P., Shubin, A.A. "The application of current pulses to fabrication of superconducting alloys" KRATKIE SOOBSHCH. PO FIZIKE FIAN, 1981, no. 6, pp. 12-15.

11. Pernik, A.D. "Problemy kavitatsii" [Problems of cavitation] Leningrad: Sudostroenie, 1966.

12. Merkulova, S.P., Shelepin, L.A. "The features of pulsed action on solids" INZH.-FIZ. ZHURN., 1985, vol. 48, no. 5, pp. 812-815.

THE INFLUENCE OF STOICHIOMETRY IN A_2B_6 MONOCRYSTAL COMPOUNDS ON THE CHARACTERISTICS OF A SEMICONDUCTOR ELECTRON-BEAM PUMPED LASER

I.V. Akimova, V.I. Kozlovskiy, Yu.V. Korostelin, A.S. Nasibov, A.N. Pechenov, P.V. Reznikov, V.I. Reshetov, Ya.K. Skasyrskiy, P.V. Shapkin

ABSTRACT

Results are given from electron-microscope, photoluminescence and cathodoluminescence investigations of CdS, ZnSe, ZnO, and ZnS monocrystals obtained from the gas phase and exposed to a variety of high-temperature annealings together with the characteristics of longitudinal electron-beam pumped semiconductor lasers based on these monocrystals. It is demonstrated that the degradation to laser performance characteristics observed as the crystal composition deviates from a stoichiometric composition can be attributed to the increase in the density of metallic microprecipitates in the case of excess metal and the metal vacancy concentration in the event of a shortage of the metal. In accordance with the phase diagrams for CdS and ZnO crystals the laser characteristics are determined largely by the microprecipitates, while in the case of ZnSe and ZnS crystals they are determined by the metal vacancies.

Introduction

Lasers longitudinally pumped by a sharply-focused electron beam are a class of semiconductor lasers of important practical interest. These lasers have a short single-pass optical gain length of the cavity and a limited depth of penetration of electrons into the semiconductor layer (5-10 μm). With such a short gain length it is possible to compensate total cavity losses with an optical gain of the order of g = 100-1000 cm^{-1}. Such values of g can be obtained only by the band-to-band transitions with a nonequilibrium carrier concentration n_0 = 10^{18}-10^{19} cm^{-3} [1]. All other recombination channels associated with the impurities and intrinsic defects are competing in this case and they increase the threshold pump density of the laser responsible for the necessary value of n_0.[1] Therefore crystal quality can be characterized by the internal quantum efficiency of the radiative band-to-band recombination at pump levels where it is necessary to achieve the lasing threshold (for example, at an electron energy E_0 = 75 keV the threshold current density of an electron beam j will not exceed 10 A/cm^2).

Undesirable impurities and intrinsic defects always exist in actual crystals. By optimizing the growing conditions and using appropriate annealing it is possible to significantly reduce the concentration of a number of impurities and defects, although this is often done at the cost of increasing the concentration of other defects. Therefore the important problem is to determine which of the defects create centers that significantly reduce the band to band recombination efficiency and to eliminate such defects first. It is also necessary to reduce the concentration of defects that, although they do not directly form effective recombination centers, they nonetheless aid their formation during the operation of the laser by association among themselves and with other defects, thereby contributing to laser degradation.

The present study covers the primary experimental results on the influence of A^2B^6 monocrystal fabrication conditions on the characteristics of longitudinal electron beam-pumped semiconductor lasers in which such crystals are used as the active medium. Since the majority of experiments were carried out on rather pure crystals grown from the gas phase the observed regularities are attributable to the intrinsic defects of nonstoichiometry. Section 1 examines the energy and degradation of cadmium sulfide lasers as a function of the monocrystal growing conditions. Section 2 gives results from an investigation of the influence of various annealings on the luminescence and laser characteristics of zinc selenide. Results from analogous investigations on zinc oxide monocrystals are examined in Section 3. The cathodoluminescence of zinc sulfide monocrystals is described as a function of the sample fabrication conditions in Section 4. In the concluding section (Section 5) the primary conclusions of the study are formulated.

---

[1] Longitudinally electron-beam pumped GaAs, InP crystal lasers normally employ n-type crystals with a majority carrier concentration $n \approx 10^{18}$ cm^{-3} [2]. This is the primary means for achieving the reduction in the absorption coefficient of the passive regions of the cavity at the lasing wavelength. Doping the A^2B^6 compounds with Cl, In-type donors or Li, Na acceptors has not yet yielded positive results in longitudinally pumped lasers, although according to [3] a reduction in the lasing threshold and improvements in radiation homogeneity have been achieved in transversely-pumped lasers.

1. Cadmium Sulfide

Even the early studies [4] on electron-pumped semiconductor lasers noted that the performance characteristics of a transversely pumped plastic cadmium sulfide crystal laser with the crystal fabricated by resublimation in an inert gas flow are significantly dependent on the composition of the vapor phase in the growth zone. It was noted that the energy characteristics of the laser are degraded while the efficiency approaches the theoretical limit with crystals fabricated at an excess cadmium pressure. Study [5] established that the lasing threshold drops when this method is employed with the reduction in crystal growth temperature, while the laser efficiency improves. However no reasonable explanation for these observations was found.

Below we provide results from electron microscope investigations of original and electron-beam irradiated crystals fabricated by various techniques. These investigations revealed that cadmium sulfide crystals obtained from the vapor phase with minimum total vapor pressure contains excess cadmium. This excess cadmium normally appears as interstitial Cd_i cadmium or V_S sulfur vacancies dissolved in the crystal; in *n*-type material which is normally the type of CdS, no deep levels are formed and there is no harmful influence on the laser characteristics, since as second phase precipitates they significantly enhance the radiative properties of the crystal. The second phase can precipitate both due to cooling of the monocrystal grown at a high temperature and by electron-beam irradiation of the crystal.

We then provide results from an investigation of the influence of excess sulfur vapor pressure in growing CdS crystals on the laser characteristics. A nonmonotonic change in these characteristics is observed with variations in sulfur pressure over a certain range. A range of sulfur vapor pressures is found where the laser performance characteristics are optimum. If at lower sulfur vapor pressures the degradation of the laser performance is associated with cadmium precipitation in the second phase, at higher pressures it produces an increase in the cadmium vacancy V_{Cd} concentration which form effective nonradiative recombination centers.

Gas-phase grown CdS crystals (by resublimation in both an argon flow (layered) [6] and in a sealed vessel using the technique similar to that described in study [7] (bulk)) were investigated. Experiments on the influence of a controlled variation in sulfur vapor pressure p_S over the growing surface of the crystal were carried out on the assembly shown in Fig. 1 which, compared to that described in Study [7], had an additional region containing the narrow end of the silica vessel filled with sulfur. The sulfur vapor pressure was temperature-controlled in this region in accordance with the data given in study [8]. The following cooling conditions were used in growing the layered crystals: slow cooling at a rate $\upsilon = 470 \exp(-0.35\ t)\ K \cdot h^{-1}$, where t is in hours (crystal types A, B, D, E) and rapid cooling to room temperature at $\upsilon = 1800\ K \cdot h^{-1}$ (crystal type C). Crystal types A, B, and C were grown from a blend consisting of 99.999% purity CdS powder;

types D and E crystals contained 2% by weight of sulfur and cadmium, respectively.

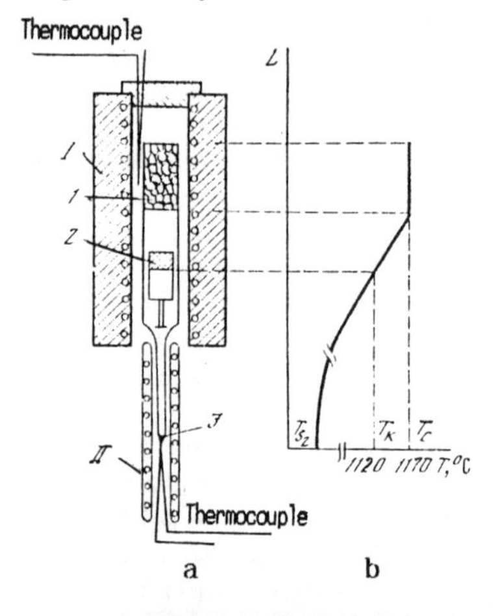

Fig. 1. Set up for crystal growing (a) and the vertical temperature distribution in the oven (b)
I - growing oven; II - sulfur vapor pressure control oven; 1 - CdS blend; 2 - seed; 3 - sulfur blend; T_C - sublimation temperature; T_K - crystallization temperature; T_{S_2} - temperature in sulfur region

Fig. 2. Electron microscope photographs of precipitates in CdS crystals
a - type D layered crystal, plane (11̄20); the arrow indicates the traces of dislocation slip on the (0001) planes; b - bulk crystal (0001) plane

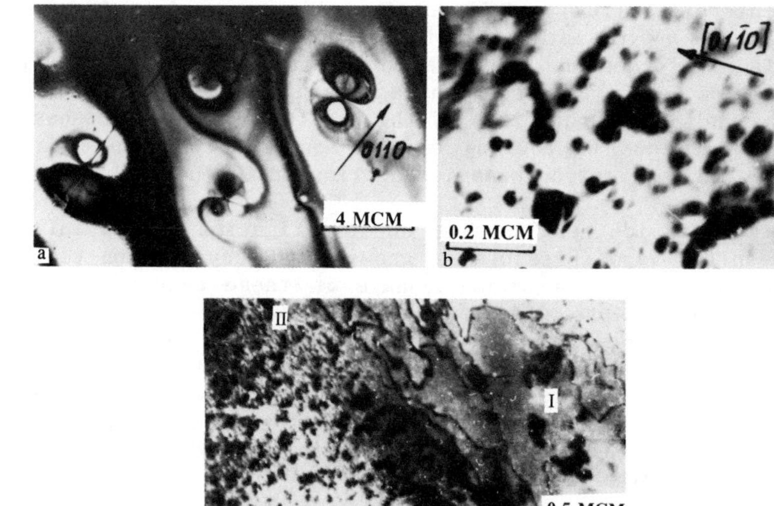

Fig. 3. Electron microscope photograph of layered CdS crystal, (1120 plane)
I - region with elevated dislocation density; II - region with elevated precipitation density

Electron microscope and cathodoluminescence investigations of layered crystals were carried out on samples obtained in a single growing process, while in the case of bulk crystals the crystals are taken from the same blank. Samples from the bulk crystals for transition electron microscopy were prepared by chemical polishing [9], while the layered samples were selected for the necessary width by the characteristic green interference color corresponding to a 100-200 nm thickness. The investigations were carried out on the JEM-100U electron microscope.

The active element of the laser (henceforth called the laser screen (LS)) was a 30-50 μm thick semiconductor monocrystal wafer with reflection coatings attached to a sapphire disk on both surfaces. The method of fabricating the LS from bulk crystals is analogous to that described in [10]. Chemical polishing in an etch was used for final surface polishing of the semiconductor wafer.

The cathodoluminescent and laser characteristics were investigated at T = 800 K on the ELT-1 assembly described in study [11] in a pulsed scanning mode at an electron energy E_0 = 75 keV, an electron beam diameter $pd_э$ = 15 μm, current densities of j = 0-300 $A \cdot cm^{-2}$ scanning rates υ_{CK} = 10^5 $cm \cdot s^{-1}$ and LS excitation frequency ν = 200 Hz. The low-temperature (4.2 K) photoluminescence spectra of the derived crystals under excitation by a DRSh-250 type mercury lamp were also investigated.

The samples were either excited in the assembly described above or in the microscope column which made it possible to observe defect formation directly during irradiation. In this case no condenser diaphragm was used in the electron-optical section of the microscope which made it possible to obtain current densities of the order of 1 $A \cdot cm^{-2}$. Estimates of the heating ΔT of cadmium sulfide crystals under irradiation in the microscope column in accordance with [12] yielded a value of less than $0.5 t_K$, where t_K is the thickness in nanometers, ΔT in degrees Kelvin.

Results from electron microscope investigations of cadmium sulfide crystals are illustrated by the photographs in Figs. 2-4. Fig. 2a shows a type D layered crystal grown using a blend with an excess sulfur concentration. Sparse precipitates with a surface density $\sim 10^6$ cm^{-2} are visible in the figure; these produce strong local elastic deformation fields. A significantly higher precipitation density is observed in a bulk crystal (Fig. 2) grown at a minimum vapor pressure over the growing surface and cooled in the oven (slowly). The precipitate density here is of the order of 10^{10} cm^{-2}. The characteristic dimensions of the precipitates is 30 nm. The contrast from these precipitates makes it possible to conclude that they are coherent producing near-symmetrical deformation fields [13]. We note that the dimensions of the inclusions grow with a reduction in their density.

Fig. 3 shows a crystal whose image contains two qualitatively different regions. Region I has an elevated dislocation density and a reduced precipitation density compared to region II. This indicates that the dislocations in CdS crystals function as effective drains for the defects producing the precipitates.

The photographs in Fig. 4 illustrate the structural changes occurring in a type C layered crystal under irradiation in an electron microscope column and after long-term exposure to light. There are no crystal precipitates immediately after the growing process (Fig. 4a). However as the radiation dosage increases (Fig. 4b, c) and the storage time grows (Fig. 4d) isolated precipitates first appear and then their concentration grows. Crystals with a high precipitation concentration have made it possible for us to record an electron diffraction pattern revealing the existence of polycrystalline cadmium in the crystals. One such electron diffraction pattern is shown in Fig. 5.

The observed precipitates will clearly be manifest in the characteristics of lasers by, on the one hand, increasing the scattering and absorption of generated radiation and, on the other, reducing the efficiency of band-to-band recombination. Fig. 6 gives the cathodoluminescence intensity plotted as a function of precipitate concentration estimated by the surface density in electron-microscope images accounting for sample thickness. It is clear from the figure that an increase in precipitation concentration significantly reduces radiation intensity when $N_B > 10^{10}$ cm^{-3}.

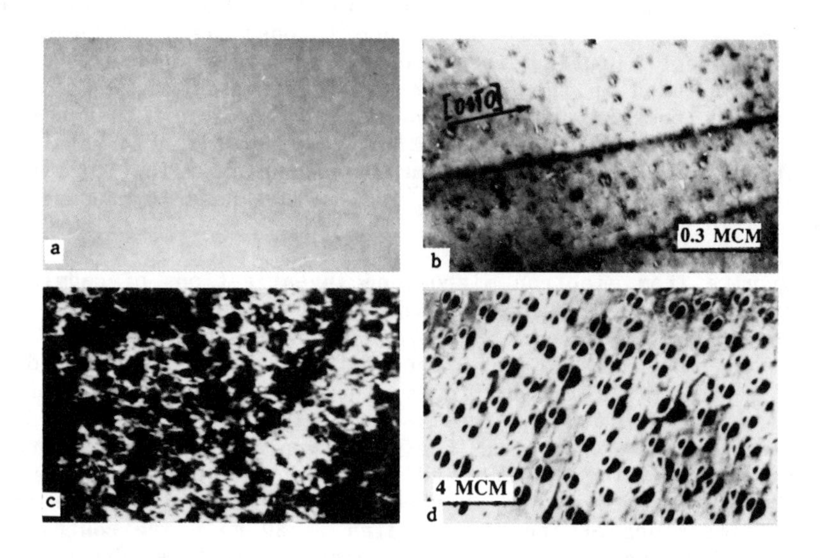

Fig. 4. Successive photographs of a type C CdS layered crystal under irradiation under a JEM-100U electron microscope (j = 1 A·cm^{-2})
a - initial crystal; b - radiation time: 13 min; c - 33 min; d - after exposure to light at room temperature for 6 months without irradiation

Changes in the precipitation concentration during irradiation and their effect on the radiative properties of the crystal reduce cathodoluminescence efficiency over time and consequently cause degradation of a laser based on the crystal. Fig. 7 shows the cathodoluminescence intensity plotted as a function of irradiation time in the EhLT-1 for type C and D layered crystals. Initially the type C crystal, as long as it has no precipitates, has a higher in-

Fig. 5. Electron diffraction pattern of type C CdS crystal after long-term irradiation in transition electron microscope
The rings represent Cd precipitation; the points correspond to the CdS monocrystal

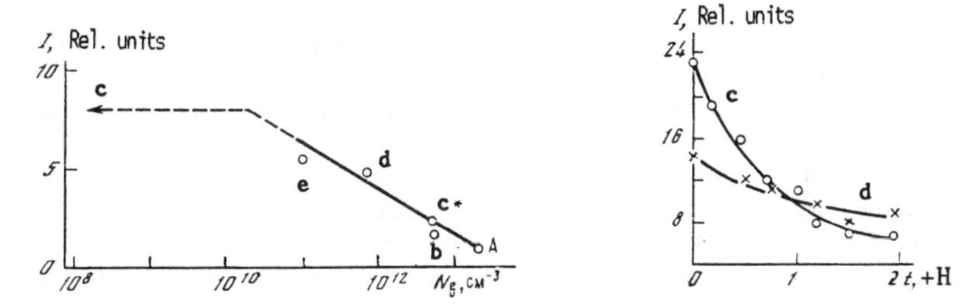

Fig. 6. The cathodoluminescence intensity I plotted as a function of precipitation concentration N_B in CdS layered crystals
C*: type C crystal after photoaging; j = 300 $A \cdot cm^{-2}$

Fig. 7. Cathodoluminescence intensity plotted against irradiation time in the EhLT-1 for two CdS layered crystals
j = 300 $A \cdot cm^{-2}$, E_0 = 75 keV; ν = 200 Hz, $d_э$ = 15 μm, υ_{CK} = 10^5 $cm \cdot s^{-1}$

tensity. Fewer cadmium precipitates form in a type D crystal after long-term irradiation. It is possible that this is related to the fact that the blend contained excess sulfur in the growing of the crystal. Therefore the cathodoluminescence intensity of a type C crystal after one hour of irradiation drops below that of a type D crystal.

The results given above from electron microscope and cathodoluminescence investigations revealed that when growing from the vapor phase in conditions similar to minimum pressure over the growing surface, CdS crystals are obtained with a significant excess cadmium level. The maximum deviation from stoichiometry observed in our experiments on a type A crystal (see Fig. 6) can be estimated by the total precipitate volume V_B. Taking a value D_B = 30 nm for the average precipitate dimensions and for a concentration $N_B = 2 \cdot 10^{13}$ cm^{-3} we obtain $V_B = 3 \cdot 10^{-4}$ cm^{-3} per 1 cm^3 of the crystal corresponding to an excess cadmium concentration $N_{Cd} \approx 6 \cdot 10^{18}$ cm^{-3}. This estimate does not exceed the maximum possible deviation from stoichiometry on the excess cadmium side of 10^{19} cm^{-3} obtained from an analysis of the phase diagram [14].

The excess cadmium "dissolves" in the growing process in the crystal as either interstitial atoms or sulfur vacancies. As the temperature drops the composition stability range of the CdS crystal narrows. Therefore if a cadmium-rich crystal is cooled slowly, the cadmium will precipitate out in a second phase from the association of the elementary Cd_i and V_S point defects. With sufficiently rapid cooling the point defects are frozen since their diffusion is rather insignificant at room temperature or below. Therefore precipitate formation is observed only over a long time period (type C crystal) or in the case of electron beam irradiation which significantly stimulates the point defect diffusion processes.

It is then reasonable to assume that the nonequilibrium carriers will recombine nonradiatively on the cadmium precipitate surface. The cathodoluminescence efficiency then drops in the bulk surrounding inclusions at distances less than the diffusion length L_D of the nonequilibrium carriers. Bearing in mind that the diffusion length of the nonequilibrium carriers in the cadmium sulfide at sufficiently high pump levels reaches values of the order of 1 μm [15] we obtain an estimate of the cadmium precipitate concentration N_B where the radiation intensity drops by 10%:

$$N_B = 0{,}1\,(^4/_3\pi L_D^3)^{-1} \approx 2{,}5 \cdot 10^{10}\ \text{см}^{-3},$$

which agrees with experimental results (see Fig. 6). We note that N_B is not proportional to the concentration of excess cadmium N_{Cd} since, as noted above, the characteristic dimensions of the precipitates grow with a reduction in their concentration. This fact can be explained within the framework of the diffusion-limiting model of the second phase precipitation process [16, 17].

The existence of precipitates even in a small quantity in the crystal can degrade laser performance characteristics due to radiation scattering. Consequently it is necessary to change the growing conditions so that the concentration of the superstoichiometric cadmium is below a specific level. Experiments with altering the composition of the blend during growing in an argon flow have revealed that this method has not been successful in altering the crystal composition in a controlled manner (the precipitate concentration does not correlate with the composition of the blend: see Fig. 6 and its legend). However it could be achieved using a static free growing technique [7] on the assembly described above (see Fig. 1).

Fig. 8 gives the threshold current density values j_{Π} (curve 1) for lasers manufactured using crystals fabricated with different values of p_S in the form of dependencies on the partial sulfur pressure p_S. This figure also gives the dependencies on p_S of the I_1 (curve 2) and I_2 (curve 3) line intensities in the photoluminescence spectra of these crystals at 4.2 K caused by the radiative annihilation of excitons by the neutral acceptor and the neutral donor, respectively. At low values of p_S the threshold current density is sufficiently high while the laser differential efficiency is low. This corresponds to low intensities of the I_1 and I_2 lines in the photoluminescence spectra. With p_S increasing to values of 10^{-3} atm a noticeable drop of j_{Π} is observed (the differential laser efficiency also improves). Here the intensity of I_1 remains unchanged while I_2 grows. The growth of P_S above 10^{-2} atm reduces I_2 line intensity and increases I_1 line intensity and j_{Π}. Electrical conductivity measurements reveal that crystals grown at $p_S < 10^{-2}$ atm are low-resistance crystals, while at 10^{-2} atm $\leqslant p_S \leqslant 10^{-1}$ atm they are high resistance crystals ($\rho \sim 10^9$ ohm·cm).

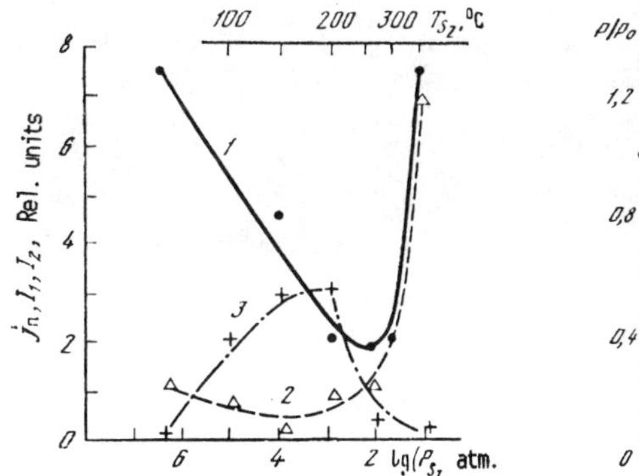

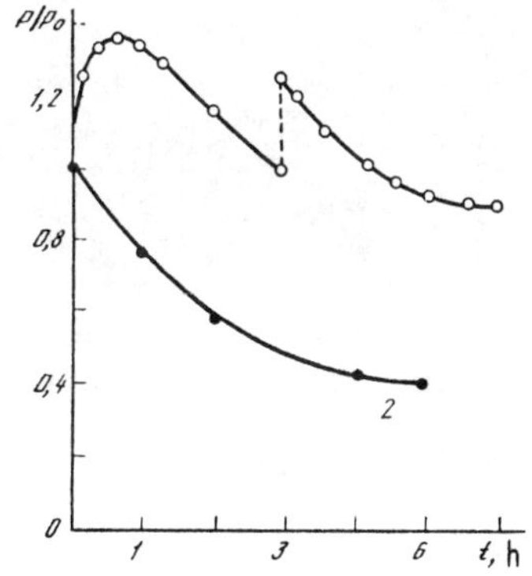

Fig. 8. The threshold current density j_{Π} (1) and the intensity of the radiation lines of the excitons bound to the neutral acceptor I_1 (2) and the neutral donor I_2 (3) in the low-temperature photoluminescence spectrum plotted as a function of sulfur vapor pressure p_{S_2} (and the temperature in the sulfur blend region T_{S_2})

Fig. 9. The laser screen lasing power P normalized to the initial value P_0 plotted as a function of electron beam irradiation time t Crystals grown at $p_{S_2} = 10^{-2}$ atm (1) and 10^{-5} atm (2); after three hours of irradiation from the LEh-1 (laser screen) a one hour break was introduced in the irradiation process; E_0 = 75 keV, ν = 400 Hz, j = 300 A·cm^{-2}, υ_{CK} = 10^5 cm·s^{-1}

If the degradation to the laser characteristics with a reduction in p_S below 10^{-3} atm is caused by the precipitates, crystals grown at $p_S > 10^{-2}$ atm are sufficiently homogeneous and the observed growth of $j_п$ can be attributed only to a growth in the concentration of intrinsic defects that form fast nonradiative recombination channels. V_{Cd} and S_i are the primary defects in a CdS crystal as its composition moves from stoichiometric to excess S. Since it is generally assumed that due to the large atomic radius of sulfur and the low diffusion coefficient a S_i defect is not likely, the V_{Cd} defect is most likely responsible for the nonradiative recombination. It is then possible to interpret the regularities shown in Fig. 8 in the following manner.

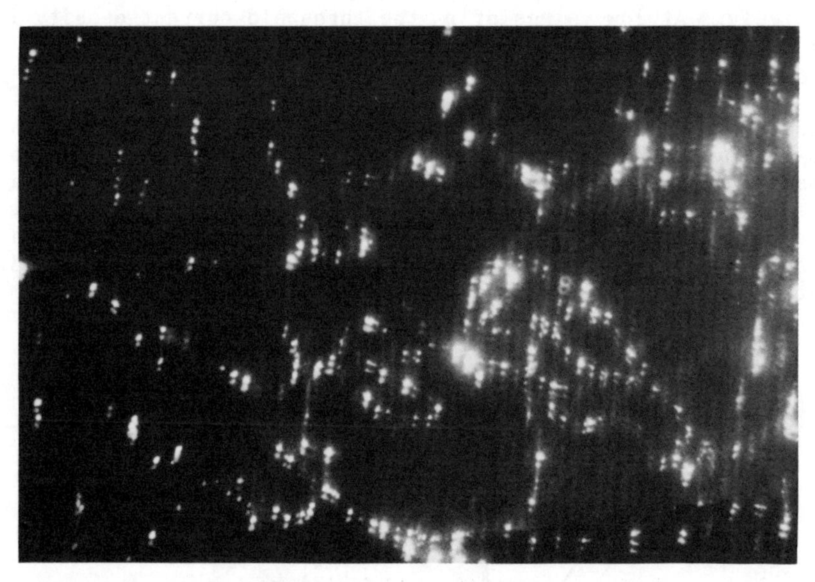

Fig. 10. Near field photograph of laser screen irradiation from a liquid cadmium annealed CdS crystal under electron beam pumping (television scanning)
E_0 = 75 keV, j = 200 A·cm^{-2}, υ_{CK} = 0.8·10^5 cm·s^{-1}, T = 80 K

With small values of p_S the crystal is saturated with cadmium and the concentration of Li_{Cd} or Na_{Cd} substitution acceptors to which the I_1 lines are normally related is low. In this case the Li and Na atoms will most likely be distributed among the interstitial sites. The concentration of donors, including the Cd_i and V_S defects, on the other hand, is so high that the intensity of the I_2 line generated by trapping by a shallow isolated donor is also low. In this case the donors form associates with the deeper levels in the bandgap. Therefore it is reasonable to relate the growth of I_2 to growth of p_S up to 10^{-3} atm to an increase in the concentration of isolated shallow donors as the concentration of Cd_i of V_S drops to values where precipitates or associates are not likely to form. Then as $p_S > 10^{-2}$ atm a reduction in I_2 can be attributed to a reduction in the total number of donors as indicated by a drop in the electrical conductivity of the crystals grown at 10^{-2} atm $\leqslant p_S \leqslant 10^{-1}$ atm. The noticeable growth of I_1 coinciding with the growth of j_Π at $p_S > 10^{-2}$ atm is the result of an increase in the V_{Cd} concentration when sites are freed up for Li and Na and nonradiative recombination centers appear.

It turned out that lasers employing crystals grown in the $p_S = 10^{-3}$-10^{-2} atm range had a higher degradation resistance than lasers employing crystals fabricated at lower values of p_S. Fig. 9 shows a sample relative change in laser radiation power under intense pulse-scanning pumping for crystal laser screens. The rapid degradation of LEh-2 (curve 2) is, as noted above, the result of the precipitation of cadmium in a second phase or from the formation of associates. The following is one possible explanation for the initial increase in LEh-1 power. The crystal used to fabricate the LEh-1 evidently has a near-stoichiometric composition. However the Cd_i and V_{Cd} type defects that do not interact evidently froze in the crystal during the cooling process in a rather high concentration. Under electron beam irradiation the mobility of these defects increases and some of the interstitial cadmium atoms occupy the free sites which reduces the nonradiative recombination centers associated with V_{Cd}. The drop in lasing power from LEh-1 following its initial growth can be attributed to the generation of new defects near the surface or large associates as well as the radiation-stimulated diffusion of these defects in the crystal bulk.

The stoichiometry of a CdS crystal therefore has a significant influence on the characteristics of a laser based on this crystal. A very simple example of this influence is shown in Fig. 10. This figure gives the near radiation field of an electron-beam pumped laser screen scanning in the television mode. The photograph was made through a sapphire substrate that was deliberately out of orientation which split the image at 30 μm. A liquid cadmium annealed CdS crystal was used. It is clear that lasing is observed only near the small-angle grain boundaries (SAGB), i.e., where the crystal had no cadmium precipitates due to the drainage of excess cadmium through the small-angle grain boundaries.

2. Zinc Selenide

Until recently large zinc selenide monocrystals for practical applications were melt-grown [18]. The first semiconductor laser designs were based on "melt-grown" crystals. However the power and efficiency of such lasers were significantly lower than the corresponding parameters of cadmium sulfide lasers. The reason for this was the low quantum efficiency of band-to-band recombination in the "melt" zinc selenide crystals compared to the best CdS crystals. The low band-to-band recombination efficiency of these crystals can be attributed to the strong contamination by uncontrolled impurities, significant deviation of the composition from stoichiometry and a significant concentration of structural defects. Twin-type and dislocation-type structural defects are present in a large concentration due to the fact that in cooling from T_m = 1520°C to room temperature the melt-grown crystals undergo a phase transition at T_f = 1425°C from the stable hexagonal phase at a high temperature to a stable cubic phase at low temperature.

In order to determine the relative role of the factors outlined above in the formation of the radiative properties of zinc selenide and, in the final analysis, its laser characteristics, we carried out a series of annealings of melt-grown crystals in liquid zinc, and zinc and selenium vapors. Gas-phase grown, purer crystals were also annealed.

In the initial annealing experiments in liquid zinc we determined the self-diffusion coefficient of zinc in the "melt-grown" crystals. Several blocks were cleaved from a monocrystal blank and were polished; one of the block faces was oriented parallel to the twinning plane. Prior to annealing the blocks were treated in a polishing etchant based on a CrO_3 solution in HCl [19]. The liquid zinc annealing process took place in a sealed silica vessel at $T_{ANN.}$ = 980°C for several hours. The liquid zinc was then drained out through the other end of the vessel and the crystals were cooled in the oven at $\upsilon \approx 150°C \cdot h^{-1}$. The blocks were then cut into slabs that were then mechanically polished and chemically treated in a polishing etchant. Fig. 11 contains photographs of the luminescence from two slabs under scanning electron beam excitation with a current density j = 1 $A \cdot cm^{-2}$ (a) and j = 300 $A \cdot cm^{-2}$ (b). The photograph in Fig. 11b was obtained by means of a filter that filtered out the longwave radiation and transmitted the band-to-band radiation. It is clear from Fig. 11 that there are three regions in the slab that appear at low excitation levels: a rim, a central region and a weakly luminescing region between them. At a high excitation level (see Fig. 11b) the intensity of band-to-band recombination radiation is significantly greater than that of the rim.

The inhomogeneity in the luminance of the wafer fabricated by the method described above can be attributed to the two-way diffusion of the impurities from the crystal bulk to liquid zinc and from the liquid phase zinc to the crystal. Evidently in the initial state the "melt-grown" crystals have insufficient zinc which aids in the formation of Li_{Zn}, Na_{Zn}, Cu_{Zn} substitution centers and associates involving V_{Zn} and the impurity atoms responsible for longwave radiation from zinc selenide. Under annealing in liquid zinc the impurity atoms gradually are released from the crystal beginning at the surface, thereby breaking down the longwave radiation centers and forming high concentration V_{Zn}. The zinc diffusion among the interstitial sites in the crystal bulk gradually occupies the liberated sites. Therefore the weak luminescence region most probably corresponds to a region with an elevated zinc vacancy concentration which we will henceforth identify with the formation of nonradiative recombination centers. In this case the slab rim has a minimum concentration of longwave radiation centers and nonradiative recombination centers, i.e., the centers presenting competition to band-to-band radiative recombination.

We obtain as a result of these experiments on annealing in liquid zinc the following estimate for the self-diffusion coefficient of zinc in zinc selenide at $T_{ANN.}$ = 980°C by the formula

$$D = L^2_{ANN.} t^{-1}_{AAN.},$$

where $L_{ANN.}$ is the rim width; while $t_{ANN.}$ is the annealing time. Substituting values of $L_{ANN.}$ = 350 μm, $t_{ANN.}$ = 12.5 h we obtain D = $3 \cdot 10^8$ $cm^2 \cdot s^{-1}$. We note, however, that the zinc diffuses in different crystallographic directions at different rates and is significantly accelerated near structural defects (see Fig. 11b).

The change in the low-temperature photoluminescence spectra from annealing of one of the melt-grown zinc selenide crystals in liquid

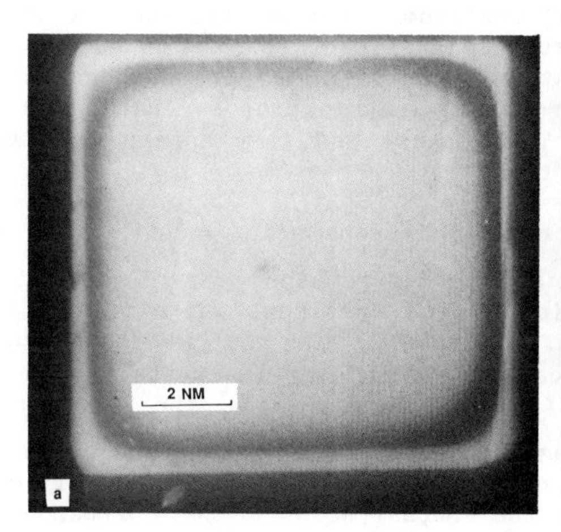

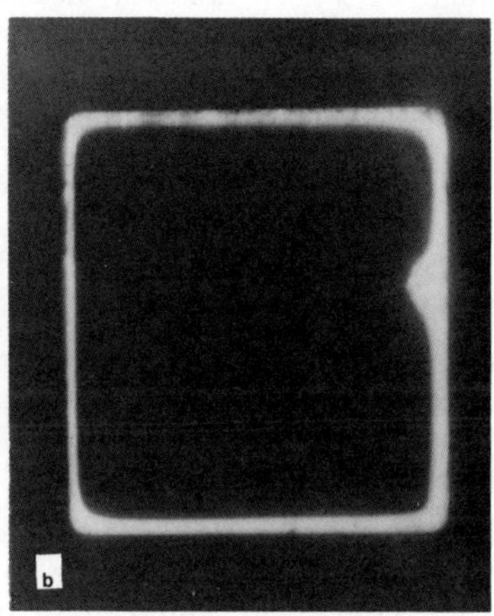

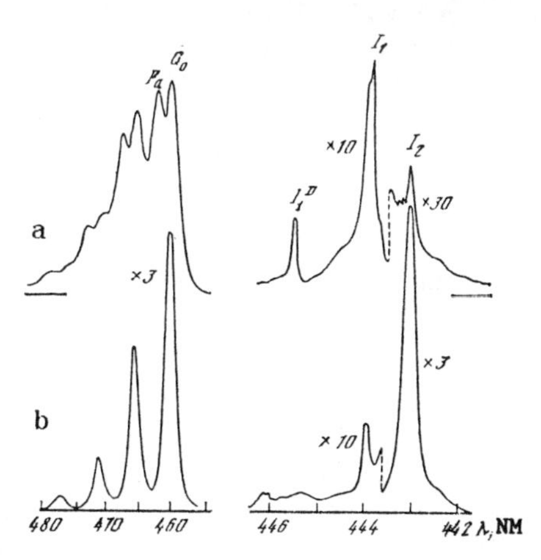

Fig. 11. Luminance of two slabs cleaved from liquid zinc-annealed zinc selenide blocks for the case of electron beam excitation (television beam scanning)
a – j = 1 $A \cdot cm^{-2}$, $t_{ANN.}$ = 12.5 h; b – j = 300 $A \cdot Cm^{-2}$, $t_{ANN.}$ = 4 h; a filter was employed to eliminate the longwave radiation at λ > 470 nm

Fig. 12. Photoluminescence spectrum at T = 4.2 K of ZnSe crystal melt-grown prior to (a) and after (b) annealing in liquid zinc.

zinc are given in Table 12. Two series of intense "edge" radiation with zero-phonon lines Q_0 and P_0 were observed prior to annealing; in study [20] these are attributed to transitions from the shallow donor to two different acceptors Li_{Zn} and Na_{Zn}. The exciton radiation intensity in this case is significantly lower and the I_1 and I_1^D lines predominate; these lines are from the annihilation of excitons by the relatively shallow and deep acceptor in the neutral state. The same acceptors as in the "edge" radiation can be the shallow acceptor, while the deep acceptor is assigned to either copper Cu_{Zn} [21, 22] or the neutral zinc vacancy $V_{Zn}^{\times}$ [23-25]. Study [26] has claimed that both Cu_{Zn} and $V_{Zn}^{\times}$ yield in the exciton spectrum the I_1^D lines that are similar in shape and spectral position and therefore are difficult to differentiate in regular conditions. An analysis of the study reveals that V_{Zn} produces the I_1^D line in the low-temperature luminescence spectra, although the appearance of this line does not infer the presence of V_{Zn}. However deviations from stoichiometry towards excess zinc in our experiments are rather significant (of the order of 10^{18} cm^{-3} in the "melt-grown" crystals and 10^{16} cm^{-3} in the "gas-phase grown" initial crystals) and are sufficient to propose that there are significantly fewer Cu_{Zn} than V_{Zn} centers and they appear much less intensely in the radiation spectra.

The intensity of "edge" radiation drops after annealing in liquid zinc and only the shorter wave series remains. The I_1^D drops significantly in the exciton region of the spectrum, while the I_2 line produced by the annihilation of excitons by the neutral donor grows; the I_1 line diminishes significantly and in certain crystals even increases. The growth of the I_2 line is related to the increase in shallow donor concentration; the $Zn_i^{\times}$ interstitial zinc atoms can serve as the shallow donors. The minor change in the intensity of line I_1 does not represent constancy of the concentration of the corresponding centers. The concentration of these centers will drop after annealing, although the concentration of other acceptors and the nonradiative recombination centers will also drop, thereby causing an increase in the free exciton lifetime and an increase in the probability of capture by the shallow acceptor. The fact that the I_1^D line always diminishes significantly after annealing while the I_1 line may even grow confirms our assumption that the zinc vacancy in our case is the deep acceptor determining the I_1^D line. Otherwise it would be necessary to assume that the copper exits the crystal due to annealing more rapidly than the Li and Na shallow acceptors, which contradicts diffusion experiments (for ZnO see study [27]).

The influence of intrinsic point defects on the radiative properties of zinc selenide and of any other material will of course be clearer in crystals of greater purity. We obtained significantly purer, large-scale monocrystals (up to 25 cm^2 in volume) by gas-phase growing in sealed silica vessels using a method similar to that described in study [7]. The assembly shown in Fig. 1 was used and auxiliary oven II was not activated. Chemical transport in hydrogen was used to enhance the mass transfer of the low volatility zinc selenide material [28]. The vessel was filled with a hydrogen/argon gas mix rather than simply with argon as in the case of growing cadmium sulfide for this purpose [7]. We should note that for growing in hydrogen the vapor composition over the growing crystal surface has a higher partial pressure of the metal compared to growing in an inert gas [29] which in turn increases the percentage of metal in the crystal.

The low-temperature photoluminescence spectra of one of the zinc selenide crystals grown from the gas phase in a hydrogen atmosphere before and after annealing in liquid zinc are shown in Fig. 13. The most intense line in the exciton region of the initial sample is the I_1^D line with its phonon repetitions. No "edge" radiation is observed against the background of these lines or a broad longwave line at $\lambda_м$ = 495 nm. The I_1^D line vanishes with growth of temperature due to the temperature dissociation of the corresponding exciton-impurity complexes, and weak "edge" radiation then becomes visible at the shortwave edge of the line at $\lambda_м$ = 495 nm. After annealing in liquid zinc the lines at $\lambda_м$ = 495 nm and I_1^D virtually vanish from the spectra.

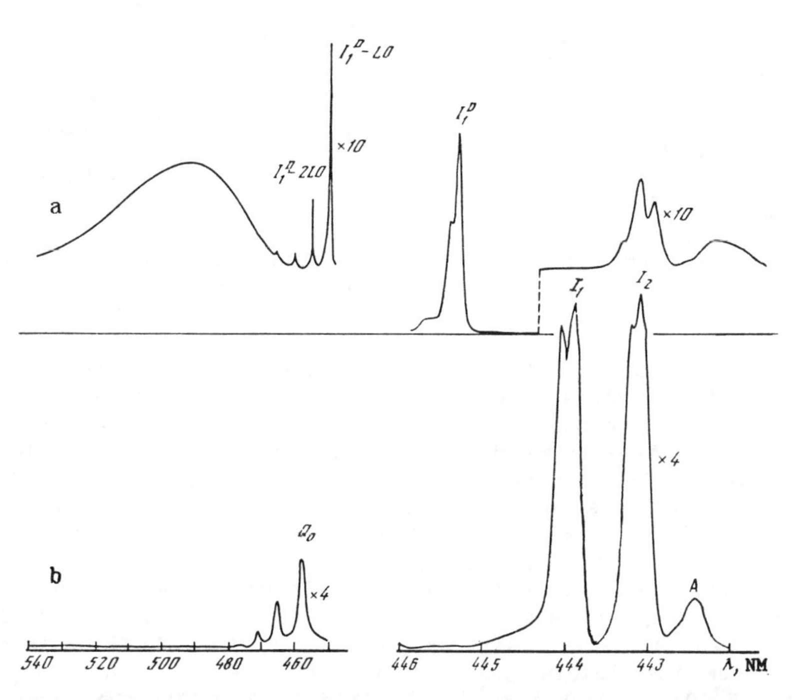

Fig. 13. Photoluminescence spectrum at T = 4.2 K of gas-phase grown ZnSe crystal prior to (a) and after (b) liquid zinc annealing

The I_1 and I_2 lines appear together with the free exciton line (A) and weak edge radiation with the zero-phonon line Q_0. The growth in intensity of the A, I_1, I_2, and Q_0 lines can be attributed to the increased lifetime of the free exciton after the vanishing of centers responsible for radiation from the I_1^D and the line at $\lambda_М$ = 495 nm. We note that the longwave line at $\lambda_М$ = 495 nm correlates with the I_1^D line. This supports their assignment to the same defect (the zinc vacancy.

The influence of annealing in zinc and selenium vapor on the radiative characteristics of zinc selenide was also investigated. Zinc vapor annealing was carried out at $T_{ANN.}$ = 1000°C and p_{Zn} = 1 atm, while selenium vapor annealing was performed at $T_{ANN.}$ = 900°C and p_{Se} = 1 atm. After zinc vapor annealing qualitatively identical changes were observed in the zinc selenide crystals as in liquid zinc annealing. The intensity of all lines diminished from selenium vapor annealing.

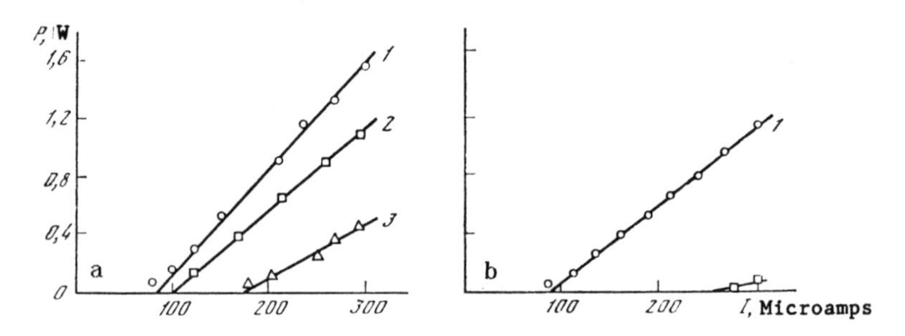

Fig. 14. Lasing power P at E_0 = 75 keV, $d_Э$ = 15 μm, T = 80 K plotted as a function of electron beam current J for laser screens fabricated from ZnSe crystals.
a - gas-phase grown; b - melt-grown; 1 - after liquid zinc annealing; 2 - before annealing; 3 - after selenium vapor annealing

We should note that commonly observed changes in the radiation spectra at low excitation levels do not always cause significant changes in the band-to-band radiation efficiency η at high excitation levels ($j \geqslant 10$ A·cm^{-2}) which determines in the lasing threshold ($j_П \sim \eta^{-1}$). This is because as the excitation level grows many radiative recombination centers become saturated and the nonradiative recombination deep centers represent the primary competition to the band-to-band transition. However, an investigation of many zinc selenide crystals grown by various techniques (melt-grown and gas-phase grown in different regimes) has demonstrated that annealing in liquid zinc always increases the intensity of exciton radiation I_E (by a factor of 2-60) at low pump levels ($j = 10^{-5}$ A·cm^{-2}) as well as in η (by a factor of 1.5-10) at high pump levels (j = 50 A·cm^{-2}); the greater the increase in intensity the lower the values of η and I_E in the initial samples. Selenium vapor annealing reduces I_E by a factor of two on the average and reduces η by a factor of 1.5 on the average.

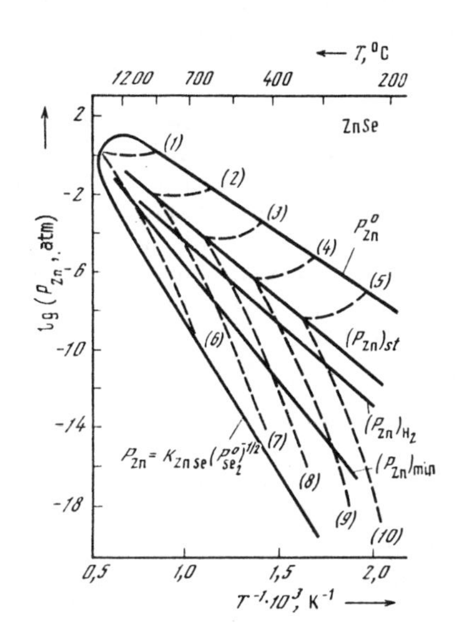

Fig. 15. Composition stability range of ZnSe compound in p–T coordinates [31]
The dashed lines represent an identical composition corresponding to given deviation from stoichiometry: 1 - $[V_{S3}]$ - $[V_{Zn}]$ = 10^{16} cm^{-3}; 2- 10^{14}; 3 - 10^{12}; 4 - 10^{10}; 5 - 108; 6 - $[V_{Zn}]$ - $[V_{Se}]$ = 10^{16}; 7 - 10^{14}; 8 - 10^{12}; 9 - 10^{10}; 10 - 10^{8}; $(p_{Zn})_{st}$ is zinc pressure over the stoichiometric compound; p^0_{Zn} and p^0_{Se} are zinc and selenium pressures at the boundary of the composition stability range; $(p_{Zn})_{min}$ is zinc pressure with minimum vapor pressure over the crystal; $(p_{Zn})_{H2}$ is the same pressure in the case of hydrogen growing [32]

The influence of liquid zinc and selenium vapor annealing on laser screen energy characteristics is shown in Fig. 14. It is clear from this figure that the increase in the laser screen power and efficiency after liquid zinc annealing is significantly greater for a "melt-grown" crystal having a lower internal quantum efficiency prior to annealing at high excitation levels. Se vapor annealing degrades lasing characteristics.

The results given above indicate that the normal melt- and gas-phase grown zinc selenide crystals contain a greater concentration of zinc vacancies that form centers that compete with the band-to-band radiative transitions. A high V_{Zn} concentration in the "melt-grown" crystals also follows from study [30] where the zinc shortage in these crystals was estimated at $2\cdot 10^{18}$ cm^{-3}. According to study [31] the gas-phase-grown zinc selenide crystals with minimum vapor pressure will also have a zinc shortage. Results from this study are given in Fig. 15 where the composition stability range of the crystal is plotted in logarithmic coordinates of the partial pressure of the zinc and the inverse growing temperature; this figure also gives the curve corresponding to crystal growing in an inert gas when $p_{Zn} + p_{Se2} = p_{min}$; the curve corresponding to stoichiometry calculated assuming a vacancy defect formation model and dashed curves corresponding to a given deviation of the composition from a stoichiometric separation. This figure also contains a curve corresponding to hydrogen growing calculated by data from study [32]. It is clear from Fig. 15 that the zinc vacancy concentration in hydrogen-grown crystals is estimated at 10^{16} cm^{-3} when T_p = 1150°C. This is significantly lower than in "melt-grown" crystals and therefore the characteristics of lasers employing the initial "gas-phase grown" crystals are much better than those based on "melt-grown" crystals (see Fig. 14). Nonetheless this same concentration of V_{Zn} is insufficient to determine radiation from "gas-phase grown" crystals at low pump levels.

Unlike cadmium sulfide for which the improvement in laser characteristics is achieved primarily by a decrease in the excess cadmium normally present in the crystal to concentrations that make it possible to precipitate the cadmium out in a second phase, in the case of zinc selenide such improvement is achieved by zinc saturation. The primary difference is the significant differential of the phase diagrams of zinc selenide (excess chalcogenide) and cadmium sulfide (excess metal).

3. Zinc Oxide

Zinc oxide is a promising material for obtaining UV radiation (λ_L = 375-395 nm) [33]. The melting point of ZnO and the vapor pressure at this temperature are quite high (1975°C and 50 atm). Therefore until recently large-scale ZnO monocrystals could be fabricated by the hydrothermal method only [34]. A gas-phase fabrication technology was recently developed for producing zinc oxide slabs 25 mm in diameter and approximately 10 mm in height by using hydrogen transport [35]. The present study focuses on crystals fabricated by these two techniques.

Fig. 16 shows the cathodoluminescence spectra of ZnO crystals fabricated by different techniques at low pump levels. The radiation spectra of all samples consisted of a series of exciton lines at 80 K and a line at λ_M = 385 nm at 300 K within which lasing is achieved as well as longwave radiation with a peak in the 490-500 nm range; however, the relative intensity of these lines varies significantly in the different samples. The most intense UV and the weakest longwave radiation are observed in crystal 1 which is a gas-phase grown crystal. As the current density of the electron beam is increased the UV intensity I_{UV} becomes comparable for crystals 1 and 2 shown in Fig. 16. Nonetheless for the working pump levels corresponding to the shaded region in the insert in the upper left corner of Fig. 16, I_{UV} is still significantly greater for crystal 1 than for crystal 2, particularly at T = 300 K. Consequently the threshold current density for a laser screen made from crystal 1 will be significantly lower than for a similar screen made from crystal 2.

The significant difference discussed above between the "gas-phase grown" and "hydrothermally-grown" crystals in favor of the former is not a chance occurrence. An investigation of several dozen different crystals has revealed that the "gas-phase grown" crystals produce more intense UV radiation, while the threshold current densities for laser screens fabricated from these crystals are lower (by a factor of two or more) than when crystals grown in hydrothermal

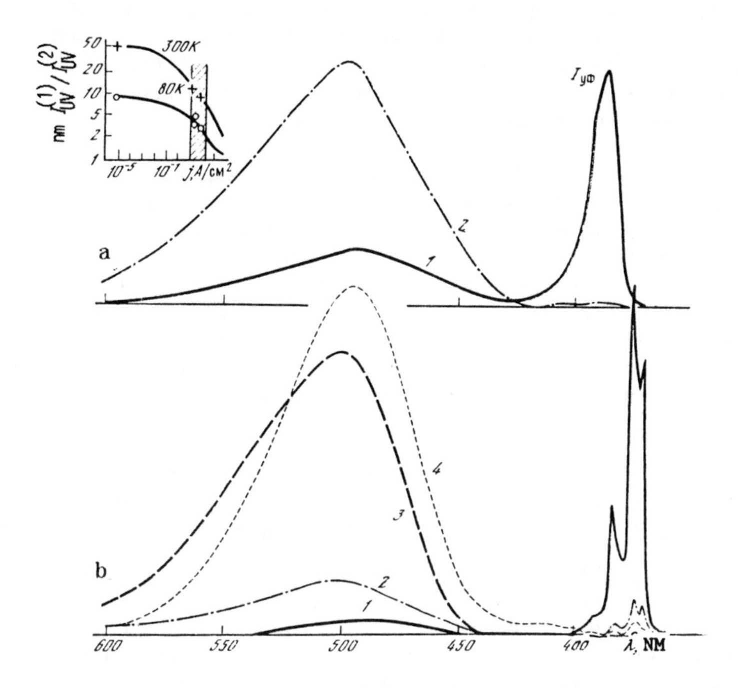

Fig. 16. Cathodoluminescence spectra of different ZnO crystals E_0 = 50 keV, $j = 10^{-5}$ A·cm^{-2}; a - T = 300 K; b - T = 800 K. 1 - gas-phase grown crystal [35]; 2 - crystal grown in hydrothermal conditions; 3 - fabrication by technique from study [37]; 4 - epitaxial film grown [38]; the insert shows the ratio of UV intensities $I_{UV}^{(1)}$ - crystal 1 and $I_{UV}^{(2)}$ - crystal 2 plotted as a function of electron beam current density j; the hatched region represents the range of laser working levels j

conditions are employed. In our opinion the primary reason for this condition is the low level of impurities in gas-phase-grown crystals in cold-end silica vessels.

The changes in the longwave radiation spectrum after different annealings of one of the "hydrothermally-grown" crystals and one of the "gas-phase grown" crystals are given in are given in Fig. 17. Liquid zinc annealing of the "hydrothermally-grown" crystal was carried out at T = 960°C for 43 hours while the oxygen annealing was carried out at T = 900°C, a pressure p_{O_2} = 0.25 atm for 65 hours (the annealed wafers 1 mm in thickness). The vacuum annealing of the "gas-phase grown" crystal was carried out at a fixed vacuum level ($p \approx 10^{-6}$ atm) while oxygen annealing employed an O_2 flow and zinc vapor annealing took place at a pressure of p_{Zn} = 1-1.5 atm. All three annealing processes were performed at an identical temperature $T_{ANN.}$ = 800°C for 24 hours followed by cooling for 10 hours. The "hydrothermally-grown" crystal samples had an $(11\bar{2}0)$ orientation; after annealing they were polished mechanically and etched in a hot CrO_2 solution in H_3PO_4. The (0001) base orientation was used for the "gas-phase grown" crystal samples (wafer thickness: 1.5 mm) and after annealing they were etched in the solution discussed above.

It is clear from Fig. 17 that both oxygen annealing and liquid zinc annealing serve to reduce the total longwave radiation intensity in hydrothermally-grown crystals, at the same time that the longwave radiation intensity of the "gas-phase grown" crystal increases significantly after vacuum annealing and zinc vapor annealing and changes insignificantly after oxygen annealing. Moreover, in the case of the "hydrothermally-grown" crystal oxygen annealing eliminates the yellow-orange line which zinc annealing eliminates the green line; the longwave radiation curve of the "gas-phase grown" crystal does not change after these annealings.

At present there is no unambiguous interpretation of the green or yellow-orange radiation lines [36]. They have been attributed to both the V_0' and V_{Zn}' intrinsic defects as well as copper and lithium impurities, respectively. Evidently as in the case of the I_1^D line in ZnSe the longwave radiation in ZnO cannot unambiguously be attributed to a specific center. The results from annealing initial crystals of different quality given above reveal that in the "hydrothermally-grown" samples the impurities eliminated from the crystal during annealing (particularly in liquid zinc annealing) play the primary role in the formation of longwave radiation, which serves to reduce longwave radiation intensity. On the other hand the longwave green radiation (yellow-orange radiation was not observed) in the "gas-phase grown" crystals is generated by the intrinsic defects whose intensity can jump significantly after annealing (zinc annealing).

The table gives a comparison of the UV intensities at high pump levels (E_0 = 50 keV, j = 150 A·cm^{-2}) for the initial samples and those annealed by various techniques. The table clearly reveals that zinc annealing clearly reduces UV radiation at the same time that after oxygen annealing the samples radiate nonuniformly and even a certain growth of I_{UV} in certain regions at 80 K is observed for a "hydrothermally-grown" crystal. However at 300 K oxygen annealing also degrades I_{UV} for "hydrothermally-grown" and "gas-phase grown" crystals. It is interesting that annealing at T = 890°C for 22 hours in a cold-ended evacuated and sealed silica vessel increases the UV radiation at 80 K.

The cathodoluminescence spectra in the exciton spectral range of the "gas-phase grown", initial, and variously annealed crystals are given in Fig. 18. The arrows indicate the energy positions of the bottom of the exciton band incorporating (A) and neglecting (*A-LO, A-2LO*) the longitudinal optical phonon energy and the more intense radiation line of the exciton bound to the shallow neutral center (I). Aside from the intense radiation of the bound exciton line at λ_M = 368.5 nm radiation in the 370-372 nm range resulting from exciton

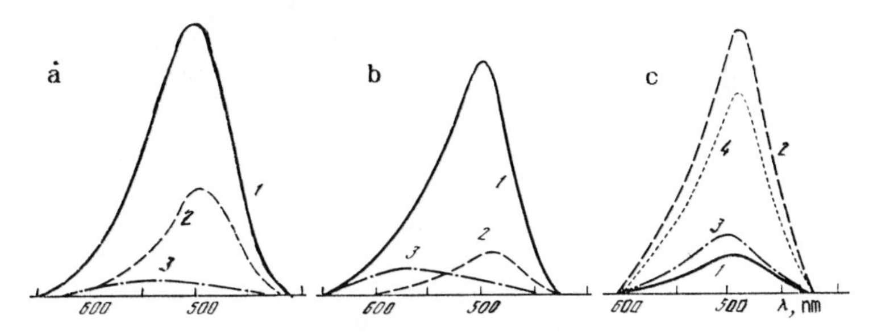

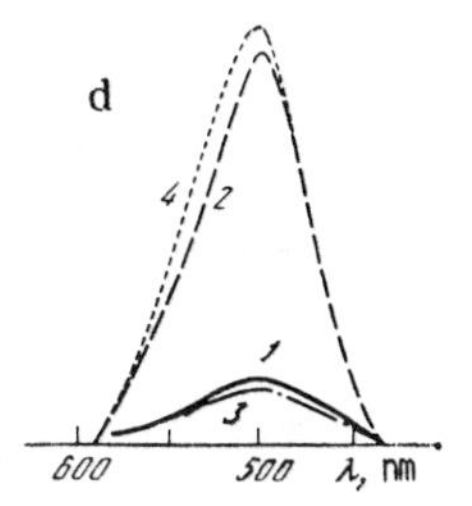

Fig. 17. Longwave radiation spectra of "hydrothermally-grown" (a, b) and "gas-phase grown" (c, d) zinc oxide crystals at two temperatures: 300 K (a, c) and 80 K (b, d) and E_0 = 50 keV, j = 10^{-5} A·cm^{-2}
1 - initial crystal; 2 - annealing in liquid zinc (a, b) in zinc vapor (c, d); 3 - oxygen annealing in a sealed vessel (a, b), oxygen flow annealing (c, d); 4 - vacuum annealing

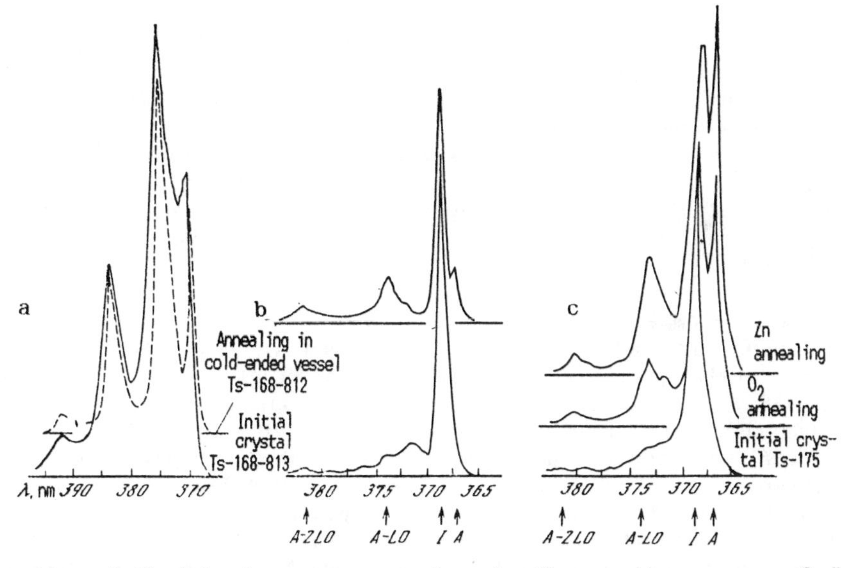

Fig. 18. Cathodoluminescence spectra in the exciton range of "gas-phase grown" ZnO crystals
a - initial crystal, E_0 = 50 keV, T = 80 K; b - crystal annealed in a cold-ended vessel, E_0 = 10 keV, T = 45 K; c - initial crystal and crystal annealed in zinc vapor and an oxygen flow at E_0 = 10 keV; T = 45 K

capture by the deep levels is observed in the cathodoluminescence spectrum of the initial sample at T = 45 K. After annealing in the cold ended vessel the intensity of these lines drops significantly and free exciton lines are clearly visible in the spectrum together with the exciton *LO*-phonon repetitions. The radiation spectrum also empties out at 80 K. Analogous spectral changes are observed even in the case of oxygen flow annealing, vacuum annealing, and zinc vapor annealing.

Table

Crystal growing technique	Measurement temperature K	UV cathodoluminescence intensity, rel. units						
			After annealing					
		Initial sample	Liquid zinc annealing	Zinc vapor annealing	Oxygen annealing	Annealing in oxygen flow	Vacuum Annealing	Annealing in a cold-ended vessel
"hydrothermally-grown" crystal	80	1	0.33	-	0.25-1.8	-	-	-
	300	1	0.045	-	0.045-0.4	-	-	-
"gas-phase grown" crystal	80	1	-	0.2	-	0.25	1	1.2
	300	1	-	0.065	-	0.15	0.54	-

The influence of the various annealings on the low-temperature photoluminescence spectra (T = 4.2 K) of both the "hydrothermally-grown" and the "gas-phase grown" crystals were investigated under excitation by a DRSh-250 lamp at λ_M = 365 nm detected by means of the MDR-3 spectrograph. The spectrally-integrated radiation intensity of the "gas-phase grown" crystal samples was significantly greater (by a factor of 10-100) than in the "hydrothermally-grown" crystal samples. The highest radiation intensity was obtained from the zinc vapor annealed sample.

The low-temperature photoluminescence spectra are given in Table 19. A weak free exciton line is observed in the spectrum at λ = 367.3 nm together with the bound exciton lines λ = 367.5, 368.0, 368.5, 368.7, 369.0, 369.2 nm, a group of lines in the 370-374 nm range together with lines with maxima at λ = 376.8 and 385.5 nm in the vicinity of the spectral position of the *LO*-phonon repetitions of the most intense group of bound excitons (λ = 368.5-369.0 nm). After vacuum and oxygen flow annealing lines appeared at 368.0 and 367.5 nm, the longer wavelength lines among the exciton lines in the 369.2-368.5 nm range were amplified and, consequently, a thin intense line at 371.95 nm appeared; this line predominated at lower excitation levels (see dashed curve in Fig. 19).

Annealing in the cold-ended vessel produced approximately the same changes although these were manifest to a lesser degree. After annealing in zinc vapors the 371.95 nm line is visible against the background of the 372.3 nm line which evidently is the primary criterion of zinc annealing. The 372.3 nm line appears clearly in the photoluminescence spectrum of the "hydrothermally-grown" crystal sample (liquid zinc-annealed). After zinc annealing the short wave length lines in the 369.2-368.5 nm range grew as well. The 371.95 nm line does not appear clearly after oxygen annealing in the "hydrothermally-grown" crystals.

An optical microscope investigation of the optical homogeneity of the samples revealed that if 1-10 μm brown sparse precipitates are observed in the initial gas-phase grown crystal samples, such precipitates are not found in the crystals vacuum-annealed or annealed in the cold-ended vessel, at the same time that the number of such precipitates increases substantially in crystals annealed in the zinc vapor or in liquid zinc (the crystal turns red after zinc annealing). Their total volume can reach $(1\text{-}5)\cdot10^{-4}$ of the sample volume.

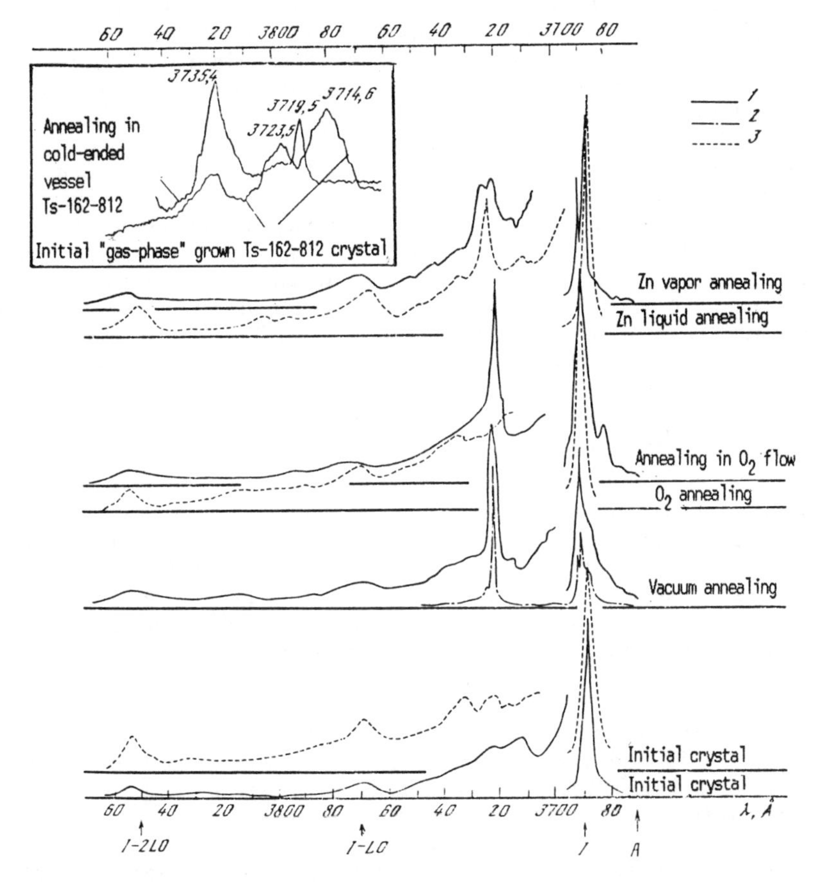

Fig. 19. Low-temperature (4.2 K) photoluminescence spectra of the initial and variously-annealed gas-phase grown and hydrothermally-grown ZnO crystals
1 - "gas-phase grown"; 2 - "gas-phase grown" at lower excitation level; 3 - "hydrothermally-grown"

The electrical measurements carried out on "gas-phase grown" crystal samples revealed that zinc vapor annealing increases the equilibrium electron concentration to $n = (1\text{-}3)\cdot 10^{19}$ cm^{-3} and reduces the resistivity to $\rho = 10^{-3}$ ohm·cm (T = 300 K), changing n and ρ compared to the initial values by more than two orders of magnitude. Vacuum annealing or oxygen flow annealing will produce a sharper temperature dependence of $\rho(T)$ and $n(T)$. Growth of T from 77 K to 300 K increases n by two orders of magnitude while the mobility μ drops by an order of magnitude at the same time that ρ and n change insignificantly in the initial sample.

These measurements reveal that the concentration of donor intrinsic defects with an ionization energy ~0.05 eV increases with growth of excess zinc in the crystals while at high excess zinc levels metallic impregnations are formed. We can assume as with cadmium sulfide that the observed precipitates are the excess zinc precipitated out in a second phase. Based on the total precipitate volume the excess zinc concentration in the zinc-annealed crystals is estimated at $2\cdot 10^{18}$–10^{19} cm^{-3}.

Accounting for the rather thin photoluminescence structure in the exciton range we can expect that the shallow donor concentration is not significant between the precipitates (less than 10^{17} cm^{-3}). The growth of precipitates observed after zinc annealing reduces cathodoluminescence intensity for excitation to a depth of several microns (E_0 > 10 keV). This is not observed under photoexcitation due to the fact that such precipitates are less likely to form near the surface at a depth of the order of 0.1 μm since the defects will drain out onto the crystal surface. The drop in UV radiation intensity after oxygen annealing can be attributed to zinc release from the crystal and the formation of zinc vacancies. The line at λ = 371.95 nm is evidently an analog of the I_1^D line in ZnSe. As with ZnSe (and ZnS) we can expect that V_{Zn} forms a fast nonradiative recombination center in ZnO as well. We can then attribute the growth in radiation after short-term annealing in the cold-ended vessel to the elimination of second phase precipitates as well as the insignificant growth in zinc vacancy concentration. Further research is required to interpret other details of the photoluminescence and cathodoluminescence spectra.

Zinc oxide crystals that are gas-phase grown by the technique from study [35] (and representing today the most promising media for lasers) have a significant deviation from stoichiometry on the excess zinc side which is in agreement with the phase diagram [39]. It is possible to achieve some improvement in the radiative properties of these crystals by employing annealing to remove the excess zinc precipitates; these processes do not significantly increase the concentration of V_{Zn}.

4. Zinc Sulfide

Today zinc sulfide is used as the medium for the semiconductor laser with the shortest radiation wavelength of λ_L = 330 nm [40]. The initial attempts to develop laser electron beam devices based on this material employed pressure-melt-grown crystals [41]. Zinc sulfide is a higher temperature material than zinc selenide (T_M = 1830°C for ZnS) while the phase transition from the high-temperature hexagonal structure to the low-temperature cubic structure occurs at a lower temperature (T_F = 1020°C). Therefore the melt-grown crystals contain many uncontrolled impurities forming multiple associate complexes with the intrinsic defects; these crystals also have a high concentration of stacking faults which made it impossible to achieve high lasing power and efficiency in study [41].

In the present study we investigated crystals of greater purity (gas-phase grown crystals using the same technique as for the ZnSe and ZnO crystals in study [35]). The crystals were grown at temperatures 1100-1200°C for 80-100 h in an argon/hydrogen mixture with partial pressures p_{H_2} = 0.3 atm, p_{Ar} = 0.1 atm employing "melt-grown" seeds and intrinsic "gas-phase grown" seeds. The derived monocrystal blanks were 30-50 mm in diameter and up to 10 mm in height. The portion of the hexagonal phase α_H produced by the stacking faults in the "gas-phase grown" crystals was two to five times smaller than in the "melt-grown" crystals and was less than 3%. α_H was determined by measuring the birefringence ΔN by the interference pattern obtained from crossed polarization filters in a POLAM-type polarizing microscope from the

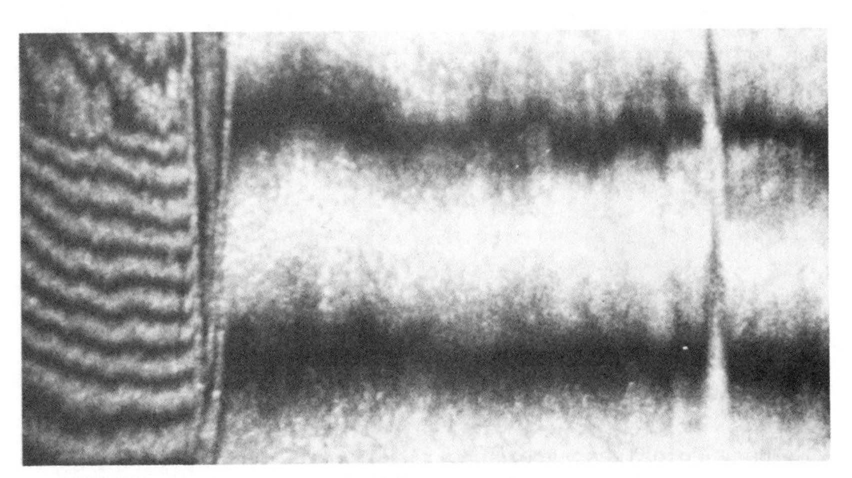

Fig. 20. Interference pattern in monochromatic light resulting from a ZnS sample wedge containing a "melt-grown" substrate and a gas-phase grown layer when placed between crossed polarization filters

transmission of polarized radiation through a wedge sample with the wedge edge along the <111> direction coinciding with the optical axis of the crystal (the former <0001> direction of the hexagonal phase). A linear dependence of α_H on ΔN as determined in study [42] was used. Such an interference pattern appearing on a sample consisting of a "melt-grown" substrate and a grown layer is shown in Fig. 20.

For comparison Fig. 21 gives the spectra of the purest melt-grown [49] and gas-phase grown (our crystals) zinc sulfide crystals on a gas-phase grown substrate. It is clear that the exciton luminescence intensity in the gas-phase grown crystal is significantly greater and therefore we will present results primarily for the gas-phase grown samples below.

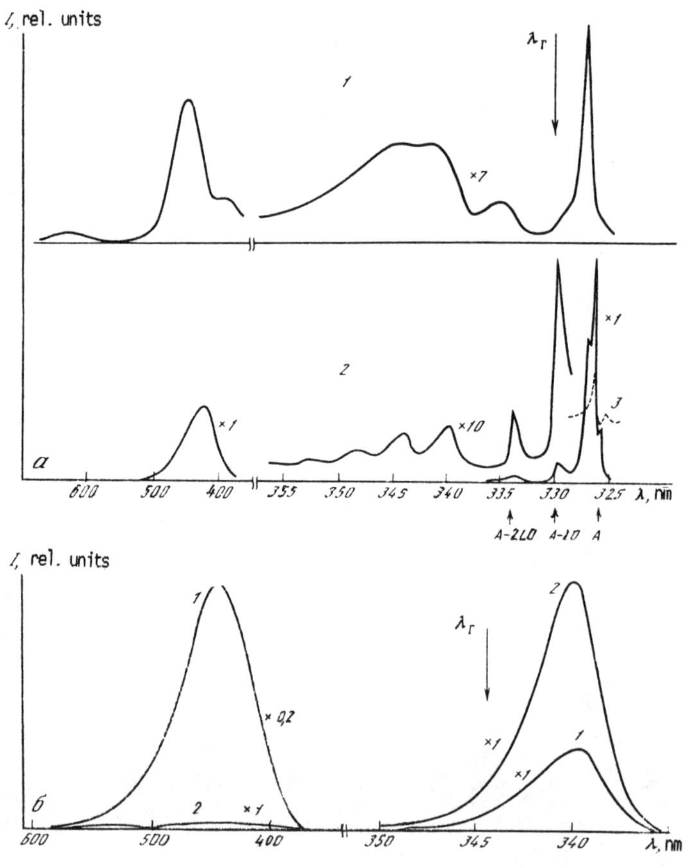

Fig. 21. Cathodoluminescence spectra of melt-grown (1) and gas-phase grown on a "gas-phase grown" seed (2) ZnS crystals at two temperatures: 30 K (a) and 300 K (b); E_0 = 10 keV, $j = 10^{-5}$ A$\cdot$cm^{-2}
Dashed curve (3) represents the reflection spectrum of the "gas-phase grown" crystal. The arrows indicate the position of the free exciton A; its LO are the phonon repetitions and lasing wavelength λ_L of a ZnS laser screen

Annealing was performed in sulfur vapor at p_{S_2} = 1 atm and zinc vapor at p_{Zn} = 0.2-1 atm at temperatures $T_{ANN.}$ = 950°C and 1150°C, i.e., both below and above the phase transition temperature T_P. When $T_{ANN.}$ = 950°C liquid zinc annealing was also performed. Two cooling states were used: air cooling of the vessel (slow cooling) and rapid cooling in water (quenching). ZnS slabs 1 mm in thickness were used for slow cooling and annealing was performed for 70 hours. After cooling the samples were polished to remove at least 200 μm and were then treated in a polishing etchant based on a CrO_3 solution in HCl. Thin platelets 100 μm in thickness were used for quenching and the annealing duration was 15 hours. After quenching the samples were treated in the polishing etchant only.

The cathodoluminescence was investigated on the assembly described in study [43] in a temperature range T = 30-300 K, an electron beam current density j = $10^{-6/102}$ A·cm^{-2} and an electron energy E_0 = 10-75 keV. A PGS-2 spectrograph and a FEU-100 photomultiplier were used to record the spectra. The radiation efficiency of the various samples was compared at high pump levels by means of the FEK-22 photocell of the SPU-M.

Fig. 22 gives results from a comparison of the radiation intensity I of different samples (both initial samples and variously-annealed samples) at high pump levels (E_0 = 75 keV, j = 100 A·cm^{-2}). It is clear that annealing of the "gas-phase grown" crystal in liquid zinc followed by slow cooling increases somewhat their radiation intensity, while sulfur vapor annealing significantly reduces this intensity.

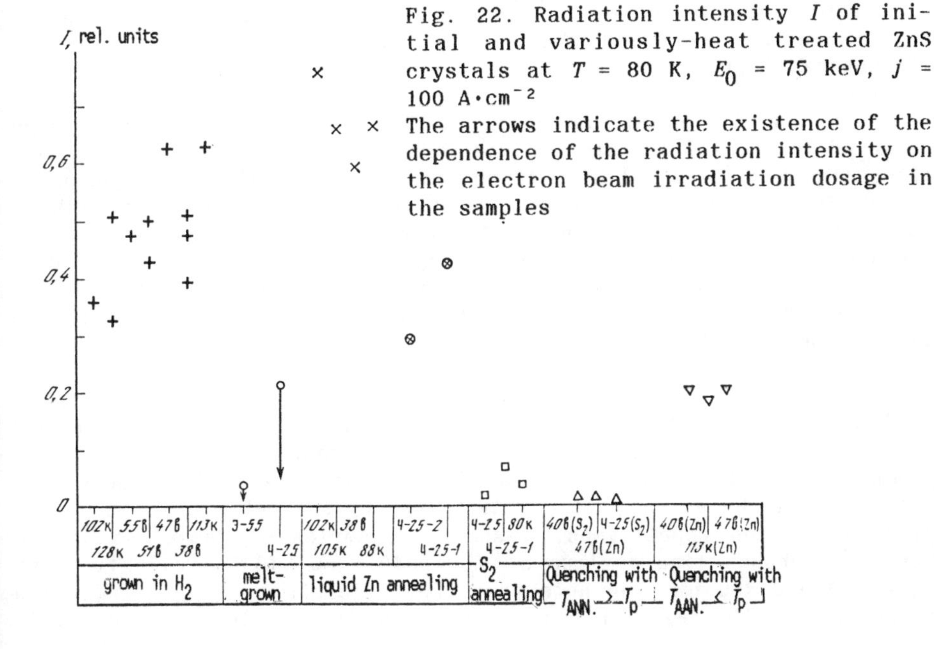

Fig. 22. Radiation intensity I of initial and variously-heat treated ZnS crystals at T = 80 K, E_0 = 75 keV, j = 100 A·cm^{-2}
The arrows indicate the existence of the dependence of the radiation intensity on the electron beam irradiation dosage in the samples

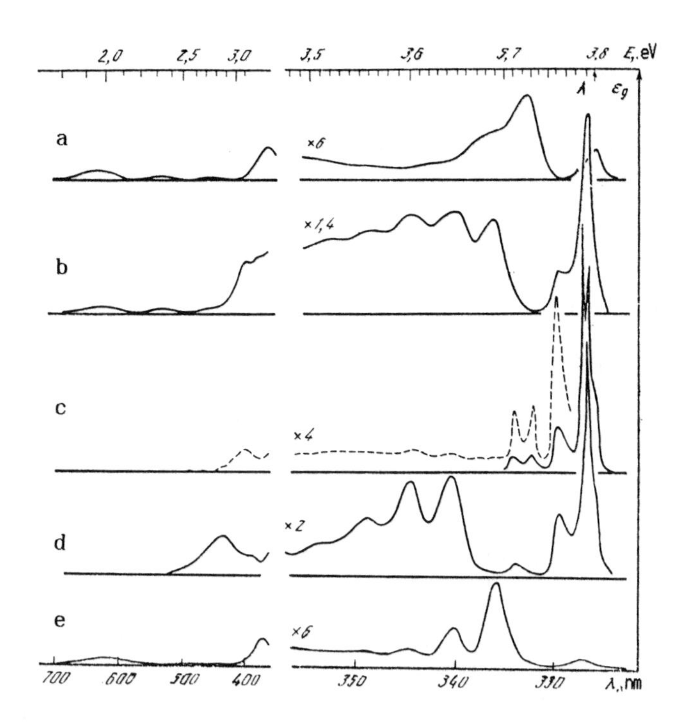

Fig. 23. Cathodoluminescence spectra of ZnS at T = 30 K, E_0 = 10 keV, $j = 10^{-5}$ A·cm^{-2}
a - annealed in zinc vapor for $T_{ANN.} > T_P$ followed by quenching; b- annealed in zinc vapor for $T_{ANN.} < T_P$ followed by quenching; c - annealed in liquid zinc; d - initial crystal; e - sulfur vapor annealing; the arrows represent the energy position of the exciton (a) and the bandgap (ε_g)

The rapid drop in radiation intensity from the radiation dosage observed in the initial samples vanishes for the "melt-grown" crystals after liquid zinc annealing. After zinc vapor annealing and quenching the radiation intensity depends on whether or not annealing is performed above or below T_P; in the first case ($T_{ANN.} > T_P$) the crystals radiated approximately an order of magnitude lower. The portion of hexagonal phase in samples quenched after annealing when $T_{ANN.} > T_P$ exceeded 50% while in the remaining samples it was approximately the same as in the initial sample.

The cathodoluminescence spectra at low excitation levels (E_0 = 10 keV, $j = 10^{-5}$ A·cm^{-2}) and T = 30 K of the initial sample and variously annealed samples from the "gas-phase grown" crystal are given in Fig. 23. The spectrum of the initial sample contains the free exciton radiation line (a) distorted by self-absorption together with its phonon repetitions, a shortwave series of "edge" radiation with zero-phonon line maximum at λ_M = 340.5 nm and longwave lines at λ_M = 430 and 400 nm. We note that the ratio of the "edge" radiation intensity to the exciton radiation intensity is lower in crystals grown by "gas-phase grown" seeds compared to "melt-grown" seeds (compare the corresponding spectra in Figs. 23 and 22).

In the dirtier crystals a second longwave series is observed in the "edge" radiation with a zero-phonon line maximum at λ_M = 343 nm. Sulfur annealing produces a radical change in the spectrum. The spectrum contains a weak exciton radiation line as well as new radiation similar to the "edge" radiation with a zero-phonon line maximum at λ_M = 336 nm; and a longwave line adjacent to this radiation at λ_M = 370 nm together with red radiation at λ_M = 630 nm. Liquid zinc annealing suppresses the "edge" and longwave radiation and increases the exciton radiation intensity. The exciton radiation line (bound to the shallow neutral donor with binding energy ε_B = 15 meV) appears clearly in the exciton region of the spectrum together with the line at λ = 332 nm. The spectrum of the sample quenched after zinc vapor annealing at $T_{AAN.} > T_P$ is similar to that of the sulfur vapor-annealed sample. However the zero-phonon line of the new series is shifted towards shorter wavelengths by approximately 35 meV while the exciton radiation line is shifted by 15 meV. The lines are broadened and the phonon structure is blurred. The spectrum of the sample quenched after annealing in zinc vapors at $T_{ANN} < T_{op}$ represents an intermediate point between the spectrum of the initial sample and the sample annealed in sulfur vapor.

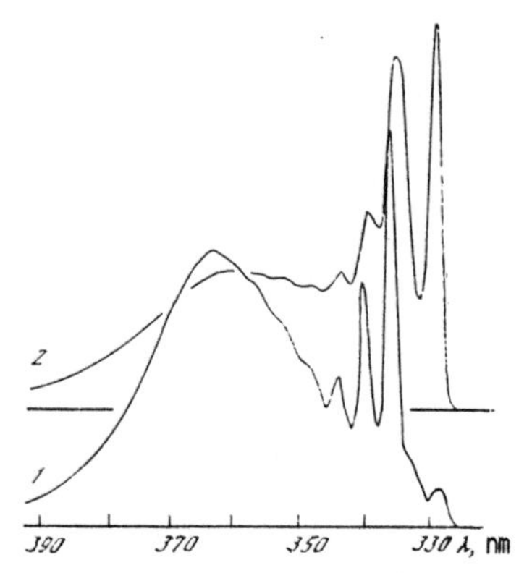

Fig. 24. Cathodoluminescence spectra of S_2 annealed ZnS crystal at j = 10^{-5} A·cm^{-2} (1) and 20 A·cm^{-2} (2), E_0 = 50 keV, T = 35 K

Changes in the radiation spectra of samples annealed in liquid zinc and sulfur vapors followed by slow cooling with an increase in the electron beam current density are shown in Figs. 24 and 25, respectively. Fig. 24 indicates that the new series of "edge" radiation shifts towards shorter wavelengths with growth of j. Such behavior is characteristic of transitions in donor-acceptor pairs.

The exciton radiation between the first and second repetitions of the free exciton radiation line observed in all samples annealed in liquid zinc and slowly cooled turned out to have a fine structure (Fig. 25). With growth of j this radiation line broadens due to the increased intensity of the longer wavelength component.

Laser screens were fabricated from the "gas-phase grown" crystals as well as the initial crystals and crystals annealed in liquid zinc or sulfur vapor. The best energy characteristics were obtained by laser screens fabricated from crystals annealed in liquid zinc.

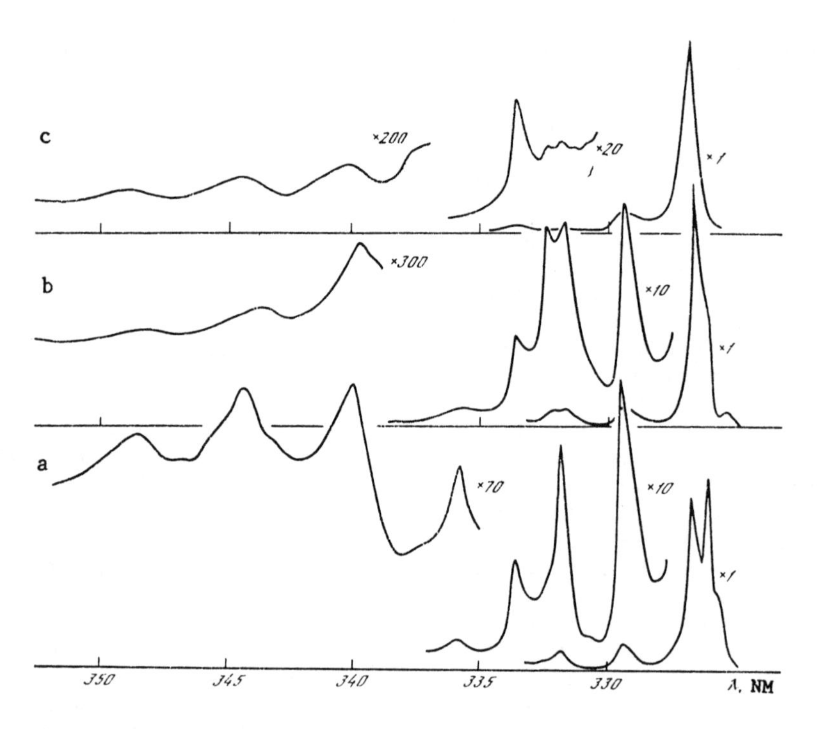

Fig. 25. Cathodoluminescence spectra of liquid zinc-annealed ZnS crystal
at $j = 10^{-5}$ A·cm^{-2} (a), $j = 0.3$ A·cm^{-2} (b), $E_0 = 10$ keV and $j = 10$ A·cm^{-2}, $E_0 = 50$ keV (c), $T = 35$ K

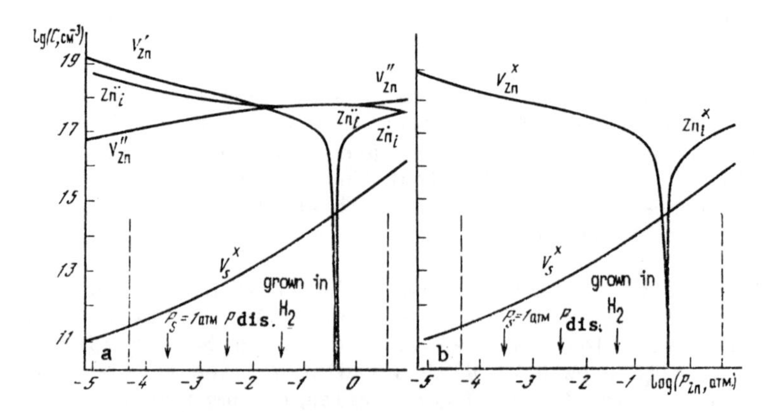

Fig. 26. Phase diagram of intrinsic point defects in ZnS crystals in thermodynamic equilibrium with the vapor at $T = 1150$°C
a - instantaneously cooled to low temperatures; b - cooled so that the defects interact with one another but not with the gas medium. The dotted line represents the composition stability limits.

The radiation power at T = 80 K in the scanning pump mode was 1 w with an electron beam-generated radiation energy conversion efficiency of 3%. The characteristics of the laser screens fabricated from the initial crystals were somewhat lower. Crystals annealed in sulfur vapor did not produce lasing.

These results indicate that in terms of radiative properties zinc sulfide is similar to zinc selenide. In these crystals the zinc is not precipitated out even after high-temperature annealing in liquid zinc, and the radiative properties of the pure samples are determined by the intrinsic point defects and the zinc vacancies form nonradiative recombination centers.

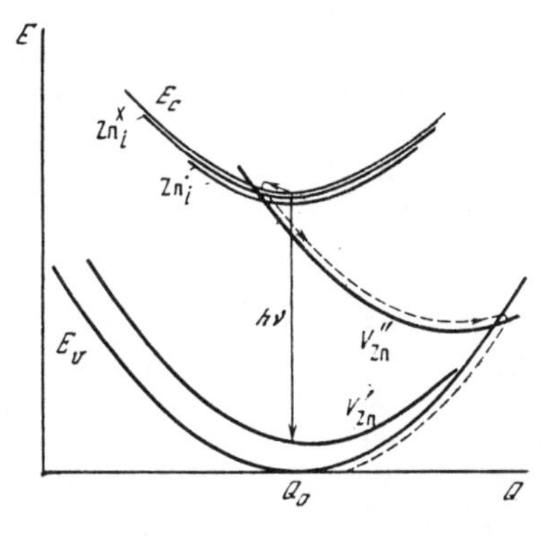

Fig. 27. Configurational-coordinate diagram
E_c - bottom of the conduction band, E_v - top of the valence band; V'_{Zn}, V''_{Zn}, $Zn_i^{\times}$ and $Zn_i^{\bullet}$ - electron levels in the zinc vacancy and interstitial zinc atom; Q - configurational coordinate. The vertical arrow shows the radiative capture of an electron from the bottom of the conduction band or of the $Zn_i^{\times}$, $Zn_i^{\bullet}$ levels by the neutral vacancy. The dotted line represents the nonradiative transition involving the V''_{Zn} level

Calculation of the intrinsic defects for zinc sulfide in good agreement with existing experimental data was carried out in study [44]. One of the results of this calculation is given in Fig. 26a. This figure plots the point defect concentration against the zinc pressure over the crystal in thermodynamic equilibrium with the vapor at T = 1150°C and instantaneously cooled to low temperatures (T = 0 K). We see that the primary defects include the zinc vacancy and the interstitial zinc in different charge states. However the case of instantaneous cooling is not achieved in practice. With this cooling state the value of the defect concentrations in the crystal will depend on the dimensions of the cooled sample as well as the coefficients of diffusion of these defects. Annealing experiments on ZnSe in liquid zinc have revealed that the self-diffusion coefficient of zinc is rather high $D = 10^{-7}$-10^{-8} cm^2 s^{-1}. If we take the self-diffusion coefficient of zinc in ZnS to be of the same order when cooling the sample after annealing for several hours the defect concentration will follow the change in the vapor phase only in the first 100 μm surface layer. In the remaining portion of the sample the concentration and charge state of the defects will change due solely to the interaction between the defects themselves. The interaction of V_{Zn} with the highly-mobile Zn_i defect will cause the concentration of these defects to drop significantly in the slowly-cooled samples, and depending on the zinc pressure either $V_{Zn}^{\times}$ or $Zn_i^{\times}$ will be the primary defect. Such a situation is shown in Fig. 26b.

It is also clear from Fig. 26 that the composition of the crystal grown in a hydrogen atmosphere is near-stoichiometric. A small shortage of zinc may also exist in this crystal. After sulfur vapor annealing the V_{Zn} concentration in the sample grows by at least an order of magnitude compared to the initial sample and reaches $\sim 10^{18}$ cm^{-3}. Zinc vapor annealing causes a significant change in the V_{Zn} concentration in the slowly cooled crystals, while the Zn_i concentration does not exceed 10^{17} cm^{-3}. Such a deviation of the composition towards excess zinc does not result in its precipitation.

We will now consider a possible model of the nonradiative center associated with V_{Zn}. It was demonstrated in studies [45, 46] that the nonradiative capture cross-section of the charge carrier by the deep level could be greater (up to 10^{-14} cm^{-2}) if the defect or impurity atom forming this level significantly deforms the lattice. It turned out that V_{Ga} was such a nonradiative recombination center in GaAs [47].

By analogy to these studies and based on data on the V_{Zn}' and V_{Zn}'' energy levels from study [44] we proposed the following nonradiative recombination model in ZnS shown schematically in the configurational-coordinate diagram in Fig. 27. The conduction band electron with low activation energy is captured by the vibrationally-excited level of the negatively-charged zinc vacancy. After contributing its energy to the lattice vibrations it first drops down following the vibrational branch of the center and then jumps over to the vibrational branch of the upper energy state of the valence band after recombining with the hole.

After sulfur annealing followed by slow cooling the crystal has a high concentration of V_{Zn} in the ground and neutral state. However due to the existence of the uncontrolled donor impurity in the crystal as well as a possible small Zn_i concentration the initial V_{Zn}' concentration may be significant. With growth of the electron beam current density the $V_{Zn}^{\times}$ part will assume a negatively charged state by capturing a nonequilibrium electron. It is clear from the preceding discussion and Fig. 27 that in addition to nonradiative recombination through V_{Zn}' there are also other radiative transitions: 1) radiative capture of a band electron by a neutral zinc vacancy; 2) radiative annihilation of the exciton bound to the neutral V_{Zn}; 3) radiative transitions in the far donor-acceptor pairs (V_{Zn}'; D') or (V_{Zn}', Zn_i); 4) emission of an exciton captured by the near donor-acceptor pair. The line at λ_M = 370 nm whose shortwave edge is approximately 0.4 eV below the energy position of the bandgap (compare to the energy position of the level attributed to V_{Zn}') can be attributed to the first and third processes. Evidently the fourth process corresponds to radiation with a zero-phonon line maximum λ_m = 336 nm. The radiation line of the exciton coupled to $V_{Zn}^{\times}$ in the spectrum of the sulfur vapor annealed sample evidently does not appear clearly due to the high $V_{Zn}^{\times}$ concentration. Occasionally in the initial samples a narrow line at λ_M = 328 nm is observed; it is possible that this is an analog of the I_1^D line in ZnSe. On the other hand the line at λ_M = 495 nm in ZnSe can possibly be analogous to the line at λ_M = 370 nm in zinc sulfide.

In samples annealed in liquid zinc and slowly cooled the zinc vacancy concentration is low and the primary defect is interstitial zinc. The line at λ = 372 nm is likely to be caused by radiation of the exciton bound to $Zn_i^{\times}$. Indeed, in this case the exciton binding energy is equal to approximately 14 meV which according to Hayne's rule gives an estimate for the donor ionization energy of ε_{ION} = 70 meV close to the accepted value of 0.1 eV [44]. (It is necessary to account for the fact that with a high Zn_1 concentration the ionization energy drops.) The radiation line near λ = 332 nm in this case can also be related to exciton annihilation by $Zn_i^{\times}$, although this results in simultaneous excitation of this center (two-electron transition [48]). The basis for this assumption is that the energy distance between these lines - 70 meV - approximately corresponds to 3/4 of ε_{ION} of the donor which will occur from the transition of the electron at the $Zn_i^{\times}$ center from the ground state to the first excited level. The longer wavelength line amplified from growth of the pump level can be attributed to additional energy transfer to the exciton captured by the neighboring center.

The phase transition from the high-temperature hexagonal phase to the low-temperature cubic phase occurs by dislocation movement [49]. In motion the dislocations can collect the interstitial zinc atoms. The results obtained for CdS crystals confirm this process (see Figs. 3 and 10). Consequently after sudden cooling of a sample from $T_{ANN.} > T_P$ to low temperatures (an increase in the dislocation density is observed here) the concentration of minority point defects V_{Zn} that have not interacted with Zn_i will be greater and will appear in the spectra in the same way as in samples annealed in sulfur vapor.

The shortwave shift of the radiation spectrum in samples quenched after annealing in zinc vapor with $T_{ANN.} > T_P$ correlates with the increase in the portion of hexagonal phase α_H in these samples. Since with growth of α_H the bandgap ε_g of the zinc sulfide crystal grows linearly [42], the observed shortwave shift can be attributed to growth of ε_g. The minor shift of exciton radiation is probably related to localization of the exciton at the bandgap fluctuations [50].

After quenching from $T_{ANN.} < T_P$ there are few dislocations and they do not hinder interaction of the Zn_i and V_{Zn} defects, which results in the establishment of a final V_{Zn} concentration that is small compared to the case of quenching from $T_{ANN.} > T_P$ and is greater than the case of slow quenching.

5. Conclusion

These investigations have revealed that the characteristics of longitudinal electron-pumped semiconductor lasers employing A^2B^6 compounds are significantly dependent on the type and concentration of intrinsic point defects. Degradation of the radiative characteristics is observed when the crystals contain a high concentration of metal vacancies that function as fast nonradiative recombination centers as well as metallic precipitates. The role of these factors depends on the phase diagram of the specific crystal. In the case of CdS and ZnO crystals normally grown with significant excess metal the laser characteristics are determined by the precipitate concentration. And, vice versa, in ZnSe and ZnS crystals normally grown with a metal shortage the primary influence on laser characteristics comes from the metal vacancy concentration.

In conclusion the authors wish to express their gratitude to E.V. Markov and V.V. Smirnov for providing the gas-phase grown zinc oxide crystals and for performing the electrical measurements on these crystals and to S.A. Pendyur for providing the layered cadmium sulfide crystals, and G.V. Bushueva for assistance in performing the transition electron microscopy investigations.

References

1. Kozlovskiy, V.I., Nasibov, A.S., Pechenov, A.N., et al. "The lasing mechanism in laser screens fabricated from A^2B^6 semiconductor compounds" KVANTOVAYA ELEKTRON., 1979, vol. 6, no. 1, pp. 189-197.

2. Kozlovskiy, V.I., Nasibov, A.S., Reznikov, P.V. "Investigation of CW lasing in an electron beam pumped GaAs laser" KVANTOVAYA ELEKTRON., 1982, vol. 9, no. 11, pp. 2211-2216.

3. Kryukova, I.V., Kuprishina, E.S., Prokof'eva, S.P. "The lasing mechanism in uncooled doped cadmium sulfide lasers" PIS'MA V ZhTF, 1979, vol. 5, no. 9, pp. 525-531.

4. Hurwitz, C.E. "High power and efficiency in CdS electron beam pumped lasers" APPL. PHYS. LETT., 1966, vol. 9, no. 12, pp. 420-423.

5. Bogdankevich, O.V., Borisov, N.A., Zverev, M.M., et al. "Cadmium sulfide monocrystals for electron beam pumped lasers" KVANTOVAYA ELEKTRON., 1972, no. 7, pp. 44-47.

6. "Fizika i khimiya soedineniy $A^{II}B^{VI}$" [The specifics and chemistry of A^2B^6 compounds] Ed. by: S.A. Medvedev, Moscow: Mir, 1970, 75 pp.

7. Markov, E.V., Davydov, A.A. "Sublimation of CdS crystals" IZV. AN SSSR. NEORGAN. MATERIALY, 1971, vol. 7, no. 4, pp. 575-579.

8. "Svoystva neorganicheskikh soedineniy: Spravochnik" [The properties of inorganic compounds: A handbook] Ed. by: A.N. Efimov, L.P. Belorukova, I.V. Vasil'kova, et al., Leningrad: Khimiya, 1983, 264 pp.

9. Smirnova, I.A., Strukova, G.K., Osip'yan, Yu.A. "Detection of dislocations in cadmium sulfide by selective etching" FIZIKA I KHIMIYA OBRABOTKI MATERIALOV, 1974, no. 6, pp. 145-149.

10. Kozlovskiy, V.I., Nasibov, A.S., Pechenov, A.N., et al. "Laser screens fabricated from CdS, CdS_xSe_{1-x}, ZnSe monocrystal wafers" KVANTOVAYA ELEKTRON., 1977,m vol. 4, no. 2, pp. 351-354.

11. Nasibov, A.S., Kozlovskiy, V.I., Papusha, V.P. "Investigation of the characteristics of a laser screen cathode ray tube" RADIOTEKHNIKA I ELEKTRONIKA, 1973, vol. 17, no. 10, pp. 2151-2157.

12. Pugachev, A.T., Volkov, Yu.A. "Determining electron energy losses from transmission through thin films" FTT, 1979, vol. 21, no. 9, pp. 2637-2640.

13. "Elektronnaya mikroskopiya tonkikh kristallov" [Electron microscopy of thin crystals] Ed. by: P. Hirsch, A. Hovy, R. Nicholson, et al., Moscow: Mir, 1968, 340 pp.

14. Shapkin, P.V. "The technology of growing A^2B^6 semiconductor compound monocrystals for electron-pumped lasers": Dissertation for Candidates Degree of Technical Sciences, Moscow, 1982, 181 pp.

15. Kozlovskiy, V.I., Nasibov, A.S., Reznikov, P.V. "The influence of dislocations on CdS laser screen characteristics" KVANTOVAYA ELEKTRON., 1981, vol. 8, pp. 745-750.

16. Ham, F.S. "Theory of diffusion limited precipitation" J. PHYS. AND CHEM. SOLIDS, 1958, vol. 6, pp. 335-351.

17. Podo, M., Shneyder, M., T'erri-Mitt, et al. "Phase diagram and kinetics of excess component precipitation in nonstoichiometric IV-VI compounds" In: Rost i legirovanie poluprovodnikovykh kristallov i plenok" [Growing and doping semiconductor crystals and films] Novosibirsk: Nauka, 1977, vol. 4, no. 1, pp. 239-248.

18. Kulakov, M.P., Kulakovskiy, V.D., Fadeev, A.V. "Twinning in ZnSe pressurized melt-grown crystals" IZV. AN SSSR. NEORGAN. MATERIALY, 1976, vol. 12, no. 10, pp. 1867-1969.

19. Strukov*α*, G.K. "Chemical etching and transition electron microscope investigation of dislocations in A^5B^6 semiconductors": Dis-

sertation for Candidates Degree of Physics and Mathematics, Chernogolovka, 1982, 161 pp.

20. Merz, J.L., Nasau, K., Shiever, J.W. "Pair spectra and the shallow acceptors in ZnSe" PHYS. REV. B, 1973, vol. 8, no. 4, pp. 1444-1452.

21. Dean, P.J., Pitt, A.D., Skolnick, M.S., et al. "Optical properties of undoped organometallic grown ZnSe and ZnS" J. CRYST. GROWTH, 1982, vol. 59, pp. 301-306.

22. Dean, P.J., Herbert, D.C., Werkh, C.J., ven, et al. "Donor bound-exciton excited states in zinc selenide" PHYS. REV. B, 1981, vol. 23, no. 10, pp. 4888-4901.

23. Röppischer, H., Jacobs, J., Novikov, B.V. "The influence of Zn and Se heat treatment on the exciton spectra of ZnSe single crystals" PHYS. STATUS SOLIDI (a), 1975, vol. 27, no. 1, pp. 123-127.

24. Satch, S., Igaki, K. "Bound exciton emission of zinc selenide" JAP. J. APPL. PHYS., 1981, vol. 20, no. 10, pp. 1889-1895.

25. Jiang, X.J., Hisamune, T., Nozue, Y. et al. "Optical properties of the I_I^{deep} bound exciton in ZnSe" J. PHYS. SOC. JAP., 1983, vol. 52, no. 11, pp. 4008-4013.

26. Huang, S.M., Nozue, Y., Igaki, K. "Bound exciton luminescence of Cu-doped ZnSe" JAP. J. APPL. PHYS., 1983, vol. 52, no. 11, pp. 4008-4013.

27. Neumann, G. "Diffusion and transport process" In: Current topics in materials science, Ed. E. Kaldis, North-Holland, 1981, vol. 7, pp. 279-303.

28. Berezina, T.I., Martovitskiy, V.P., Talenskiy, O.N., et al. "The perfection of gas-phase grown A^2B^6 compound crystals on a crystal substrate" In: 6-ya Mezhdunar. konf. po rostu kristallov (10-16 sentyabrya 1980): [The Sixth International Conference on Crystal Growing (10-16 September 1980)] Expanded topic papers, Moscow: VINITI, 1980, vol. 1, p. 207.

29. Flögel, P. "Zur Kristallzüchtung von Cadmiumsulfid und anderen II-VI-Verbindungen. III: Gleichgewichtsdrucke über den Sulfiden in wasserstoffenthaltenden Gasen" KRIST. UND TECHN., 1971, vol. 6, no. 4, pp. 499-515.

30. Kulakov, M.P., Fadeev, A.V. "Stoichiometry of melt-grown zinc selenide crystals" IZV. AN SSSR. NEORGAN. MATERIALY, 1976, vol. 12, no. 10, pp. 1867-1971.

31. Testova, N.A., Osipova, G.E., Tkhim, T.V., et al. In: Rostovye protsessy v poluprovodnikovykh kristallakh i sloyakh" [Growing

processes in semiconductor crystals and layers] Ed. F.A. Kuznetsov, Novosibirsk, Nauka, 181, p. 9.

32. Hartmann, H., Mach, R., Selle, B. "Wide gap II-VI compounds as electronic materials" In: Current topics in materials science" Ed. E. Kaldis, North-Holland, 1982, vol. 9, p. 373.

33. Kozlovskiy, V.I., Markov, E.M., Nasibov, A.S., et al. "Longitudinal electron beam pumped ultraviolet ZnO semiconductor laser" PIS'MA V ZhTF, 1983, vol. 9, no. 14, pp. 873-876.

34. Kuz'mina, I.P., Nikitenko, V.A., Teretsenko, A.I., et al. "The influence of growing and doping conditions on select optical properties of zinc oxide monocrystals" In: Gidrotermal'nyy sintez i vyrashchivanie monokristallov" [Hydrothermal synthesis and growing of monocrystals] Moscow: Nauka, 1982, pp. 40-68.

35. Markov, E.V., Smirnov, V.V., Khryapov, V.T. "Gas phase growing of large-scale pure zinc oxide monocrystals by oriented seed" In: V Vsesoyuz. sobeshch. "Fizika i tekhnicheskoe primenenie poluprovodnikov $A^{II}B^{VI}$ (1-2 dek. 1983)" [The Fifth All Union Conference "The Physics and Technical Application of A^2B^6 semiconductors (1-2 December, 1983] Tez. dokl. Vil'nys: VGU, 1983, vol. 3, p. 131.

36. Kuz'mina, I.P., Nikitenko, V.A. "Okis' tsinka. Poluchenie i opticheskie svoystva" [Zinc oxide. Fabrication and optical properties] Moscow: Nauka, 2984, pp. 111-119.

37. Helbig, R. "Über die Züchtung von grösseren reinen und dotierten ZnO-Kristallen aus der Gasphase" J. CRYST. GROWTH, 1972, vol. 15, no. 1, pp. 25-31.

38. Abduev, A.Kh., Adukhov, A.D., Azaev, B.M., et al. "Recombination radiation from epitaxial zinc oxide layers under powerful one-photon excitation" PIS'MA V ZhTF, 1979, vol. 5, no. 3, pp. 149-152.

39. Hagemark, K.J., Toren, P.E. "Determination of excess Zn in ZnO" J. ELECTROCHEM. SOC., 1975, vol. 122, pp. 992-994.

40. Hurwitz, C.E. "Efficient ultraviolet laser emission in electron beam-excited ZnS" APPL. PHYS. LETT., 1966, vol. 9, no. 3, pp. 116-118.

41. Kozlovskiy, V.I., Korastelin, Yu.V., Nasibov, A.S., et al., "UV longitudinal electron beam pumped ZnS semiconductor laser" KVANTOVAYA ELEKTRON., 1984, vol. 11, no. 3, pp. 618-621.

42. Brafman, O., Steinberger, I.T., "Optical bang gap and birefringence of ZnS polytypes" PHYS. REV., 1966, vol. 143, pp. 501-505.

43. Dinov, Yu.S., Zykov, V.M., Kozlovskiy, V.I., et al. "Laser CRT with independent cryogenic cooling of the active medium" PIS'MA V ZhTF, 1984, vol. 10, no. 22, pp. 1373-1376.

44. Morozova, N.K., Morozova, O.N. "Phase equilibrium diagram of point defects and deviation from stoichiometry of zinc sulfide" IZV. AN SSSR. NEORGAN. MATERIALY, 1981, vol. 17, no. 8, pp. 1335-1340.

45. Lang, D.V., Henry, C.H. "Nonradiative recombination of deep levels in GaAs and GaP by lattice-relaxation multiphonon emission" PHYS. REV. LETT., 1975, vol. 35, no. 22, pp. 1525-1528.

46. Henry, C.H., Lang, D.V. "Nonradiative capture and recombination by multiphonon emission in GaAs and GaP" PHYS. REV. B, 1977, vol. 15, no. 2, pp. 989-1016.

47. Lang, D.V., Logan, R.A., Kimerling, L.C. "Identification of the defect state associated with a gallium vacancy in GaAs and $Al_xGa_{1-x}As$" PHYS. REV. B, 1977, vol. 15, no. 10, pp. 4874-4882.

48. Reynolds, D.C., Collins, T.C. "Excited terminal states of a bound exciton donor complex in ZnO" PHYS. REV., 1969, vol. 185, no. 3, pp. 1099-1103.

49. Kulakov, M.P., Shmurak, S.Z. "Structural changes in ZnS crystals on account of partial dislocation movement" PHYS. STATUS SOLIDI (a), 1980, vol. 59, pp. 147-153.

50. Maslov, A.Yu., Suslina, L.G. "The influence of one-dimensional disorder on the exciton states in semiconductors" FTT, 1982, vol. 24, no. 11, pp. 3394-3405.

PIEZOOPTIC EFFECTS AND THE INFLUENCE OF ANISOTROPIC DEFORMATION ON GaInPAs/InP HETEROLASERS

P.G. Eliseev, B.N. Sverdlov, N. Shokhudzhaev

ABSTRACT

The piezooptical effects and the anisotropic deformation of the active medium in GaInAsP/InP heterolasers arising either due to external pressure or lattice mismatch in the heterojunctions are investigated. Their influence on the radiative properties of heterolasers is investigated in the 1.0-1.6 μm range. It is found that the threshold, differential efficiency, and temperature parameter T_0 are sensitive to deformation. It is demonstrated that the degradation rate is virtually independent of lattice mismatch within the coherency range of the heterojunction. Piezogain is investigated and the deformation potential constants *b*, *b'* and *a* in lasers corresponding to a composition with $E_g \sim 0.95$ eV are estimated. A strong predominance of *TE*-polarization was observed in samples with a thin active region ($\leqslant 0.2$ μm). Spectral tuning by uniaxial pressure is investigated. The mechanisms responsible for deformation influence are discussed.

Introduction

The energy spectral parameters of a semiconductor are sensitive to lattice deformations which is the principal reason for a number of piezooptic effects, i.e., changes in optical properties under the influence of mechanical stress (shift of the natural absorption edge, induced birefringence and dichroism, spectral tuning of luminescence, etc.). Stresses are important in several respects for radiation sources (LED's, injection lasers, etc.) based on semiconductor heterostructures.

First in actual operating conditions the active (radiating) region of the device contains residual mechanical stresses due to the mismatch of the lattice periods at the heteroboundaries and due to the crystal assembly features (the use of holding clips, thermal compression contacts, soldering to a substrate with a different coefficient of thermal expansion (CTE), the use of dielectric coatings with different CTE's, etc.). As a result the radiative characteristics of the devices vary from sample to sample due to the deviation in the actual stressed state of the active regions. In lasers the stressed state can predetermine the modal composition of radiation, particularly radiation polarization and transverse mode indices. The high level of mechanical stresses in AlGaAs-lasers has a critical influence on service life, as has been demonstrated by direct measurements [1]. As soon as the initial stresses become undesirable in the devices, special mounting techniques are used (elimination of clamping devices, soldering by plastic solder, etc.) and the heterostructure lattice periods are carefully matched.

Second, external stresses can be used to successfully control radiation particularly for polarization switching and spectral tuning. We know that hydrostatic compression is one widely employed technique for tuning semiconductor lasers in IR spectroscopy. Regarding uniaxial compression, its practical application is limited due to the low pressure level and, consequently, the narrow tuning range. Another sample useful application of the piezooptic effect in semiconductor lasers is ultrasonic or hypersonic frequency modulation of radiation [2, 3]. Proposals and experiments on the application of acoustic waves for fabricating dynamic periodic feedback structures are well known [4, 5].

Third, the piezooptic effects represent a nonlinearity factor in the behavior of the radiation source, since the stressed state is sensitive to the pump level, the dissipative power and the radiation intensity. An example of such a nonlinearity is the anomalous change in radiation power accompanying the steps (kinks) in the P-I characteristic in select GaInPAs/InP stripe lasers [6, 7]. The mechanism of this phenomena has not been entirely established, although an interrelationship has been identified between this anomaly and residual stresses. The appearance of induced mechanical stresses and resulting deformations in the operating conditions can be caused by several generally identifiable processes such as: 1) thermoelastic stresses due to spatially-inhomogeneous dissipation of pump power; 2) electrostriction; 3) acoustoelectronic and acoustooptic effects, including Brillouin scattering.

The observation of so-called current injection induced acoustic signals (CIA) in semiconductor lasers has a history [8, 9]. The CIA signal itself is detected as acoustic noise in a laser diode from the flow of pump current saturated at a current somewhat below the threshold current [8]. According to the conclusions in study [9] this signal is primarily an acoustic "bell" due to pulsed excitation of vibrations by current pulses.

The pulsed state is normally rich in attendant processes since it is normally accompanied by nonstationary heating. The influence of a nonstationary temperature waveguide in stripe lasers is one example [10]. In this case a nonstationary waveguide appears in a laser where there is no internal lateral optical confinement and eliminates the nonlinearities in the P-I characteristic typical of this state and even improves radiation power (in spite of the slight increase in active medium temperature which normally degrades radiative properties). Naturally the temperature waveguide is accompanied by a three-dimensional field of thermoelastic stresses whose influence has not been empirically identified.

Of all the phenomena associated with the piezooptics of semiconductor lasers the present study has selected the uniaxial compression cases (normal to the active layer plane) most accessible to experimental investigation and theoretical modeling together with the influence of lattice period mismatch at the heteroboundaries of the active layer. Both uniaxial compression and mismatch produce qualitatively identical anisotropic deformation of the cubic lattice in the active region, which can be characterized as tetragonal deformation with a slight deviation from cubic deformation. This deformation eliminates the degeneration of electron terms, particularly energy splitting of the valence band when $k = 0$ by Δ. The primary consequences of deformation result in the following changes:

- changes in the symmetry of the isoenergetic surfaces in k-space;

- changes in the state density spectrum due to the energy splitting Δ and the redistribution of carriers among the subbands;

- changes in the effective carrier masses, the matrix elements of the transitions, and other characteristics of the electronic subsystem;

- changes in the optical length of the diode cavity.

These changes occur with little inertia and their appearance in static conditions closely corresponds to their manifestation at frequencies of interest in the dynamic mode of a laser and an LED, say, up to frequencies $\sim 10^9$ Hz, i.e., including the hypersonic range. The present study investigates the statistical characteristics of piezoeffects in GaInPAs/InP heterolasers radiating between 1.08 and 1.61 μm. Select preliminary results from this research were presented in studies [11, 12].

Given the urgency of the problem of lattice mismatch and the deformations resulting from lattice mismatch Section 1 of the present study is devoted to a description of tetragonal deformation in an actual heterojunction. Without analyzing the initial deformed state it is impossible to interpret the piezoeffect in a laser since the measurement results of, say, the influence of pressure on the lasing threshold depend on whether or not the deformation pressure increases or decreases lattice deformation, i.e., on the sign of the initial deformation. Section 2 presents a brief survey of the influence of deformation on the energy spectrum and radiation from semiconductors, while Section 3 covers the influence on laser radiation.

The experimental techniques used in the present study are described in Sections 4 and 5. The method of determining deformations (particularly the mismatch of the lattice periods) by the polarization of spontaneous radiation is important.

The experimental results are given in Sections 6-10. An interesting qualitative result is the change in the approach to the role of anisotropic deformation thereby making possible a redistribution of radiation to the desired polarization at the expense of orthogonal polarization and a certain reduction in threshold current, including the threshold current at an elevated temperature. The piezooptic phenomena in GaInPAs/InP lasers are rather varied, primarily due to the significant variety of initial stressed states and the more complex relation between the radiative characteristics and the diagram of bands. A discussion of the results can be found in Section 11.

1. Coherency Range and Tetragonal Lattice Distortions in a Heterojunction

As we know an important requirement on an ideal heterojunction is the requirement for a close correlation between the lattice period of the emitter layers and the substrate. The general solution to this problem is the so-called principle of isoperiodic substitution in multicomponent solid solution systems. The GaInPAs/InP system is an example. The lattice period of InP at room temperature is 0.687 nm and the $Ga_xIn_{1-x}P_{1-y}As_y$ quaternary compositions have the same lattice period; these satisfy the "isoperiodicity" condition [13]

$$y = 2.2x/(1 + 0.07x). \quad (1)$$

Minor violations of this condition will not necessarily make the heterostructure unsuitable. A "coherency" range of the lattices exists within which the lattices, under elastic deformation will match in the longitudinal direction and the heterojunction will remain dislocation-free. The coherency range depends on the thickness of the epitaxial layer and has been investigated in many studies. Within its range elastic deformation will reduce lattice symmetry from cubic to tetragonal.

With a limited epitaxial layer thickness the formation of a misfit dislocation is advantageous only if this thickness exceeds a critical value approximated as [14]

$$n_{cr} \approx b/2\varepsilon, \quad (2)$$

where b are the components of the Burger's vector compensating the lattice misfit; ε is the deformation caused by the misfit. The time required for formation of misfit dislocations (MD) diminishes with an abundance of seed dislocations. Therefore, for example, in a AlGaAs/GaAs heterojunction grown at a temperature close to correlation of the lattice periods no MD's are observed at various thicknesses h

and at room temperature $\varepsilon \approx 1.3 \cdot 10^{-3}$ and $n_{c4} \approx 0.16$ μm. The entire deformation is elastic in nature since the stresses occur from cooling (due to the discrepancy in the coefficients of thermal expansion) and during the period within the plasticity temperature (above 500°C) which amounts to a few seconds MD's cannot form. Therefore with a given active layer thickness, critical misfits of both signs triggering MD formation depend on the temperature progression of the misfit and the cooling conditions.

In a misfit GaInAs/GaAs heterojunction the critical deformation is $4 \cdot 10^{-3}$ [14]. If the epitaxial layer is extended along the substrate cracking is observed even at $\varepsilon \geqslant 3 \cdot 10^{-3}$ in the InGaP/GaAs heterojunction, while the critical deformations are $\sim 1.5 \cdot 10^{-4}$. When $\varepsilon \approx 2 \cdot 10^{-4}$ MD's appear in one [011] direction, while an orthogonal MD grid has been observed at $3 \cdot 10^{-4} < \varepsilon < 10^{-2}$. Under greater deformation the MD grid is very dense and the dislocations extend into the layer body.

Consistent with study [15] in the test GaInPAs/InP heterojunctions the MD's appear in heterostructures with an active layer thickness of 0.4 μm at $(\Delta a/a)_{\perp} > 5 \cdot 10^{-3}$. Here $h_{cr} \approx 0.055$ μm, i.e., seven times smaller than the thickness used. Many studies have occasionally reported no misfit dislocations in layers with thicknesses greater than h_{cr}. With growth of the lattice mismatch at $\Delta a/a \geqslant 2 \cdot 10^{-3}$ by Vegard's law morphologic degradation of the layer surface has been observed [16] which evidently causes irregularities in the active layer. Study [17] has noted that the threshold current density normalized to unity thickness in lasers (1.3 μm) remains approximately constant within the range $\Delta a/a \leqslant 1.5 \cdot 10^{-3}$, but precipitously diminishes outside this range. Study [18] has also noted that the coherency range corresponds to $(\Delta a/a)_{\perp} \leqslant 1.5 \cdot 10^{-3}$; the degree of plastic deformation outside this range grows to 0.2 when $(\Delta a/a)_{\perp} = \pm 4.5 \cdot 10^{-3}$ (for layers of thickness ~ 1 μm radiating at 1.3 μm). We note that the strong longitudinal distension often forms cracks in the layer. Thus, cracking has been observed in the GaInAs/InP heterostructure when $\Delta a/a = -2.2 \cdot 10^{-3}$ [19].

Completely matched heterojunctions at the growing temperature (~650°C) due to the differential in the CTE's are in a stressed state at room temperature with $\varepsilon \approx -6 \cdot 10^{-4}$ (the active layer is distended in longitudinal directions). Based on measurements carried out in study [20] the CTE differential for a composition with $x = 0.26$, $y = 0.60$ is $\approx 0.9 \cdot 10^{-6}$ K^{-1} in the 0-400°C range.

Tetragonal distortions to the cubic lattice are due to elastic deformation of one sign in two tangential directions (in the heteroboundary plane) and of opposite sign in the normal direction (due to the Poisson effect). According to study [14] the relation between the tangential deformation $\varepsilon_{\parallel}$ and the normal deformation $\varepsilon_{\perp}$ can be expressed as

$$\varepsilon_{\perp} = -\frac{2\nu}{1-\nu}\varepsilon_{\parallel}, \tag{3}$$

where ν is the Poisson coefficient equal to approximately 1/3 in these materials. Numerically we obtain

$$\varepsilon_{\perp} \approx \varepsilon_{\parallel}. \tag{4}$$

Thus, if the chemically-induced difference in the periods is Δa, when $\Delta a \ll a$

$$|\varepsilon_{\perp}| \approx |\varepsilon_{\parallel}| = \Delta a/a, \tag{5}$$

while the jump $\Delta a_{\perp}$ in the normal direction will consist of the chemically-induced quantity Δa and the Poisson deformation which is also equal to Δa. Therefore in the coherency range we have

$$(\Delta a/a)_{\perp} \approx 2\Delta a/a. \tag{6}$$

More strictly consistent with [21] we can obtain for the heteroboundary in the (100) plane

$$\frac{\Delta a}{a} = \frac{1-\nu}{1+\nu}\left[\left(\frac{\Delta a}{a}\right)_{\perp} - \left(\frac{\Delta a}{a}\right)_{\parallel}\right] + \left(\frac{\Delta a}{a}\right)_{\parallel}, \tag{7}$$

which yields the same relation as formula (6) in the coherency range when $\nu \approx 0.35$. Here we remember that virtually all the misfit deformation is concentrated in the thin active layer whose thickness d is much less than the substrate thickness t.

External pressure perpendicular to the heteroboundary plane produces an analogous effect and when the lattice period of the quaternary layer is greater than the InP period and, consequently, is distended in the normal direction within the coherency period external compression along the normal will compensate this effect at an elastic deformation of $\Delta a/a$. Here the symmetry of the quaternary layer is recovered, i.e., it will have a cubic lattice. This does not mean the vanishing of $(\Delta a/a)_{\perp}$, since the tetragonal deformation is transferred to the substrate, i.e., $(\Delta a/a)_{\perp} = 2\Delta a/a$ in this case as well.

2. Influence of Anisotropic Deformation on the Energy Spectrum and Luminescence of Semiconductors

Tetragonal lattice distortion under anisotropic deformation modifies the diagram of bands of the semiconductor. In cubic A^3B^5 compound crystals there is, first, an energy shift of the band extrema so that the bandgap changes. Second, since this displacement depends on the quantum numbers of the corresponding states the degenerate states are split. This primarily effects the top of the valence band as well as the impurity acceptor levels. Changes in the electron spectra in different conditions are examined in study [22]. The magnitude of valence band top splitting at the point $k = 0$ [21] is

$$\Delta = \delta E_{1,2} = E_1 - E_2 = 2\mathscr{E}_{\varepsilon}^{1/2}, \tag{8}$$

where under distension along the [001] axis

$$\Delta = 2|b\,\varepsilon'_{zz}|, \tag{9}$$

where $\varepsilon'_{zz} = \varepsilon_{zz} - \varepsilon_{xx} = \varepsilon_{zz} - \varepsilon_{yy}$ is the relative deformation on the [001] axis, b is the shear constant of the deformation potential.

Occasionally it is easier to interpret valence band splitting as splitting into bands of heavy and light holes between which an energy gap Δ arises according to (8) when $k = 0$. This representation is incorrect as it does not correspond to the Kane model. Splitting of the valence subbands occurs such that the heavy mass of the hole remains along two axes in one of the subbands (the "semiheavy" subband) and on one axis in the other subband (the "semilight" subband). The light mass remains on two axes (one in the first case and two in the second). The elements of the tensor of effective mass when $k = 0$ also change as a function of pressure. The isoenergetic surfaces in the k-space in the valence band become ellipsoidal (an elongated ellipsoid of revolution for the "semiheavy" subband and a compressed ellipsoid of revolution for the "semilight" subband). From the viewpoint of radiation polarization with band-to-band transitions it is significant that the "semiheavy" subband is characterized by a total moment 3/2, while the "semilight" subband is characterized by a moment 1/2. According to the selection rules linearly-polarized radiation in lateral directions corresponding to the *TE*-mode arises for optical transitions from the conduction band to a valence band with a total moment of 3/2, at the same time that with analogous transitions to a band with a total moment of 1/2 partially polarized radiation with a predominantly *TM*-mode arises. The intensities of the corresponding radiation components depend not only on these selection rules but also on changes in the matrix elements of the transitions and the occupancy probabilities of the corresponding working states (due to changes in their energy). Under compressive strain along the normal the 1/2 subband occupies the upper energy level while under tensile strain the 3/2 subband occupies this level [23].

Internal elastic stresses arising in structures due to mismatched lattice periods can cause anisotropic deformation. If the active region has a single period less than that of the substrate, it is distended in tangential directions and $(\Delta a/a)_\perp < 0$, which also corresponds to external compression. Here the holes occupy more of the "semiheavy" band with a moment of 1/2 and the *TM*-mode predominates in the radiation. In the opposite case the *TE*-mode will predominate. This can be detected by the luminescence polarization and is used for optical determination of internal stresses in the active region and the lattice period misfits. According to study [24] the degree of linear polarization of spontaneous radiation ρ in terms of the spectral density of radiation $I(\hbar\omega)$ is determined as

$$\rho = {}^3/_8\Delta\left[\frac{1}{kT} - \gamma + \frac{d(\ln I(\hbar\omega))}{d(\hbar\omega)}\right], \tag{10}$$

where Δ is the energy splitting of the valence band when $k = 0$, corresponding to expression (8), k is Boltzmann's constant, γ is the constant characterizing the quantum working levels. This formula is also applicable to the band-to-band transitions in p-type material which normally corresponds to actual laser heterostructures. The change in ρ along the contour of the luminescence band makes it possible to calculate the quantities Δ and γ in the same cases when $\Delta \ll kT$ and the energy splitting is difficult to directly determine by the difference of the photon energies at the spectral peaks of orthogonal polarizations.

Within the coherency range of the heteroboundaries it is possible to unambiguously attribute energy splitting to the relative lattice period mismatch (RLPM)

$$(\Delta a/a)_\perp = \Delta_{100}/2b \text{ for the (100)-heteroboundary}$$

$$(\Delta a/a)_\perp = \Delta_{111}\sqrt{3/2d} \text{ for the (111)-heteroboundary}$$

where b and d are the shear constants of the deformation potential for the corresponding orientations. An investigation of the polarization technique using GaInPAs/InP heterostructures in study [18] revealed good agreement of the results from determining $(\Delta a/a)_\perp$ and internal stresses from spectral splitting Δ to results form X-ray techniques where the sensitivity of the polarization technique is significantly better and it directly characterizes the radiating volume (at the same time that the X-ray technique gives an averaged result). According to data from study [18] the spectral splitting rate with uniaxial pressure on the (100) axis is ~4.7 meV/kbar. Below we will show that such a simple picture is typical only for some samples since often in addition to band-to-band radiation there is noticeable impurity radiation (with a different splitting coefficient) overlapping the edge band.

3. Influence of Anisotropic Deformation on Laser Radiation

The influence of unidirectional compression on the radiative characteristics of laser diodes has been investigated in many studies [25-31] primarily with respect to GaAs and AlGaAs. A theoretical analysis of the optical gain effect was given in studies [27, 30]. Studies [32, 33] focus on the fact that in GaInPAs/InP heterostructures the laser radiation polarization depends on the mismatch of the lattice periods.

Uniaxial elastic deformation from uniaxial compression creates a state of the crystal lattice in a particular pressure range that arises in the active layer of a two-sided heterostructure if the lattice period in the substrate is greater than in the active layer (i.e., when $\Delta a/a > 0$). The active layer lattice is distended in two tangential directions while the period diminishes in the normal direction. This makes it possible to manipulate the deformation of the active layer; specifically, to compare the effect produced by external compression to the effect of internal residual stresses. Both deformations (residual deformation and deformation induced by external compression) are summed algebraically. Unfortunately from the technical viewpoint it is easy to implement only compression on the axis normal to the active plane, i.e., only on one (100) axis and of a single sign only (compression). This is insufficient, however, to identify the primary regularities associated with the influence of lattice mismatch.

It is comparatively easy to achieve unidirectional compression experimentally by, for example, using adjustable spring-loaded clamps. A simple technique for controlled application of loads to a holder whose electrodes are press punches has also been used. The majority of experiments have employed wide contact laser samples that generate simple and easily-interpreted deformation geometries. The source of errors in this case is the load nonuniformity due to the nonparallel contact surfaces of the diode. Generally such samples can withstand a lower load than normal samples without failure. Their influence on the measurement results is reduced by including only reproducible measurement data in the statistics of reliable tests. The irreversible degradation of the sample during the measurement process served as a criterion for rejection of measurements.

In the case of stripe structures the loading and deformation geometry is more complex. First the samples (particularly those containing grown mesa stripes) often have an irregular contact surface and this causes nonuniform loading or premature failure of the sample. Second the stripe lasers contain additional sources of stress that are added to the residual mismatch stresses in the active layer. Thermoelastic stresses between the semiconductor and the oxide layer concentrated along the edges of the oxide layer are known to have an effect in oxide insulation lasers. These stresses depend on the temperature at which the oxide layer is applied as well as its thickness, the width of the stripe window (since the stresses from both edges are summed) and the depth of the active layer. The lattice period mismatch between the grown material an the laser mesa stripe itself plays a role in grown mesa stripe structures. These stresses depend on the magnitude of the mismatch as well as the width and shape of the mesa stripe. Additional stresses in stripe lasers produce a three-dimensional homogeneous deformation pattern in which these deformations are not necessarily predominant. This must be accounted for when comparing measurements on different types of samples.

The influence of unidirectional compression on the radiative characteristics was examined previously in GaAs lasers (homolasers) and AlGaAs/GaAs lasers (heterolasers). The initial research [25, 26] identified the possibility of reducing the threshold current of the homolasers including those operating at room temperature by 10-20% as well as the possibility for spectral tuning. A change in the predominant linear polarization and a reduction in the *TM*-mode threshold (with an increase in the *TE*-mode threshold) as well as spectral tuning have been observed in heterolasers [28, 29]. Since *TE*-modes normally predominate in the initial state of AlGaAs/GaAs heterolasers when pressure is applied to the diode its threshold grows until at pressure P^* it switches to the *TM*-mode whose threshold diminishes with pressure. According to different measurements P^* in AlGaAs/GaAs lasers is ~300 bar [29].

Study [31] has noted that the development of elliptical polarization under uniaxial compression in GaAs homolasers eliminates the reduction in the threshold, since both linearly-polarized modes (the *TE*- and *TM*-modes) participate in lasing. The threshold of such a mode complex is determined by averaging the gain and loss parameters in both linearly-polarized modes and the threshold remains near constant by virtue of their mutually opposite variation. Such an interrelationship of the orthogonal polarizations assumes that the frequencies of the corresponding modes are identical. They are in fact close in homolasers. However in heterolasers waveguiding causes significant spectral splitting of the polarizations thereby hindering their locking. This splitting is due to the influence of polarization on the longitudinal propagation constant which with a given laser cavity length will cause a differential in the mode eigenfrequencies. Therefore as long as the orthogonal modes have different frequencies they behave quite independently and compete. This results in the excitation of linear polarization with a lower lasing threshold. Specifically under unidirectional compression competition of the *TM*-mode results in its predominance at $P > P^*$.

It is worth noting that the observation of orthogonal mode frequencies in an AlGaAs/GaAs laser in study [30] has revealed that these frequencies are pressure-dependent and, moreover, spectral splitting is reduced, since it can be diminished to the limit where the elliptically-polarized mode complex is formed. According to study [34] coincidence of the orthogonal mode frequencies is achieved in AlGaAs/GaAs stripe lasers at $P = 1.1$ kbar; further growth of P was not accompanied by spectral splitting. This indirectly reveals a probable mutual locking effect of the orthogonal modes. We know of no more detailed investigations of this issue. The mechanisms responsible for this effect can include depolarization of radiation within the cavity and Faraday rotation of the polarization plane in the magnetic field induced by the pump current. Since we are discussing the Faraday effect at the operating frequencies of the laser, i.e., in direct proximity to the natural absorption edge, Verdet's constant can be sufficient to produce noticeable rotation of the polarization plane even with relatively weak magnetic fields of the pump current.

4. Technique for Measuring the Radiative and Polarization Characteristics of Laser and Spontaneous Radiation. Methods of Achieving Uniaxial Compression of Diodes

The radiative polarization properties of GaAs/GaAs, GaInPAs-/InP laser diodes were investigated. The majority of samples had a broad (nonstripe) contact. A DFS-12 spectrometer and MDR-2, MDR-23 monochromators were used to record the spectra. Uncooled FEU-22, FE-U22, and FEU-63 photomultipliers were used as the photodetectors together with a FDG-3 germanium photodiode. The signal taken from the detector output was amplified by selective amplifier V6-4 and injected to synchronous detector SD-1. The reference signal injected to SD-1 was generated by pulse generator G5-54 triggered synchronously by a G5-15 pulse generator and the reference signal was amplified by a U2-6 selective amplifier. A 232V nanovoltmeter was used as the selective amplifier in certain cases.

An emitter follower fired by a G5-15 pulse generator was used for laser pumping in pulsed operation. The amplitude of the pump pulses was measured by an S1-72, S1-17, or S7-8 oscillograph. A B5-21 d.c. power supply was used for the laser diode in CW operation and the working current level was controlled by means of a standard transistorized current regulator. The test was conducted at 77 and 300 K. The laser diodes had Fabry-Perot cavities 120-400 μm in length and 100-300 μm in width. The laser characteristics were investigated over a broad range of pump currents and pressures.

A silicon photodiode with a detection surface ~1 cm^2 was used to measure the absolute laser diode output power. The photodiode and laser diode were placed in a photometric medium ~30 cm in diameter. These absolute measurements were employed to determine the internal and external quantum efficiency. The measurement error for measuring absolute radiation power was less than 15%. The output power from the laser diodes under external pressure was measured on one mirror side based on the average current employing the FDG-3 or the silicon photodiode (for lasers radiating at <1.1 μm) in relative units.

The laser diode radiation distribution in the far field was recorded by the photoelectron multiplier (lasing wavelength below 1.1 μm). The horizontal and vertical cross-sections of the directional pattern (the *p-n*-junction plane lies in the horizontal plane) were investigated using a slit and a photodetector. The diode was attached to a goniometric head that made it possible to rotate the diode 360° with a recording accuracy of 0.1° by means of an RD-09 reversible motor.

The lasing threshold was determined visually (for lasers radiating at less than 1.1 μm) by means of an MIK-1 microscope by the radiation spectrum and the P-I characteristic.

In order to measure the temperature dependence of the lasing threshold current the laser samples were placed in a nitrogen cryostat whose temperature was held between 100-360 K. The temperature was controlled by means of a copper-constantan or chromal-aluminum thermocouple and measured by a Shch31 digital millivoltmeter. The FDG-3 germanium photodiode and the S7-8 oscillograph were used to monitor the light pulse. The lasing wavelength was monitored simultaneously by means of the monochromator, detector, selective amplifier and recorder; the radiation spectra were also recorded.

Uniaxial compression was achieved in the following manner. The samples were mounted in spring-loaded crystal holders with variable-tension springs or the diode was placed in a non-spring-loaded crystal holder and the compressive force was applied to the holder by means of a lever whose long arm contained variable weights for establishing the desired pressure on the sample. Using these weights it was possible to vary the pressure over a range from 10 bar to 4 kbar. Uniaxial compression was applied in the [100] direction perpendicular to the heterojunction plane. The pressure measurement error was ±10 bar. The measurement error in determining the position of the maximum of the diode radiation band was less than ±10 Å.

Uniformity of applied pressure over the sample area is important in investigating the characteristics of lasers under uniaxial compression. Cases of nonuniform distribution were observed in the experiments; these appeared as instabilities in the degree of polarization of laser radiation. For example, there were cases where the degree of linear polarization of laser radiation was unstable and varied from experiment to experiment. The data obtained in these cases were excluded from further consideration. The experiments revealed that with a uniform stress distribution across the diode area the degree of linear polarization of laser radiation (outside the switching range) is high and a sharp mode switch was observed. Certain parameters of the initial heterostructures are given in Table 1.

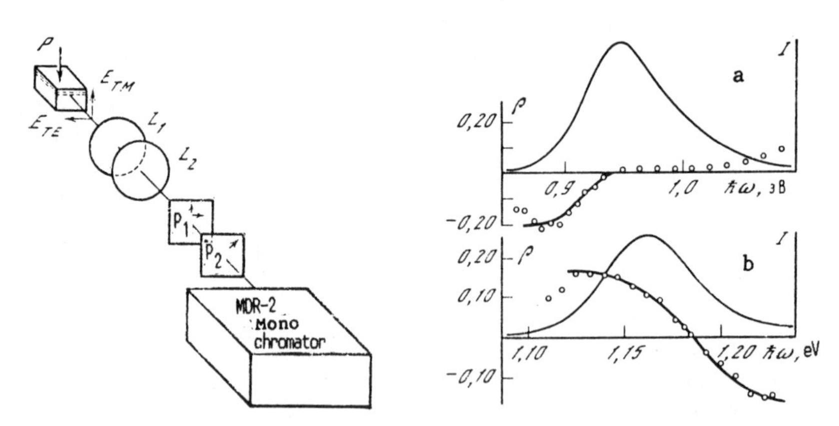

Fig. 1. Experimental set-up for investigating radiation polarization

Fig. 2. Electroluminescence spectrum of GaInPAs/InP diode heterostructure and the spectral dependencies of the degree of linear polarization ρ at 300 K without external loading for lasers with an elongated active region (a) and with a compressed active region (b)

The experimental set-up for investigating the radiation spectra and polarization characteristics is shown in Fig. 1. Diode radiation was observed in the plane perpendicular to the axis of deformation. A single polarization filter (thin film polarization filters ~ 4 cm in diameter were used) was first placed in front of the entrance slit of the monochromator to measure the degree of linear polarization of spontaneous radiation; in lasing the precise filter position where the perpendicular radiation component (i.e., the *TE*-mode) is extracted was identified; the second polarization filter was then mounted with the polarization plane 45° out of phase with respect to the first polarization filter. The *TM*-mode could be extracted by rotating the first polarization filter (that was initially tuned to the *TE*-wave) by 90°. The second fixed polarization filter made it possible to eliminate the influence of the polarization properties of the monochromator. The degree of linear polarization of radiation was determined by the formula

$$\rho = (I_{TM} - I_{TE})/(I_{TM} + I_{TE}), \qquad (11)$$

Table 1

Initial parameters of the GaInPAs/InP heterostructure

Series number	Active region thickness mcm	λ, nm (300 K)	Series number	Active region thickness mcm	λ, nm (300 K)	Series number	Active region thickness mcm	λ, nm (300 K)
1[1]	0.4	1080	11	0.5	1320	21[3]	0.075	1350
2	0.5	1235	12	0.9	1330	22	—	1375
3	0.3	1285	13	0.5	1330	23	0.10	1410
4	—	1286	14	0.7	1340	24	0.6	1516
5	0.6	1239	15	0.5	1340	25	0.4	1520
6	0.5	1288	16	0.6	1340	26	0.4	1560
7[2]	—	1290	17	0.6	1340	27	0.2—0.4	1597
8	—	1310	18	0.5	1342	28	0.5	1610
9	0.6	1312	19	0.5	1343	29[4]	0.8	835
10	0.8	1320	20[3]	0.06	1345			

[1] Stripe structure, w = 10 mcm
[2] Grown mesa structure, w ~ 5 mcm
[3] Three-layered waveguide structure
[4] AlGaAs/GaAs stripe structure, w = 10

where I_{TM} is the intensity of radiation with photon polarization parallel to the deformation axis; I_{TE} is the intensity of radiation with photon polarization perpendicular to the axis of deformation. Measuring $\rho(\hbar\omega)$ along the contour of the spontaneous radiation band made it possible to determine Δ (the magnitude of the energy split between the bands) and γ (a constant entering into formula (10)) that were used for the theoretical calculation of the degree of linear polarization.

The spectral dependencies $\rho(\hbar\omega)$ and $d\ln I/d(\hbar\omega)$ made it possible to calculate the relative lattice period mismatch $(\Delta a/a)_{\perp}$ which was then used to calculate the deformation potential constant of the bands

$$b = \Delta/2\varepsilon_{zz}, \quad (12)$$

where ε_{zz} is the deformation on the axis perpendicular to the plane of the heterojunction; i.e., in our case on the (100) axis.

Certain sample applications of this method are shown in Fig. 2. Fig. 2a shows the electroluminescence spectrum of a heterodiode (from a structure with a heavily distended active layer in the (100) direction without external stresses together with the spectral dependence of the degree of linear polarization ρ along the electroluminescence band contour (calculated by formula (10) - solid curve - and experimental - circles). As we see from the diagram on the longwave side where $dI/d(\hbar\omega) > 0$ the radiation is primarily polarized perpendicular to the axis of deformation and $\rho < 0$ corresponding to the distension state, and $\Delta < 0$. On the shortwave edge of the spectral dependence ρ heavily influences the second radiation band where these bands are accumulated. Therefore at the shortwave edge the experimental points follow the theoretical curve of ρ. The solid (theoretical) curve calculated by formula (10) is in satisfactory agreement with experimental results when $\Delta = -10.6$ meV, $\Delta \cdot \gamma = -0.04$.

We should note that the symmetric form of the dependence of ρ on $\hbar\omega$ was observed when the electroluminescence band was not influenced by additional radiation bands. For example, a symmetrical spectral dependence of the degree of linear polarization was observed in structures corresponding to the composition at 1.06 μm where the spontaneous radiation spectrum included a single band at 300 K (see Fig. 2b). It follows that the dependence of ρ on $\hbar\omega$ makes it possible to determine the number of unresolved radiation bands in the electroluminescence spectrum of the structures.

The lattice parameter mismatch $(\Delta a/a)_\perp$ was determined based on the values of Δ found for all test structures using the data from study [18].

5. Technique for Analysis of Polarization Spectra

We will consider the polarization spectrum within the spectral band produced by identical optical transitions such as band-to-band transitions. Since uniaxial deformation causes splitting of at least one of the working states (in this particular case: polarization splitting ceiling of the valence band), the spectral radiation band also splits and the energy splitting Δ of the orthogonal polarizations can be measured directly by the energy distance between the corresponding peaks. Small values of Δ can be measured by recording the polarization spectrum, i.e., the spectral dependence of the degree of linear polarization $\rho(\hbar\omega)$. The polarization spectra contain additional information on the suitability of the model on which the calculation is based; for example, the correlation between the calculated form of $\rho(\hbar\omega)$ to the experimental value is determined together with Δ

and other adjustable parameters. According to study [35] we can use the following expression for the band-to-band transitions when $\Delta < kT$:

$$\rho = \gamma_1 [\Delta/kT + \gamma_2 + \Delta d \ln I/d (\hbar\omega)], \tag{13}$$

where γ_1 and γ_2 are the constants dependent on the type of working transitions. For *p*-type materials study [29] employed the expression with two adjustable parameters Δ and γ.

It is comparatively easy to determine Δ using this formula if ρ_{max} are measured at the peak of the spectral band and $\xi_0 = d \ln I/d(\hbar\omega)$ at the point where $\rho = 0$ then $\Delta = 8\rho_{max}/3\xi_0$. The calculation of γ is a priori not reliable, although we can expect that for identical transitions γ has an identical value in different samples. We can therefore expect better reliability of the measurement results of Δ with coincident (within reasonable limits) values of γ. On the other hand variations in γ suggest changes in the predominant transitions in the spectral radiation band. In the spectra of doped samples the overall band includes the contribution of both the band-to-band transitions and the transitions involving the shallow impurities where, with identical lattice deformation, polarization splitting is characterized by a different splitting energy (Δ_A or Δ_D for the acceptors and donors, respectively). At room temperature or higher temperatures the bands are broad and it is difficult to identify their internal structure. It is therefore necessary to approach the formal analysis based on formulae (10) or (13) with some care. We will employ the phenomenological formula

$$\rho = C_1 + C_2 d \ln I/d (\hbar\omega), \tag{14}$$

in which the constants C_1 and C_2 can be interpreted in accordance with (10) in the following manner:

$$C_1 = {}^3/_8\Delta (1/kT - \gamma), \quad C_2 = {}^3/_8\Delta \tag{15}$$

and in turn the parameters Δ and γ are equal to

$$\Delta = 2{,}67 C_2, \quad \gamma = 1/kT - C_1/C_2. \tag{16}$$

From this viewpoint the interpretation above is not always suitable and requires proof. On the other hand in order to determine the limits on the applicability of the technique it is important to carry out a comparative investigation of polarization splitting by different techniques. One particle case is an analysis of the polarization spectrum with a 10.4 meV distance between the corresponding peaks (measured at 77 K) (Fig. 3). This corresponds to $\Delta > kT$, i.e., this case lies outside the range of formulae (10), (13).

The formal application yields $\Delta = -11.2$ meV which is within 7.5% of the value given above, i.e., within the limits close to the corresponding measurement error (±5%). We can carry out the analysis at room temperature where the condition $\Delta < kT$ is already satisfied, however the splitting cannot be determined or will be inaccurately determined based on the spectral peaks. Formula (10) yields -11 meV in this case. Therefore a comparatively simple case can be used to establish the compatibility of both techniques if we limit the relative measurement error to, say, 7-10%.

The situation becomes more complicated when the spectral radiation bands are superimposed; in the general case these bands have different values of Δ and γ. Evidently the most important case of this type is a mixed band of band-to-band transitions and transitions to the shallow acceptor separated by 10-20 meV depending on the carrier concentration which is less than kT at 300 K. In this case band-to-band transitions predominate on the shortwave wing of the line while impurity transitions predominate on the longwave wing (to the acceptor level or the acceptor band). Since deformation splitting of the low levels varies it turns out that the longwave and shortwave wings of the band yield formally different values of Δ. In the specific example discussed above $\Delta \approx -6$ meV at room temperature on the longwave wing, so a value $\Delta \approx -11$ meV is obtained by the band maximum and its vicinity. Therefore at room temperature the various mechanisms are combined with one mechanism predominating, corresponding to the band peak. These data revealed that in order to exactly determine $(\Delta a/a)_{\perp}$ in each series it is necessary to know the shear coefficient of the deformation potential in the valence band corresponding to the predominant mechanism.

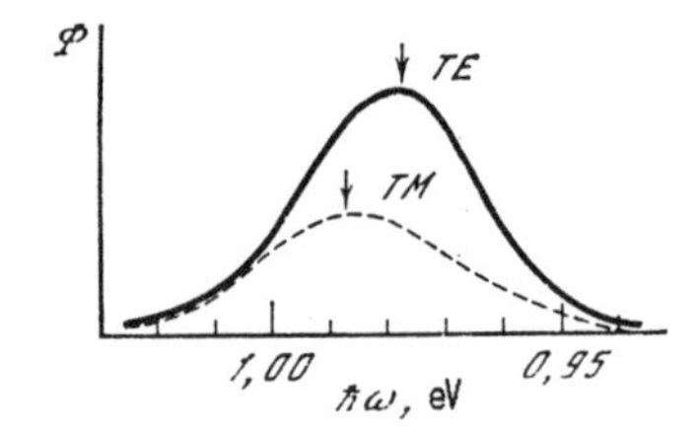

Fig. 3. Electroluminescence spectrum of orthogonal polarizations at 77 K without external loading for a GaInPAs/InP diode heterostructure laser
I = 0.65 A, $S = 9 \cdot 10^{-4}$ cm2

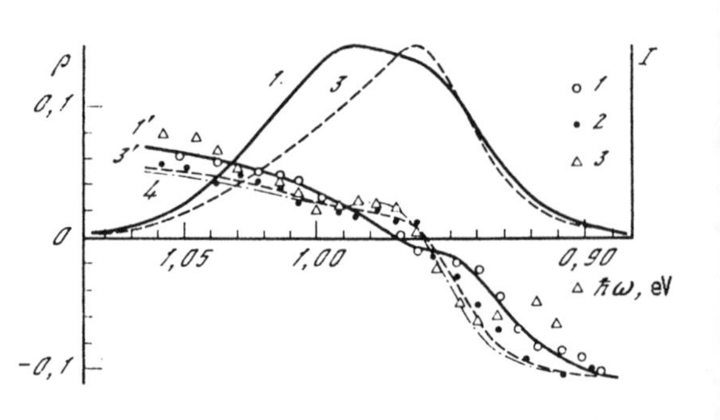

Fig. 4. Spontaneous radiation spectra (1, 3) and spectral dependencies of the degree of linear polarization ρ (1', 2, 3', 4) at 300 K plotted as a function of pump current for a GaInPAs/InP diode heterostructure

Fig. 4 shows the spontaneous radiation spectra and the linear polarization spectra of spontaneous radiation ρ as a function of the pump level for GaInPAs/InP diode heterostructures (nonplanar). Here the circles correspond to experimental data while the solid and dashed curves correspond to calculation by expression (10). It is clear from

Fig. 4 that as the pump current rises to 1.5 A (in this sample the threshold current is 3.4 A), in spite of the significant change in the form of the radiation band, the experimental points still lie on the curve calculated with identical values of Δ and γ. At currents $I \geqslant 1/2 I_{TH}$ the longwave band stands out due to the gain effect and divergences appear at the far points, i.e., at the edges of the spontaneous radiation band. However ρ is largely described by identical values of Δ and γ. With near-threshold currents the narrowband predominates in the radiation spectrum and ρ is approximately described by identical Δ and γ within limits in the proximity of the band maximum.

These results then reveal that the pump current has no influence on the degree of linear polarization of spontaneous radiation up through superluminescence.

6. Stressed State of Laser Heterostructures and its Effect on Laser Radiation Polarization

We know that initial heterostructures based on binary compound solid solutions are in the majority of cases in a stressed state. The causes of residual stresses were discussed in Section 1. An identical trend towards a change in lasing conditions in favor of the *TM*-mode with growth of pressure perpendicular to the heterojunction is also manifest in lasers employing an A^3B^5-based medium. However unlike AlGaAs/GaAs a variety of initial states are possible in the quaternary heterosystem (distension or compression) due to the possibility of violation of the isoperiodic condition on either side (by virtue of the two degrees of freedom of the chemical compound). In AlGaAs/GaAs the narrowband region is always in a distended state, since it lies between the narrowband layers with a large lattice period (the magnitude of the distension also depends on the thicknesses of the broadband layers and the substrate).

Actual heterostructures have a certain terminal mismatch that can be characterized by the quantities $(\Delta a/a)_\perp$ or Δ (in meV). The quantities $(\Delta a/a)_\perp$ and Δ obtained from an analysis of the spectral dependence of the degree of linear polarization of electroluminescence are given in Table 2. We will assume positive values of both quantities corresponding to distension along the normal (the lattice period is greater in the layer than in the substrate).

The test samples can be divided into three types based on the behavior of the radiation polarization:

1) samples radiating a *TE*-mode at all attainable pressures (*TE*-lasers);

2) samples switching from the *TE*- to the *TM*-mode with application of an external pressure P (*TE/TM*-lasers);

3) Samples radiating the *TM*-mode at all accessible pressures (*TM*-lasers).

Table 2

The quantities $(\Delta a/a)_{\perp}$ and Δ obtained from an analysis of the spectral dependence of the degree of linear polarization of electroluminescence

Series No.	λ, nm	j_p, ka/cm^2	p^*, kbar	$(\Delta a/a)_{\perp}\cdot 10^3$		Δ, meV (calculated by
				Measurement by ρ	Estimate by p^*	
1	1080	~7	—	—3,65	—	6
2	1235	6	—	0,6	—	—2,5
3	1285	2.5	—	2.0	—	—8
4	1286	6.9	—	1,95	—	—4
5	1239	8.9	0.65	0.7	—	—3.3
6	1288	3.2	—	5.45	—	—12
8	1310	—	—	3.4	—	—7.5
9	1312	4	—	—3,3	—	7
10	1320	7.7	—	4.6	—	—10,6
11	1320	4.2	—	2.3	—	—5.5
12	1330	8.7	—	5,0	—	—11
13	1330	3.5	—	2.1	—	—8
14	1340	5.4	—	3,2	—	—13
15	1340	6.2	—	0,7	—	—3,0
16	1340	8	—	1,1	—	—4.5
17	1340	4.2	—	—	—	—2,3
18	1342	3	—	3,0	—	—12
19	1343	3.1	—	3.1	—	—11,5
23	1410	1.1	—	3.0	—	—6,9
24	1516	3.2	0,3	—	1.0	—2,0
25	1520	3.5	0,4	—	1,2	—2.5
26	1560	1.2	—	—2,0	—	3,3
27	1597	2.2	—	—0.6	—	0,74
28	1610	2	0,8	—	2,0	—4,0

Clearly the *TE*-lasers include lasers heavily distended along the normal, i.e., those having a natural lattice period of the active layer greater than the InP lattice, so that an external pressure up through the failure point of the sample will not compensate the residual deformation. The *TE/TM* switchable lasers correspond to the better matched heterostructures. If the elastic stress factor does not exist we can expect that the laser operates at the *TE*-mode since it has greater Fresnel reflection at the end mirrors (i.e., a somewhat greater *Q*) than the *TM*-modes and, possibly, somewhat better optical gain. This advantage is comparatively easily canceled by the deformation effect from external or residual stresses. The dependence of P^* (switching pressure) on the initial mismatch is shown in Fig. 5. It is clear that with complete matching P^* is only approximately 0.1 kbar. The value of P^* vanishes when $(\Delta a/a)_{\perp} \approx -2.5\cdot 10^{-4}$.

We therefore have a set of samples with a broad variation of the initial stressed state of the heterostructure which is most clearly manifest in the polarization characteristics of the radiation and its behavior with application of uniaxial stress. The plane of the variables P and $(\Delta(a/a)_{\perp}$ shown in Fig. 5 is split by the line P^* into two parts corresponding to the predominance of *TM* and *TE* polarization. The left and right boundaries of the permissible range are limited by the degradation of laser characteristics due to mismatch dislocations (see Fig. 4); in terms of the pressure level the acceptable range corresponds to $0 \leqslant P \leqslant P_{cr}$, where P_{cr} is the critical pressure for samples of this type resulting in crystal failure. This figure is 2.5 kbar on the average, although in certain samples (~0.1 mm thick) measurements were performed at a pressure of 3-4 kbar.

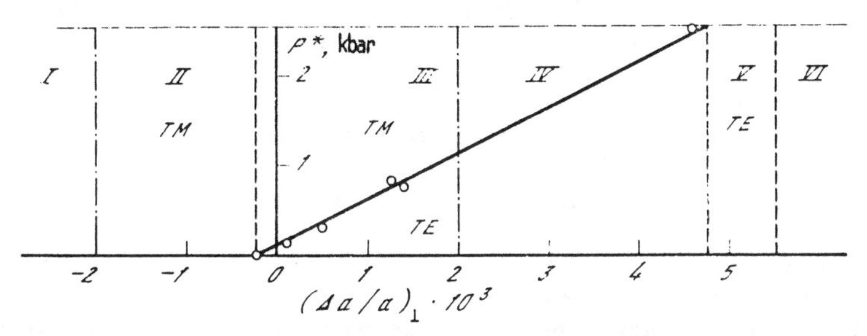

Fig. 5. Ranges of the initial deformations $(\Delta a/a)_{\perp}$ and external pressures P with different laser polarization states
P^* - uniaxial pressure at which polarization switching occurs
I - range of strong compression (along the normal) of the active layer; likely formation of mismatch dislocations (MD), i.e., noncoherent heteroboundaries; II - coherency range of heteroboundary with reduced compression (along the normal) of the active layer producing *TM*-polarization at all pressures P up to 2.5 kbar; III - coherency range with weak and zero deformation; polarization is easily controlled by uniaxial pressure up to 1 kbar; IV - coherency range with moderate distension (along the normal) to the active layer producing predominance of *TE*-polarization (at least when $P \leqslant 1$ kbar); V - distension region with *TE*-polarization for all P (up to 2.5 kbar); coherence of the heteroboundaries is under question; VI - range of strong distension (along normal) with likely formation of MDs

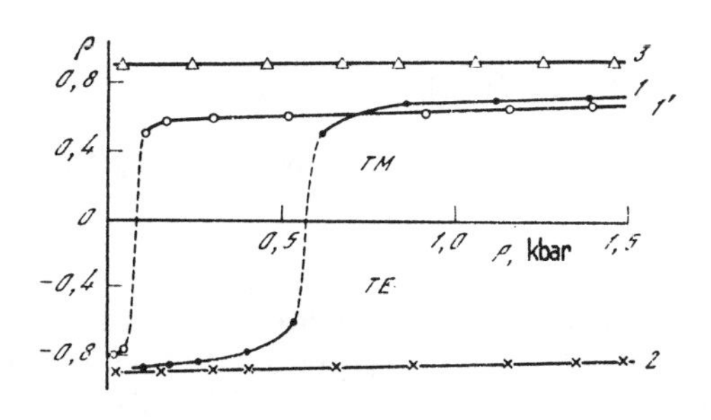

Fig. 6. The degree of polarization ρ of laser radiation plotted as a function of uniaxial pressure P in various samples at 300 k
1 - wavelength 1.24 μm, initial threshold current 2.9 A; 1' - 1.52 μm and 1.05 A; 2- 1.08 μm and 0.04 A; 3 - 1.33 μm and 2.75 A

The GaInPAs/InP heterolasers are characterized by a high degree of linear polarization of laser radiation; as we see from Fig. 6, switching occurs rather rarely. However anomalous samples have been observed with a clearly expressed polarization pattern and polarization switching. Evidently such anomalies are related to the significant spatial inhomogeneity of the polarization characteristics of radiation within a single diode which is, at least partially, related to the inhomogeneity in the distribution of elastic stresses (for example, with dislocations, etc.

Therefore the stressed state of the radiating layer (including the state produced by external uniaxial pressure) predetermines radiation polarization and its variations explain the spread of polarization characteristics. External pressure can be used (within certain limits) to control polarization.

7. Pressure-Induced Reduction in Threshold Current and Alteration of Differential Efficiency

The measurement of the piezoeffect in laser radiation (i.e., an investigation of the influence of pressure on laser radiative characteristics) using a large set of samples (~50) has revealed that in the case of GaInPAs/InP lasers we encountered two types of effects; the threshold current and the differential efficiency as well as the laser radiation wavelength are sensitive to pressure. We will first consider the "threshold piezoeffect." Here it is important to remember that in spring-loaded clamps pressure influences the thermal and electrical resistance of the contacts between the diode and the holder and, consequently, may have an indirect influence on lasing, particularly in CW operation. However if the contact resistance is not too great we can neglect their influence in pulsed operation.

We will select groups of samples classified by polarization according to the approach described in the preceding section from a wide range of samples. It turns out that the following correlation can be used:

- in the *TE*-samples the threshold rises with increasing pressure in the majority of cases;

- in the *TM*-samples the threshold diminishes with growth of P;

- in the *TE-TM*-samples the threshold rises when $P < P^*$ and diminishes when $P > P^*$.

Fig. 7 gives the P-I characteristics of a *TE/TM*-sample illustrating this regularity. Fig. 8 shows the differential efficiency and threshold plotted against the pressure, also in accordance with the general regularity. A review of the investigated parameters of the laser diodes is given in Table 3. The coefficient K_j is determined as the average value of the quantity $d(\ln j_{\Pi})/dP$ on the monotonic section of the dependence and is a characteristic of the relative change in threshold with pressure. The *TE* and *TM* indices represent the relation to the corresponding polarizations. It is clear that the coefficient K_j^{TE} is positive in the majority of cases, although occasionally it is close to zero and in two out of eighteen cases is negative. This therefore yields a behavioral anomaly of ~11%. All observed values of K_j^{TM} are negative while the maximum decrease in the threshold current achieved by virtue of the piezoeffect is 30-35%. Analogous coefficients are given for AlGaAs/GaAs are given for comparison (lasing at 835 nm). We note that absolute values of K_j are significantly greater for a shortwave laser. The P-I characteristics of this laser are shown in Fig. 9.

The temperature dependence of the threshold current is also sensitive to pressure and, as shown in Fig. 10, the threshold reduction effect is magnified with external pressure, resulting in simultaneous growth of the parameter

$$T_0 = (d \ln j_{\Pi}/dT)^{-1}, \tag{17}$$

characterizing the slope of the temperature dependence on the semi-logarithmic scale. This applies to the high-temperature part above the knee whose temperature is normally represented as T_b. Such measurements were not carried out at $T < T_b$. We note, however, that the point T_b drops with increasing pressure. This is valid for the case of *TM*-samples. More extensive investigations of the temperature dependence have revealed that the piezoeffect has a positive sign for *TE*-samples and T_0 drops, while T_b grows. Review graphs in Fig. 11 reveal that with total matching of the lattice periods the threshold current density ($j_s = j_t/d$, Fig. 11a) and T_b (Fig. 11b) pass through a maximum, while T_0 (Fig. 11c) passes through a minimum; the piezoeffect is represented by arrows.

Another feature of GaInPAs/InP lasers is the change in the slope of the P-I characteristic. Therefore in addition to the change in threshold the total piezoeffect in the radiation power is quite significant. For example, for the sample shown in Fig. 7 at a current of 4 A (40% above threshold) with a pressure jump from 410 to 1500 bar the power grows by a factor of 2.7. In the *TE/TM* samples the progression of the differential efficiency is nonmonotonic while the change in radiation power at a fixed current is also nonmonotonic. On the other hand in the shortwave laser shown in Fig. 9 the piezoeffect has no influence on the slope of the characteristics. This suggests that in the case of longwave lasers deformation influences not only the gain but also internal optical losses.

The pressure dependence of the differential efficiency is shown in Fig. 8a. The relative variation in this quantity reaches a factor of 2 if we ignore the minimum value near the polarization switching point. Qualitatively the regularities of the piezoeffects are identical for different wavelengths. The spread of the quantitative characteristics makes it impossible to identify the dependence of the piezoeffect on the laser radiation wavelength.

Table 3

Radiative and threshold characteristics of test laser diodes

Series no.	η, nm	$K_j^{TE} \cdot 10^2$, kbar^{-1}	$K_j^{TM} \cdot 10^2$ kbar^{-1}	Maximum relative change in threshold	Operational service life $T \cdot 10^{-4}$, hr.
1	1080	-	-33	-0.15	-
4	1286	17	-	0.10	-
5	1239	14	-19	-0.30	-
6	1288	35	-	0.20	-
9	1312	-	-13	-0.15	40[1]
10	1320	15	-	0.15	30[1]
11	1320	5($\leqslant$0.5 kbar)	-	0.10	-
		34(>0.5 kbar)	-		-
12	1330	14	-	0.14	-
13	1330	13	-	0.15	10[2]
14	1340	-7	-	-0.06	-
16	1340	29	-	0.35	-
19	1343	1($\leqslant$0.7 kbar)	-	0.08	10[2]
		8.5(>0.7 kbar)			
21	1350	0($\leqslant$0.5 kbar)	-	0.06	-
		30(>0.5 kbar)			
23	1410	0($\leqslant$0.3 kbar)	-	0.05	-
		28(>0.3 kbar)			
24	1516	10	-14	-0.30	-
25	1520	0	-5	-0.10	-
	1520	-0.15	-	-0.2	-
26	1560	12	-18	-0.30	-
27	1597	-	-15	-0.26	-
28	1610	16	-30	-0.35	-
29	835	268(0<P<0.05 kbar)	-67 (>0.05 kbar)	-0.15	-

[1] d.c. testing at 100°C
[2] In lasing

These results reveal that by optimizing the stressed state of the active region (by introducing a given mismatch in the lattice periods or by external pressure) it is possible to enhance the radiative properties of laser diodes, i.e., enhance differential efficiency, reduce threshold current, increase T_0, and in the final

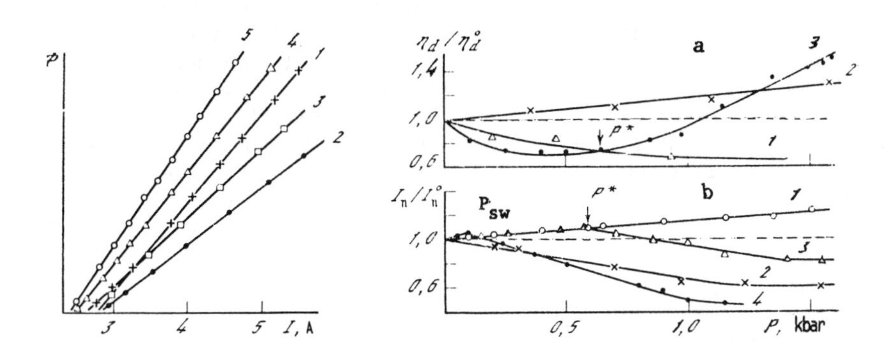

Fig. 7. P-I characteristics of a nonstripe GaInPAs/InP diode heterostructure laser operating at 1.24 μm at 300 K plotted as a function of the uniaxial pressure P
1-5: P = 10 kg/cm² (1); 410 (2); 860 (3); 1270 (4); 1500 (5)

Fig. 8. The differential efficiency (a) and threshold current (p) normalized to the initial value plotted against the uniaxial pressure P at 300 K in a variety of samples.
1 - wavelength 1.33 μm; 2 - 1.08 μm; 3 - 1.24 μm; 4 - 1.52 μm; P_{sw}- polarization switching point

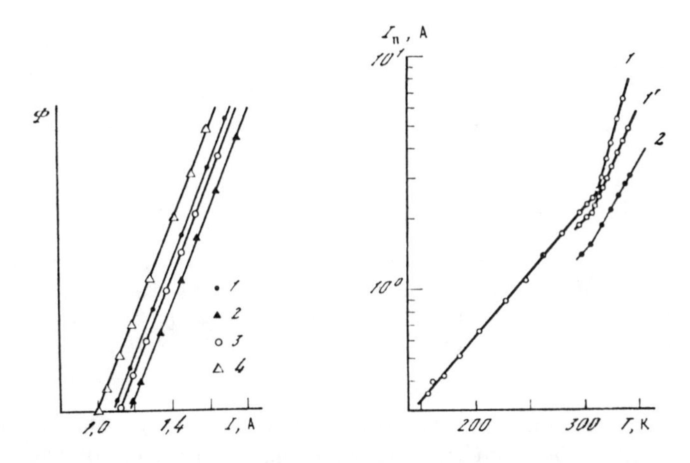

Fig. 9. P-I characteristics of a stripe AlGaAs/GaAs diode heterostructure laser operating at 0.835 μm at 300 K plotted as a function of uniaxial pressure
1-4: P = 21 kg/cm² (1); 300 (2); 610 (3); 1460 (4)

Fig. 10. The temperature dependence of the threshold current in 1.24 μm heterolasers at various uniaxial pressure levels
1, 1' - cross-sectional area of the sample $4 \cdot 10^{-4}$ cm², P = 0 (1) and ~1.6 kbar (1');
2 - 2, $2.5 \cdot 10^{-4}$ cm², P = 2.2 kbar

analysis increase radiation power. With respect to these parameters a total match of the lattices is not optimum and a deviation from this total match to either side in principle will enhance the radiative characteristics. Below we will return to the question of stressed heterostructures with respect to their degradation characteristics, since the advisability of deliberate mismatching of the lattice periods is largely dependent on the influence of stresses on operating conditions.

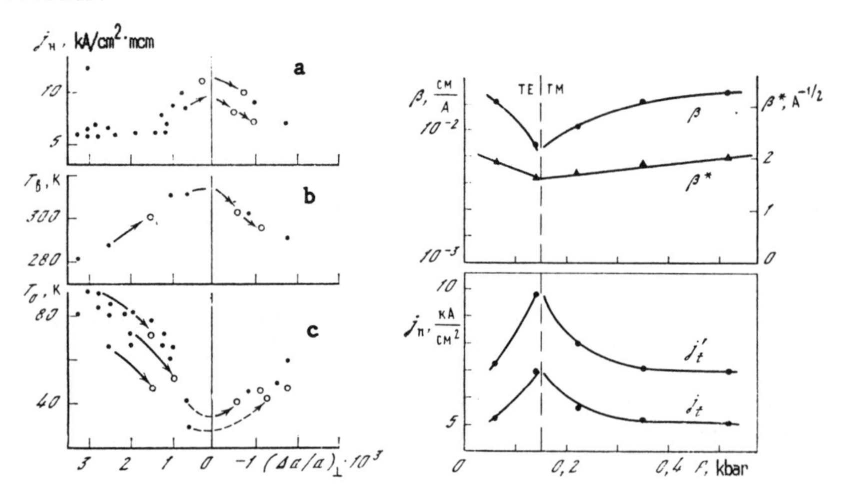

Fig. 11. The dependence of T_b, T_0 and the normalized current density j_N (at 300 K) on the initial lattice mismatch in GaInPAs/InP heterostructures.
the arrows indicate the shift in T_b and T_0 under external uniaxial pressure

Fig. 12. Investigation of the gain piezoeffect by means of immersion in benzene
j and j' are the threshold current density with and without immersion; β and β* are the differential coefficients characterizing the rate of gain increment under pumping (calculated in two hypotheses, see text). Laser diode operating at 1.56 μm; cavity length: 250 μm

Random variations in the internal stresses in the active layer can to one degree or another explain the deviation of the radiative characteristics including the spread of such parameters as the threshold current density, the differential efficiency T_0 and T_b observed from sample to sample in the same way that they explain variations in the polarization properties of laser radiation.

Additional experiments were performed to establish the influence of pressure on optical gain in laser diodes. The threshold current was measured at different pressure levels (0.06-0.52 kbar) together with its change from immersion in benzene (n = 1.467 for 1.5 μm and n = 1.47 for 1.3 μm). Immersion reduces the reflection coefficient of the cavity faces and consequently increases optical losses. The gain increment is easily calculated, $\Delta\alpha = (1/L)[\text{In}(1/R)] = \text{In}(1/R_0)$, where R and R_0 are the reflection coefficients with and without immersion. The ratio of this increment to the threshold current density increment characterizes the differential gain, i.e., the rate of increase in the optical gain factor with pump current density.

The measurements were carried out on a *TE/TM* sample at 300 K operating in a pulsed mode at 1.56 μm (cavity length L = 250 μm, diode area: $5 \cdot 10^{-4}$ cm^2, active layer thickness: 0.4 μm; initial threshold current density: 5.2 kA/cm^2). The greatest rise over threshold with immersion was 38% due to the increase in the calculated optical loss factor at 24.6 cm^{-1}. The measurement results are given in Fig. 12.

The analysis was carried out in two hypotheses: 1) the gain g grows linearly with current density j, i.e., $g = \beta(j - j_0)$, where β is the differential coefficient (the gain slope), j_0 is the current density corresponding to the inversion threshold; 2) $g(j)$ grows as $\beta^* j^{1/2} - g_0$, where β^* is the proportionality factor, j_0 is the adjustable constant. From the threshold condition (with threshold density j_T)

$$g(j_T) = \alpha_0 + (1/L)\text{In}(1/R) \tag{18}$$

for a planar cavity (where α_0 is the internal optical loss factor) we have for the first hypothesis:

$$\beta = (1/L)\text{In}(R20/R)/j_T' - j_T), \tag{19}$$

where j_T' and j_T are the threshold densities with and without immersion; β is measured in cm/A; for the second hypothesis

$$\beta^* = (1/L\text{In}(R_0/R)/(\sqrt{j_T'} - \sqrt{j_T}) \tag{20}$$

where β^* is measured in A$^{-1/2}$.

Moreover, we can easily calculate the parameters

$$\alpha = \beta j + \alpha_0 \text{ (in the first hypothesis)}$$

or

$$\alpha^* = g_0 + \alpha_0 \text{ (in the second hypothesis),}$$

which, unfortunately, are inaccessible to a spectral analysis without additional measurements such as power measurements. Fig. 12 gives calculated curves for β and β^* passing through a minimum at P^*, i.e., at the polarization switching point ($P^* \approx 0.15$ kbar). Qualitatively both hypotheses are consistent with the idea that the differential gain diminishes for the *TE*-mode while the *TM*-modes grow with application of pressure. The average rate $d\beta/dP$ for the *TM*-mode is $1.1 \cdot 10^{-2}$ cm/A$\cdot$kbar ($d\beta^*/dP \approx 1.2$ A$^{-1/2}\cdot$kbar^{-1} for the same mode). β is also measured using a *TE* sample operating at 1.34 μm (cavity length L =

tial threshold current density 5.1 kA/cm²). The average rate $d\beta/dP$ for this *TE*-sample $1\cdot 10^{-2}$ cm/A·kbar, while $d\beta^*/dP \approx 0.9\ A^{-1/2}\cdot kbar^{-1}$.

8. Spectral Piezoeffect

As noted previously anisotropic deformations cause shifts in the spectral bands that are different for different polarizations and, consequently, produce band splitting. This applies to both spontaneous radiation and optical gain; the gain values are more sensitive to deformation than luminescence for the different polarizations, since gain is critically dependent on the position of the Fermi quasi levels with respect to the corresponding energy subbands.

Table 4

Spectral coefficients in GaInPAs/InP lasers under external pressure

Series No.	λ, nm	P^*, kbar	P, kbar	meV/kbar	P, kbar	$K_{\hbar\omega}^{TM}$, meV/kbar
1	1080	—	—	—	0—1.8	—2.8
5	1239	0,7	0—1.2	8.4		
			0—0,8	2.4	0,7—1.6	—0.6
6[1]	1288	—	0—1.5	6.8	—	—
9	1312	—	—	—	0—1.5	—0.3
10	1320	—	0—2.2	7.6	—	—
11[1]	1320	—	0—0.5	1.5	—	—
			0,5—2.1	7.4		
14	1340	—	0—0.8	1.2	—	—
19	1343	—	0—0.7	3	—	—
			0,7—1.2	6.4		
21[1]	1350	—	0—0.5	0.9	—	—
			0,5—1.2	6.1		
23	1410	—	0—0.3	1	—	—
			0,3—1.2	6,5		
24	1560	0.3	0—1.5[1]	0.8	0.4—1.5	0.8[2]
			0,3—1.0	7		
26	1560	0.45	0—0.45	4.0	0.5—1.2	—0.4
			0—0.5	1.2		
27	1597	—	—	—	0—1.5	—0.6
28	1610	0.8	0—0.8	6.2	0.8—1.2	—1.0
			0—0.8	0.8		

[1] Based on the shift of the spontaneous band peak.
[2] Both *TE* and *TM* present at $P > P^*$.

Fig. 13 shows the influence of pressure on the position of the spectral peaks of the *TE*- and *TM*-polarizations in the spontaneous bands. It is clear that the photon energy grows rapidly at the *TE*-peak at the same time that it has a weak dependence on P at the *TM*-peak. The peaks converge in cases 1 and 2 and diverge in case 3. This differential reflects the influence of the initial internal stresses. We therefore cannot predict the relative behavior of bands of different polarization without incorporating the initial stresses. If we plot the photon energy at the band peak as a function of $P - P_{in}$ where P_{in} is the internal equivalent pressure, it is possible to obtain a more ordered picture, as we see in Fig. 14. The shift and splitting of the gain bands were the first spectral piezoeffects in lasers. In addition secondary piezoeffects arise from the change in band population and in the refractive index. The associated change in threshold alters the carrier concentration in the active region (at the same pump current) and causes additional spectral shifts in addition to the primary effect. Therefore measurements of the spectral piezoeffect based on laser peaks produces a large spread of spectral shift coefficients. Table 4 gives a list of such coefficients for lasers.

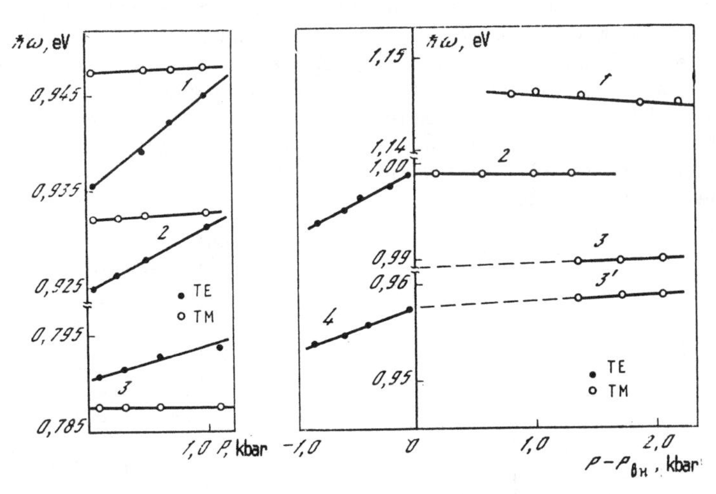

Fig. 13. The spectral peaks of orthogonal polarizations plotted as a function of external pressure in GaInPAs/InP lasers in spontaneous operation at 300 K
1 - Sample from series No. 18, $S = 1 \cdot 10^{-3}$ cm², I = 2 A, $I_n \sim 4.3$ A; 2 - no. 11, $S = 6 \cdot 10^{-4}$ cm², I = 0.6 A, $I_n \sim 2$ A; 3 - no. 28, $S = 7 \cdot 10^{-4}$ cm², I = 1 A, $I_n \sim 2.5$ A

Fig. 14. The spontaneous radiation spectral peaks in GaInPAs/InP lasers plotted as a function of $P-P_{in}$
1 - Sample from series No. 1, I = 0.1 A, $I_n \sim 0.2$ A; 2 - No. 5, $S = 4.5 \cdot 10^{-4}$ cm², I = 0.6 A; 3 = No. 9, $S = 6 \cdot 10^{-4}$ cm², I = 1 A (3'-3 is shifted on the energy scale); 4 - No. 11, $S = 7 \cdot 10^{-4}$ cm², I = 1.05 A

The spectral shift coefficient $K^{TE}_{\hbar\omega}$ is positive everywhere and is between 0.8 and 8.4 meV/kbar at the same time that $K^{TM}_{\hbar\omega}$ is largely negative (aside from case number 24 in which polarization switching at P^* occurred irregularly) and is rather small in absolute value. These data are in qualitative agreement with the behavior of spontaneous radiation. The correlation between $K_{\hbar\omega}$ and K_j can be seen in Fig. 15.

A group of samples with $K_{\hbar\omega}$ between 6 and 8 meV/kbar can be identified. These are the *TE/TM* and *TE* samples with comparatively simple behavior. Below we will discuss in detail the possibility of identifying the corresponding line as band-to-band transitions involving the valence subband V_1 (in our terminology the "semiheavy" subband). Another group has $K_{\hbar\omega}$ between 0.5 and 4 meV/kbar and purportedly corresponds to the participation of the shallow acceptor in the transitions, since the latter is characterized by a weaker energy split. The third group consists of samples with negative values of $K_{\hbar\omega}$ and K_j operating with *TM*-polarization.

In addition to influencing the laser radiation peak pressure will also have some influence on the modal composition of radiation without influencing the change in the relative level over threshold. Fig. 16 gives some picture of the nature of the spectral piezoeffect showing one-frequency spectra with piezotuning and multifrequency spectra at a high current (the relative level over threshold is fixed at two values of 1.05 and 1.20). The most significant influence on the spectral composition has been identified near the point P^* where it is possible to observe one or two groups of longitudinal modes with a large total number of laser modes. Evidently the instability of the lasing conditions of both polarizations due to competition plays a role here.

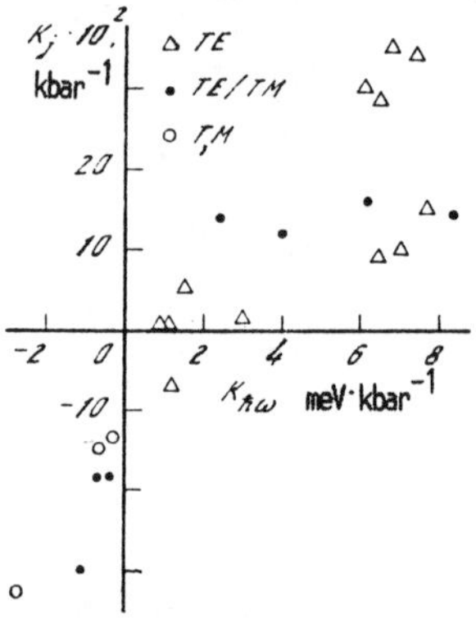

Fig. 15. The interrelationship between the coefficients $K_j(P)$ and $K_{\hbar\omega}(P)$ in GaInPAs/InP diode heterostructure lasers

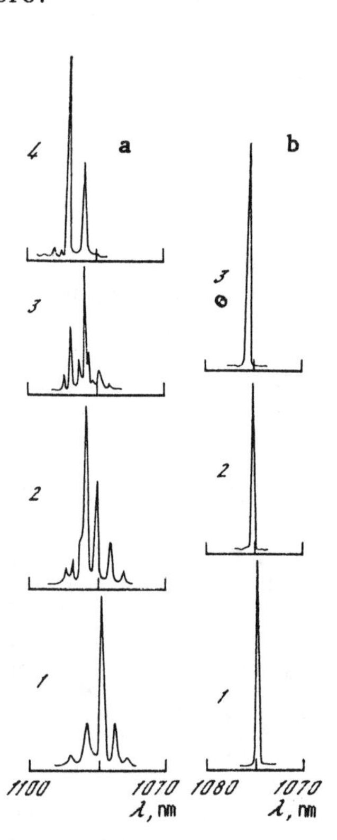

Fig. 16. Lasing spectrum of a GaInPAs/InP diode heterostructure *TM*-laser ($\lambda \sim 1.08$ μm) at 300 K plotted as a function of uniaxial pressure
a - $I = 1.2,\ I_T$; b - $I = 1.05\ I_T$, 1-4: P - 12 kg/cm² (1); 700 (2); 962 (3); 1820 (4)

9. Stressed Heterostructures. Performance Tests

The results from this study indicate that in order to optimize the characteristics of the active medium of lasers it is most advisable to have a certain specific level of elastic deformation of the active layer corresponding to the boundary of the coherency range. In a first approximation the deformation sign is not important. Clearly the discussion surrounds the relative chemical mismatch between the lattice periods on the level of $\pm 10^{-3}$. It is possible to increase the anisotropic deformation by graded or stepped mismatch of the lattice periods between the substrate and the active layer. Another method involves using variable-pressure spring-loaded clamps (this method can be used with *TM* and certain *TE/TM* samples). The magnitude and sign of the stress can be determined by the spontaneous radiation polarization.

The key issue in the suitability of this technique is the problem of the reliability and service life of the stressed heterostructures. We should note that experiments designed to test the service life of GaInPAs/InP lasers carried out in a number of studies [36-38] have revealed a greater service life than traditional AlGaAs/GaAs heterostructure lasers. It is assumed that in GaInPAs/InP lasers there is a different degradation mechanism that is not so closely related to the extending dislocations. After extended operation no abundance of "dark lines" or spots typical of AlGaAs/GaAs degraded lasers were identified. Unlike AlGaAs/GaAs-lasers there are no penetrating dislocations, misfit dislocations lying in the lane of the heteroboundary and no sites for accelerating the degradation process in GaInPAs/InP lasers. High-temperature tests subsequently revealed that at 250°C it is possible to observe the formation of dark spots due to degradation. Experimental results have therefore revealed that GaInPAs/InP lasers are virtually free of the "dark line disease" at normal operating temperatures and are potentially more reliable and the requirements on their initial substrates are not as rigid as in the case of AlGaAs/GaAs-lasers.

In the course of this study we carried out experiments on the service life of GaInPAs/InP stressed structures. We considered the power stability in the initial stage in a 1.3 μm pulsed laser under elevated mechanical loading (up to 2.5 kbar); there was no evidence of direct acceleration of degradation under loading. In an analogous experiment with an AlGaAs-laser [1] accelerated degradation was clearly identified. High-temperature tests were then carried out on rather heavily-stressed structures. The energy gap of the valence band Δ caused by internal stresses was ~10 meV at room temperature. D.c. tests were carried out at a density 1-1.5 kA/cm^2 at as temperature 100°C for 27-30 h. According to [34] the rate of relative power decay

$$v_p = \frac{d}{dt}\ln P = \frac{1}{P}\frac{dP}{dt} \tag{21}$$

can serve as an estimate of the service life

$$\tau = 1/v_p, \tag{22}$$

while τ has the following temperature dependence:

$$\tau(T) = \tau_0 \exp(E_A/kT), \tag{23}$$

where E_A is the activation energy of the degradation process equal to 0.61 eV [34]. An estimate of the extrapolated value of τ at an active region temperature of 25°C using the value of v at 100°C yielded τ = (3-4·10 h. No significant service life differential was observed for structures with a different lattice mismatch sign. There were cases of degradation due to indium solder leakage onto the diode mirror, which were not incorporated. Tests on weakly-stressed structures ($\Delta \approx$ -3-4 meV) yielded identical results. Changes in the threshold and differential efficiency were insignificant, which corresponds to the estimated data obtained on the basis of degradation rate.

Therefore test results on structures with a variable residual stress level revealed that the degradation rate is independent of the relative lattice mismatch within the coherency range. This fact can support the long-term service life of GaInPAs/InP stressed heterostructure lasers.

10. Influence of Active Layer Thickness and the Features of Compound Waveguide Structures

Structures with a three-layer waveguide (TLW) and an ultrathin active layer have the lowest threshold current density among GaInPAs-/InP heterostructures [40]. Fig. 17 shows the dependence of j_p on the active layer thickness in the 30-200 nm range whose lower boundary already corresponds to quantum dimensional structures. An investigation of the optical confinement in such structures and radiation directivity has revealed that a three-layer waveguide produces approximately the same modal profile as an optimum waveguide in a regular heterostructure, i.e., it corresponds to the case of greatest localization of optical flux within the waveguide. As noted in study [41] this corresponds to the greatest differential of the optical confinement parameters for orthogonal polarizations and according to [42] the greatest differential of the end reflection coefficients (in favor of the *TE*-mode). These circumstances, as in the case of regular optimum heterostructures ($d \approx$ 0.1-0.2 μm) represent a certain advantage for generation of the *TE*-mode in lasing with similar, equal conditions. Hence the polarization switching point P^* shifts from the point of total compensation of active layer stresses by a certain amount within the 0.1-0.2 kbar range. The polarization of spontaneous radiation is much less sensitive to waveguiding effects.

Another important fact relates to the influence of the active layer thickness. As noted in extensive research on AlGaAs quantum dimensional heterostructures [43, 44] radiation is heavily polarized in such structures (the *TE*-mode) due to the thin narrow layer and this influences the symmetry of the wave functions in the working quantum states. Calculations in study [44] reveal an analogy to the influence of the tetragonal distortion to a lattice under anisotropic deformation.

An investigation of TLW laser structures in the present study has revealed that at an active layer thickness of less than 0.2 μm, *TE*-polarization predominates. The dependencies of ρ at the band peak on the active layer thickness d are shown in Fig. 18. There are deviations in the 0.3-0.8 μm range resulting from the influence of residual stresses. When $d \leqslant 0.2$ μm the predominance of *TE*-polarization increases and the *TM*-polarization peak vanishes. In the thinner layers (0.05 μm) ρ reaches 0.6. These data reveal that in order to employ the polarization technique of analyzing internal stresses it would be desirable to use samples with a thin active layer (0.3 μm or more). A new factor is added in thin layers in favor of the *TE*-mode although we cannot isolate the contribution of internal stresses. All lasers with an ultrathin active layer ($\leqslant 0.1$ μm) are of the *TE*-type, i.e., they do not switch to the *TM*-mode upon application of pressure. Evidently the waveguide selection and quantum-dimensional factor discussed at the beginning of this section play a role.

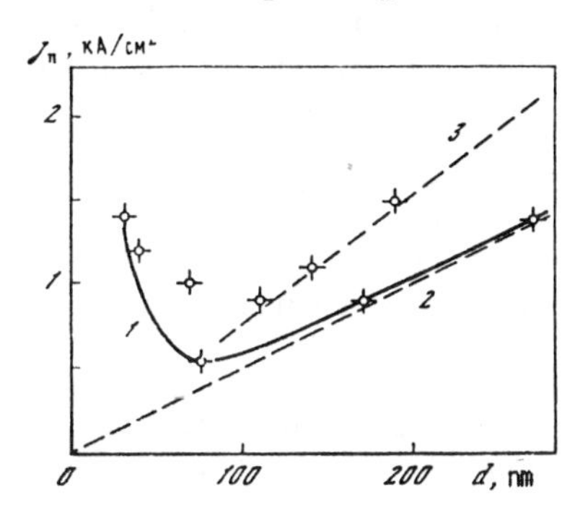

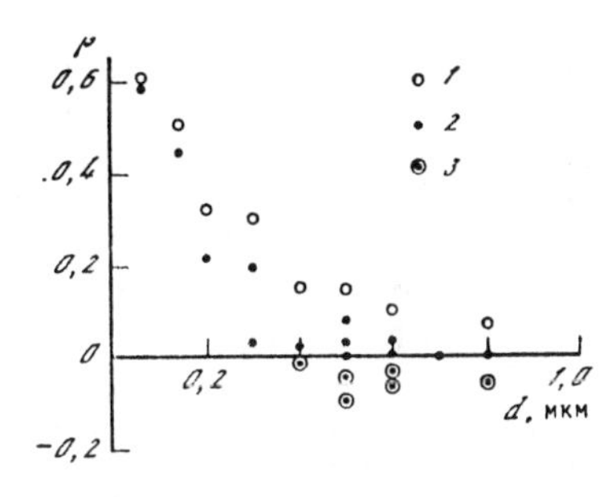

Fig. 17. The threshold current density j_T at 300 K plotted as a function of active layer thickness in GaInPAs/InP TLW diode heterostructure
1 - empirical curves; 2, 3 - linear asymptotes for large values of pd

Fig. 18. The influence of active layer thickness in GaInPAs/InP diode heterostructures and TLW diode heterostructures (1.06-1.6 μm) on the degree of linear polarization
1 - at the *TE*-polarization peak; 2 - at the integral spectral peak; 3 - at the *TM*-polarization peak

The latter, according to direct observations, plays a role at thicknesses of less than 0.05 μm, although these observations refer to the energy spectrum rather than the polarization. It is possible that the polarization perturbation is a stronger effect of the ultrathin layers than observation of the quantum levels, since this perturbation also develops and appears when inhomogeneities in the layer thickness and other spectral blurring factors make it impossible to resolve the quantum levels.

11. Discussion

Certain qualitative aspects of the influence of anisotropic deformations on laser radiation in GaInPAs/InP heterostructures can be explained analogous to the same phenomena in AlGaAs. These include:

- growth of the *TE*-mode threshold and reduction of the *TM*-mode threshold with increasing compression along the normal;

- switching from *TE*- to *TM*-polarization with certain compression (at pressure P^*) where the thresholds of both modes are comparable.

The selection rules for transitions to the valence subbands accounting for their splitting can be used as the basis for explanation. In the case of longwave GaInPAs/InP lasers these regularities appear against a background of a variety of variations in the internal stressed state of the active layer due to the additional chemical degree of freedom in the quaternary compound. Therefore observations of various types of polarization behavior and significant variations of P^* in longwave lasers can also be reduced to the regularities discussed above regarding the influence of anisotropic deformation. The central element is that it is necessary to know the initial stressed state in order to predict piezoeffects.

The following are the qualitatively new regularities characteristic of GaInPAs/InP:

- pressure influences the differential efficiency in lasing; it increases the *TM*-mode efficiency and reduces the *TE*-mode efficiency;

- pressure influences the temperature dependence of the threshold current; in the high-temperature section T_0 grows for the *TM*-mode and diminishes for the *TE*-mode;

- in ultrathin active layers *TE*-polarization predominates in spontaneous radiation ($\rho \approx 0.6$) as well as laser radiation; this predominance is noticeable even at $d \leqslant 0.2$ μm;

- there was no record of mechanical stresses having an influence on gradual degradation of laser diodes (at least for a service life of the order of ~10^4 h based on accelerated tests).

Aside from these results it is also advisable to discuss qualitative data based on measurements of the spectral piezoeffect and estimates of stresses from spectral splitting.

The influence of deformation on the differential laser efficiency indicates either a change in pump efficiency η_p or a change in internal losses. The near-threshold differential efficiency is normally described as

$$\eta = \eta_p [1 + \alpha L/\ln(1/R)]^{-1}, \qquad (24)$$

where α is the internal optical loss factor. A pressure dependence of α is quite likely, at the same time that no assumptions are made regarding the influence of pressure on η_p with any laser radiation polarizations. Assuming in a first approximation that $d\eta_p/dP = 0$, while $d\alpha/dP \neq 0$, we obtain

$$\frac{d\eta}{dP} = \eta_p f^2(\alpha) \frac{L}{\ln(1/R)} \frac{d\alpha}{dP}, \qquad (25)$$

or

$$d(\ln \eta)/dP = -(1 - f)\, d(\ln \alpha)/dP, \qquad (26)$$

where $f = [1 - \alpha L/\mathrm{In}(1/R)]^{-1}$ is the function entering into the expression for η. If $f \approx 0.5$ the relative rate of rise of η is approximately twice the rate of decay of α. This means, for example, that the fastest sections of growth of η in Fig. 8a having $d(\mathrm{Ln}\ \eta)/dP$ on a level of ~ 1 kbar^{-1} corresponds to a rate of $d(\mathrm{In}\alpha)/dP \approx -0.5\ \mathrm{kbar}^{-1}$ and when $\alpha \approx 30\ \mathrm{cm}^{-1}$ we have an estimate of $d\alpha/dP \approx -15\ \mathrm{cm}^{-1}\cdot\mathrm{kbar}^{-1}$. We note that this derivative will be positive for *TE*-modes and negative for *TE*-modes.

This hypothesis is attractive since an internal optical loss mechanism has been postulated for GaInPAs/Inp lasers; these losses are due to absorption at the transitions between the valence subbands [45]. There are experimental results confirming the existence of such losses dependent on the hole concentration. They are also temperature-dependent, since for transitions from band V_3 to the heavy hole subband (Fig. 19) free states are required on the upper levels (i.e., heavy holes). These states are shifted somewhat from the point $k = 0$ due to the law of conservation of energy $\hbar\omega = E_{V_1} - E_{V_3}$ and therefore the free state concentration is sensitive to temperature and increases with its growth.

If we take α_V to be the absorption coefficient at the intravalence transitions dependent on P it is therefore also possible to explain the change in the differential efficiency and T_0. We must therefore discuss the reasons behind the pressure dependence of α_V.

The primary reason may be that anisotropic deformation which alters the relative energy position of the valence subbands as well as their occupation by holes and the effective masses changes the phase volume in the k-space where conditions are satisfied for transitions by absorption of laser radiation quanta $\hbar\omega$. An analogous effect has been reported with respect to the influence of hydrostatic pressure [46]. Since the V_1 and V_3 subbands have a different symmetry we can expect that the absorption piezoeffect will be sensitive to radiation polarization (analogous to the piezoeffect at the band-to-band transitions).

Transitions from the V_3 subband to the heavy hole band also occur as a result of Auger recombination whose rate is temperature-dependent. In a first approximation nonradiative recombination has no influence on the differential efficiency although it can make a contribution to changing the gain and the quantity T_0.

The secondary effect is that deformation alters the threshold current, as established in experiment, and consequently influences the hole concentration at threshold. This in turn means that the intravalence absorption changes together with the factor α and η. Growth of the threshold will be accompanied by a decrease in η and vice versa as observed in experiment. This secondary cause qualitatively provides a complete explanation for experimental data and the question remains whether or not it is sufficient in the quantitative respect. Typical rates of change of the threshold with pressure amount to (see Fig. 8b) approximately $-0.4\ \mathrm{kbar}^{-1}$ which, with a threshold carrier concentration $2\cdot 10^{18}\ \mathrm{cm}^{-3}$ will correspond to a change in concentration of $8\cdot 10^{17}\ \mathrm{cm}^{-3}$. In order to explain the value of $15\ \mathrm{cm}^{-1}\cdot\mathrm{kbar}^{-1}$ obtained above we must expect an optical absorption cross-section of $\sim 2\cdot 10^{-17}\ \mathrm{cm}^2$.

For comparison in InP the intravalence absorption dependent on the hole concentration N_h is [47]

$$\alpha = 14(N_h\cdot 10^{-18}\ \mathrm{cm}^{-3})\ \mathrm{cm}^{-1} \tag{27}$$

at 1.3 μm and 297 K, corresponding to a cross-section of $1.4\cdot 10^{-17}\ \mathrm{cm}^2$ and close to within an order of magnitude of our estimate. The theoretical estimate of the intravalence absorption cross-section in InP at the band-to-band transition wavelength is $\sim 0.7\cdot 10^{-17}\ \mathrm{cm}^2$. If we remember that the energy gap between the band-to-band transition and the intravalence absorption peak at 1.3 μm is reduced by ~560 meV a growth of the calculated cross-section by a factor of 3-5 is quite likely. With these estimates we then have a quantitative explanations of variations in η for the secondary cause discussed above (i.e., due to changes in the hole concentration from a change in the threshold current). Overall the problem is self-consistent, i.e., the increase in threshold is caused by both a change in differential gain and a change in the losses dependent on the threshold increment.

It is worth noting that the primary piezoeffect is sufficient quantitatively to explain the change in the quantity α_V as well as to significantly alter the Auger recombination rate involving the subband V_3. For example, in study [12] we carried out an illustrative calculation of the influence of the change in effective mass of the heavy holes m_{hh} on the intravalence absorption between the nondegenerate subbands V_1 and V_3. The energy of heavy holes involved in the transitions with an energy $\hbar\omega$ is equal to

$$E = (\hbar\omega - \Delta_{so})\, m_{hs} / (m_{hh} - m_{hs}), \tag{28}$$

where Δ_{so} is the spin-orbit splitting energy; m_{hs} is the effective mass in the V_3 subband. For a laser operating at 1.3 μm we take Δ_{so} = 0.27 eV, m_{hh} = 0.45 m_0, m_{hs} = 0.14 m_0, and $\hbar\omega$ = 0.95 eV. Therefore a reduction of m_{hh} by 10% produces an increase in E by $2kT$ and a reduction in the rate of absorptive transitions by a factor of $e^2 \approx 7.3$. The actual change in effective masses has not yet been investigated in quaternary systems, although significant changes have been observed in other semiconductors (for example, in GaAs under compression along the [100] direction when $\Delta \gg kT$ the mass m_{hh} was ~0.2 m_0 at the same time that in undeformed material it is ~0.5 m_0). Therefore the primary absorption piezoeffect can be significant at least for the 1.3-1.6 μm range.

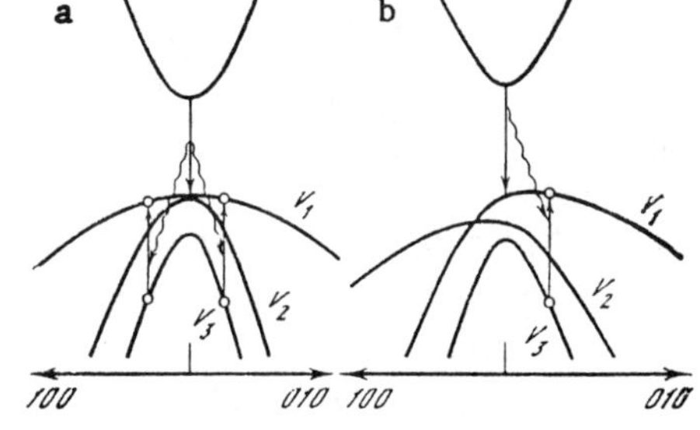

Fig. 19. Configuration of the band structure demonstrating optical transitions with band-to-band radiation and intraband absorption at the same frequency
a - zero deformation; b - anisotropic deformation (uniaxial compression)

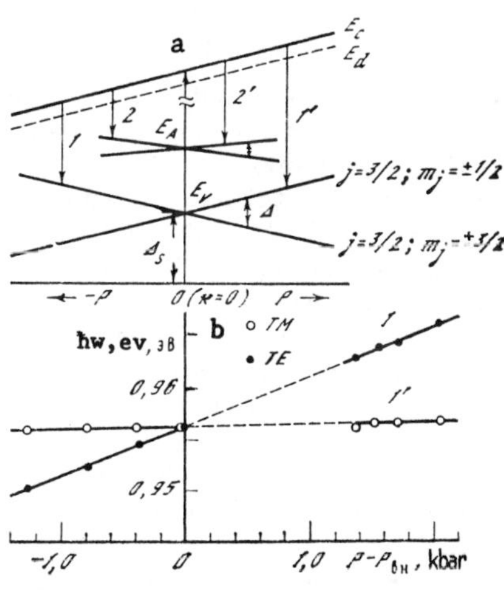

Fig. 20. Diagram of the dependence of the relative energy position of the band extrema on relative deformation (a) and the spontaneous radiation band peak in a GaInPAs/InP diode heterostructure (series No. 11, 9) at 300 K plotted as a function of $P - P_{IN}$ (b)

Spectral piezoeffects have been investigated previously in the photoluminescence of GaAs [48] and other materials. The technique for analyzing the spectral shift of the lines remains standard. Fig. 20a gives a diagram corresponding to the dependence of the relative energy position of the band maxima on relative deformation. It is clear that in the range of negative values of P (i.e., with distension along the normal) application of pressure will increase the photon energy of the band-to-band transition 1 and the conduction band/acceptor transition 2. This corresponds to the valence subband with $m_j = \pm 3/2$ and

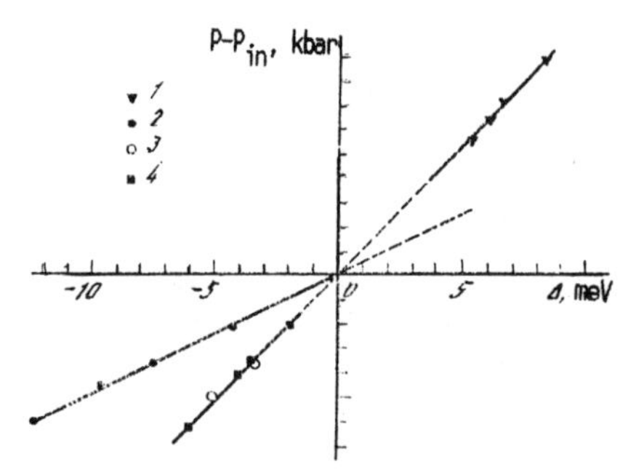

Fig. 21. The energy separation of the valence subbands Δ plotted as a function of $P - P_{IN}$ in a GaInPAs/InP diode heterostructure
Series No. 9; 2 - No. 18; 3 - No. 4; 4 - No. 11

radiation with preferential *TE* polarization. With positive values of *P* the change in the energy of band-to-band transition 1 is much less and it is clear that the difference in the spectral shift coefficients yields the quantity $d\Delta/dP$, where Δ is the splitting (separation) of the top of the valence band. Analogously for impurity transition 2 the difference of the shift coefficients yields $d\Delta 2A/dP$, where Δ_A is the splitting (separation) of the acceptor level. Fig. 20b gives an illustration of such changes where the quantity $P - P_{IN}$ is plotted on the horizontal axis with the origin corresponding to deformation compensation, while the photon energy at the peak of the spontaneous band is plotted along the vertical axis. According to [48] the pressure dependencies of the energy gaps at $k = 0$ for transitions 1 and 1' take the form

$$\delta E(P) = a(S_{11} + 2S_{12})P + b(S_{11} - S_{12})P, \tag{29}$$

$$\delta E'(P) = a(S_{11} + 2S_{12})P - b(S_{11} - S_{12})P[1 + \Delta_s^{-1} b \times (S_{11} - S_{12})P], \tag{30}$$

where *a* is the hydrostatic deformation potential constant; *b* is the shear constant of the deformation potential; S_{11} and S_{12} are the elasticity constants; Δ_S is the energy distance to subband V_3. The interrelationship between Δ and $P - P_{IN}$ is given in Fig. 21 for material with $\lambda = 1.3$ μm. We see that it is largely linear although it has at least two slopes differing by a factor of approximately two. We will interpret this as the appearance of two types of radiative transitions: band-to-band transitions (1 and 1') and impurity transitions (2 and 2') where a small coefficient $d\Delta/dP$ is characteristic of the impurity transitions.

If we ignore the quadratic term in the formula for $\delta E'(P)$ (corresponding to a linear dependence of Δ on *P*), we obtain

$$\Delta = \delta E + \delta E' = 2a(S_{11} + 2S_{12})P \tag{31}$$

and

$$d\Delta/dP = 2a(S_{11} + 2S_{12}). \tag{32}$$

We use the values given in study [18] for S_{11} and S_{12} and their dependence on the composition of the quaternary material was ignored. The derived value of the coefficient *a* for the band-to-band transitions in GaInPAs at 1.3 μm was 10.6 ±0.5 eV. An analogous coefficient for the transitions from the acceptor level was negative (5.1 ±0.5) eV. A recalculation of the data given in study [49] for InP produces a value a = -9.3 eV which is quite close to the value for the composition at 1.3 μm.

We anticipate that measurements of the spectral piezoeffect will be widely used to identify the radiative mechanism behind laser operation (band-to-band or impurity). In our experiments we had cases where the quantity $d\Delta/dP$ was conserved throughout the entire spontaneous radiation band as well as cases where the longwave part of the band was characterized by a near half value of $d\Delta/dP$. The latter case can be interpreted as the result of the overlap of the band-to-band and impurity bands whose width is comparable to or greater than the energy distance between them (the depth of the acceptor level). The significant spread in the values of $k_{\hbar\omega}$ fundamentally related to $d\Delta/dP$ is due to the influence of the composition (or the radiation wavelength) as shown in Fig. 15 as well as variations in band occupation levels causing a shift in the spectral radiation bands (including alteration of the predominant radiation mechanism). The occupation of the bands associated with the position of both quasi-Fermi levels confirms the influence of changes in threshold current and the effective masses of the density of states under the influence of deformation. A detailed description of the secondary effects is not possible due to their irregularities.

12. Conclusion

The use of the polarization technique [18] to determine internal stresses in the active region of GaInPAs/InP injection lasers has made it possible to investigate the interrelationship between stresses and radiative characteristics, and the application of external uniaxial compression has made it possible to expand this picture. In addition to analyzing data on the polarization splitting of the spectra we focus on the differential of the shift coefficients for the band-to-band and impurity bands which can be used to identify the radiative mechanism.

It has been established that composition along the normal aids in the generation of *TM*-modes (it reduces the threshold, increases differential efficiency and increases T_0) while distension generates *TE*-modes. In samples with a thin active layer a strong predominance of *TE*-polarization in both the laser and spontaneous spectrum is observed. Samples with a thin active layer of 0.3-0.8 μm are divided into *TE, TE/TM,* and *TM* groups.

Evidently in these cases the initial polarization is determined by the internal stresses caused by deviations from a strict match of the lattice periods in the heterojunctions. These stresses represent a significant cause of the spread of other radiative characteristics, including the temperature dependence of the threshold etc.

Preliminary experiments on the influence of moderate pressure (up to 2 kbar) on laser degradation have revealed that no noticeable acceleration of degradation on a service life scale of $(3-4)\cdot 10^4$ hours is observed and, consequently, the stressed state of the laser heterostructures can be selected to obtain optimum laser characteristics. The advisability of using stressed heterostructures with, for example, a given mismatch of the lattice period within the coherency range of the heteroboundaries is substantiated. We can estimate based on piezoeffect measurements that such optimization compared to total compensation of stresses reduces the threshold by 30-40%, increases differential efficiency by a factor of 1.5-2 and increases T_0 (in approximately equal conditions).

The uniaxial pressure technique makes it possible to obtain spectral tuning of laser radiation comparatively easily. This applies to the spectral envelope and occasionally to the individual mode in single-frequency lasing. In the general case deformation will cause a substantial change in the modal composition of radiation and will, for example, produce multimode lasing near the polarization instability point ($P \gtrsim P^*$). This same technique makes it possible to control polarization, i.e., obtain switching of the primary polarization type or to stabilize *TM*-polarization (in *TE/TM* samples). Further investigations of the nature of radiation polarization are required to differentiate structures with ultrathin active layers (to the advantage of the *TE*-mode). It has been suggested that the advantage of *TE*-polarization in the case of thin and ultrathin radiating layers results from the influence of thickness on the symmetry of the wave functions (the polarization-dimensional effect), i.e., an action of the same nature as the quantum dimensional effects in the spectrum yet much more visible than spectral effects. Such an effect can be expected when the de Broglie wavelength of the electrons ($\sim 10^5$ cm) is comparable to the thickness of the radiating layer and when the spectral broadening does not make it possible to observe discrete quantization lines of the energy spectrum.

References

1. Eleseev, P.G., Khaydarov, A.V. "The role of mechanical stresses in the gradual degradation of light emitting diodes and injection lasers" KVANTOVAYA ELEKTRON., 1975, vol. 2, no. 1, pp. 127-129.

2. Ripper, J.E., Pratt, G.W., Whitney, C.G. "Direct frequency modulation of a semiconductor laser by ultrasonic waves" IEEE J. QUANT. ELECTRON., 1966, vol. 2, no. 9, pp. 603-605.

3. Ripper, J.E. "Analysis of frequency modulation of junction lasers by ultrasonic waves" IEEE J. QUANT. ELECTRON., 1970, vol. 6, no. 2, pp. 129-132.

4. Gulyaev, Yu.V., Shkerdin, G.N. "Acoustic-wave-generated dis-

tributed feedback injection laser" FTP, 1975, vol. 9, no. 7, pp. 1434-1436.

5. Yamanashi, M., Ameda, M., Ishii, K., et al. "Optically pumped GaAs lasers with acoustic distributed feedback" APPL. PHYS. LETT., 1973, vol. 33, no. 3, pp. 251-253.

6. Craft, D.C., Dutta, N.K., Wagner, W.R. "Anomalous polarization characteristics of 1.3 μm InGaAsP buried heterostructure lasers" APPL. PHYS. LETT., 1984, vol. 44, no. 9, pp. 823-825.

7. Dutta, N.K., Craft, D.C. "Effects of stress on the polarization of stimulated emission from injection lasers" J. APPL. PHYS., 1984, vol. 56, no. 1, pp. 65-70.

8. Suemune, L., Yamanishi, M., Mikoshiba, N., et al. "Observation of acoustic signals from semiconductor laser" JAP. J. APPL. PHYS., 1981, vol. 20, no. 1, pp. 9-12.

9. Suemune, L., Nonomura, K., Yamanishi, M., et al. "Generation mechanism of current-injection induced acoustic (CIA) signals in semiconductor lasers" JAP. J. APPL. PHYS., 1982, vol. 21, no. 3, pp. 110-112.

10. Vu Van Lyk, Eleseev, P.G., Man'ko, M.A., et al. "The influence of a nonstationary temperature waveguide on the radiation characteristics of heterolasers" In: Tez. dokl. III Vsesoyuz. konf. po fiz. protsessam v poluprovodnikovykh geterostrukturakh" [Topic Papers of the Third All Union Conference on Physical Processes in Semiconductor Heterostructures] Odessa, 1982, vol. 2, pp. 64-65.

11. Eleseev, P.G., Sverdlov, B.N., Shokhudzhaev, N. "Reduction of the threshold current of GaInAsP/InP heterolasers under uniaxial compression" KVANTOVAYA ELEKTRON., 1984, vol. 11, no. 8, pp. 1665-1667.

12. Eleseev, P.G., Sverdlov, B.N., Shokhudzhaev, N. "The influence of anisotropic deformation on the radiative characteristics of GaIaAsP/InP lasers" Preprint. FIAN, no. 197, Moscow, 1984.

13. Dolginov, L.M., Eliseev, P.G., Mil'vidskiy, M.G. "Multicomponent semiconductor solid solutions and their application in lasers" KVANTOVAYA ELEKTRON., 1976, vol. 3, no. 7, pp. 1381-1393.

14. Olsen, G.Kh., Ettenberg, M. "The features of fabricating A^3B^5 type heteroepitaxial structures" In: Rost kristallov [Crystal growing] Moscow: Mir, 1981, no. 2, pp. 9-76.

15. Oe, K., Shinoda, Y., Sugiyama, K. "Lattice deformations and misfit dislocations in GaInPAs/InP DH layers" APPL. PHYS. LETT., 1978, vol. 33, no. 11, pp. 962-964.

tributed feedback injection laser" FTP, 1975, vol. 9, no. 7, pp. 1434-1436.

5. Yamanashi, M., Ameda, M., Ishii, K., et al. "Optically pumped GaAs lasers with acoustic distributed feedback" APPL. PHYS. LETT., 1973, vol. 33, no. 3, pp. 251-253.

6. Craft, D.C., Dutta, N.K., Wagner, W.R. "Anomalous polarization characteristics of 1.3 μm InGaAsP buried heterostructure lasers" APPL. PHYS. LETT., 1984, vol. 44, no. 9, pp. 823-825.

7. Dutta, N.K., Craft, D.C. "Effects of stress on the polarization of stimulated emission from injection lasers" J. APPL. PHYS., 1984, vol. 56, no. 1, pp. 65-70.

8. Suemune, L., Yamanishi, M., Mikoshiba, N., et al. "Observation of acoustic signals from semiconductor laser" JAP. J. APPL. PHYS., 1981, vol. 20, no. 1, pp. 9-12.

9. Suemune, L., Nonomura, K., Yamanishi, M., et al. "Generation mechanism of current-injection induced acoustic (CIA) signals in semiconductor lasers" JAP. J. APPL. PHYS., 1982, vol. 21, no. 3, pp. 110-112.

10. By Ban Luk, Eleseev, P.G., Man'ko, M.A., et al. "The influence of a nonstationary temperature waveguide on the radiation characteristics of heterolasers" In: Tez. dokl. III Vsesoyuz. konf. po fiz. protsessam v poluprovodnikovykh geterostrukturakh" [Topic Papers of the Third All Union Conference on Physical Processes in Semiconductor Heterostructures] Odessa, 1982, vol. 2, pp. 64-65.

11. Eleseev, P.G., Sverdlov, B.N., Shokhudzhaev, N. "Reduction of the threshold current of GaInAsP/InP heterolasers under uniaxial compression" KVANTOVAYA ELEKTRON., 1984, vol. 11, no. 8, pp. 1665-1667.

12. Eleseev, P.G., Sverdlov, B.N., Shokhudzhaev, N. "The influence of anisotropic deformation on the radiative characteristics of GaIaAsP/InP lasers" Preprint. FIAN, no. 197, Moscow, 1984.

13. Dolginov, L.M., Eliseev, P.G., Mil'vidskiy, M.G. "Multicomponent semiconductor solid solutions and their application in lasers" KVANTOVAYA ELEKTRON., 1976, vol. 3, no. 7, pp. 1381-1393.

14. Olsen, G.Kh., Ettenberg, M. "The features of fabricating A^3B^5 type heteroepitaxial structures" In: Rost kristallov [Crystal growing] Moscow: Mir, 1981, no. 2, pp. 9-76.

15. Oe, K., Shinoda, Y., Sugiyama, K. "Lattice deformations and misfit dislocations in GaInPAs/InP DH layers" APPL. PHYS. LETT., 1978, vol. 33, no. 11, pp. 962-964.

28. Ripper, J.E., Patel, N.B., Brosson, P. "Effect on uniaxial pressure on the threshold current of double-heterostructure GaAs lasers" APPL. PHYS. LETT., 1972, vol. 21, pp. 124-125.

29. Eleseev, P.G., Krasil'nikov, A.N., Khaydarov, A.V., Kharisov, G.G. "Controlling radiation polarization from a heterolaser by means of uniaxial compression" KVANTOVAYA ELEKTRON., 1974, vol. 1, pp. 196-197.

30. Patel, N.B., Ripper, J.E., Brosson, P. "Behavior of threshold current and polarization of stimulated emission of GaAs injection lasers under uniaxial stress" IEEE J. QUANT. ELECTRON., 1973, vol. 9, no. 2, pp. 338-341.

31. Patel, N.B., Morosini, M.B.Z., Serra, T.J., Ripper, J.R. "Pressure induced polarization ellipticity in injection lasers" REV. BRASIL. FIS., 1982, vol. 12, no. 1, pp. 51-59.

32. Akhmetov, D., Bezhan, N.P., Bert, N.A., et al. "The influence of internal deformations on radiation polarization in InP-InGaAsP heterolaser structures" PIS'MA V ZhTF, 1980, vol. 6, no. 12, pp. 705-706.

33. Gorelenok, A.G., Tarasov, I.S., Usikov, A.S. "Detection of electroluminescence polarization ($\lambda \sim 1.3\ \mu m$) in InGaAsP/InP heterostructures generated by external deformations" PIS'MA V ZhTF, 1981, vol. 7, no. 6, pp. 453-456.

34. Bogatov, A.P., Dolginov, L.M., Druzhinina, L.V., et al. "The radiative characteristics of stripe geometry symmetric heterostructure lasers in CW and pulsed operation at 300 K" FTP, 1972, vol. 6, no. 1, pp. 43-48.

35. Averkiev, I.S., Gorelenok, A.G., Tarasov, I.S. "The polarization features and deformation potential constant in n- and p-type InP" FTP, 1983, vol. 17, no. 6, pp. 997-102.

36. Yamamoto, T., Sakai, K., Akiba, S. "10,000-h CW operation of InGaAsP/InP DH lasers at room temperature" IEEE J. QUANT. ELECTRON., 1979, vol. 15, no. 8, pp. 684-687.

37. Horikoshi, Y., Kobayshi, T., Furukawa, Y. "Lifetime of InGaAsP-InP and AlGaAs-GaAs lasers estimated by the point defect generation model" JAP. J. APPL. PHYS., 1979, vol. 18, pp. 2237-2244.

38. Begotosnyy, V.P., Duraev, Eliseev, P.G., et al. "Service life characteristics of InGaAsP/InP heterostructures" KVANTOVAYA ELEKTRON., 1981, vol. 8, no. 9, pp. 1985-1987.

39. Fukuda, M., Wakita, K., Iwane, G. "Observation of dark defects related to degradation in InGaAsP/InP DH lasers under accelerated operation" JAP. J. APPL. PHYS., 1981, vol. 20, no. 2, pp. L87-L90.

40. Vasil'ev, M.G., Dolginov, L.M., Drakin, A.E., et al. "InGaAsP/InP three-layer waveguide injection lasers" KVANTOVAYA ELEKTRON., 1984, vol. 11, no. 3, pp. 631-633.

41. Eliseev, P.G., "Heterojunction injection lasers" KVANTOVAYA ELEKTRON., 1972, no. 6(12), pp. 3-28.

42. Ikegami, T. "Reflectivity of mode at facet and oscillation mode in double-heterostructure injection lasers" IEEE J. QUANT. ELECTRON., 1972, vol. 8, pp. 470-476.

43. Kobayashi, H., Iwamura, H., Saku, T., Ostuka, K. "Polarization-dependent gain-current relationship in GaAs-AlGaAs MQM laser diodes" ELECTRON. LETT., 1983, vol. 19, no. 5, pp. 166-168.

44. Asada, M., Kameyama, A., Suematsu, Y. "Gain and intervalence band absorption in quantum-well lasers" IEEE J. QUANT. ELECTRON., 1984, vol. 20, no. 7, pp. 745-753.

45. Adams, A.R., Asada, M., Suematsu, Y., Arai, S. "Temperature dependence of the efficiency and threshold current of InGaAsP lasers related to intervalance band absorption" JAP. J. APPL. PHYS., 1980, vol. 19, no. 10, pp. L621-624.

46. Adams, A.R., Patel, D., Greene, P.D., Hensball, G.D. "Influence of pressure on temperature sensitivity of GaInAsP lasers" ELECTRON. LETT., 1982, vol. 18, no. 21, pp. 919-920.

47. Casey, H.C., Carter, P.L. "Variation of intervalence band absorption with hole concentration in *p*-type InP" APPL. PHYS. LETT., 1984, vol. 44, no. 1, pp. 82-83.

48. Bhargave, R.N., Nathan, M.I. "Stress dependence of photoluminescence in GaAs" PHYS. REV., 1967, vol. 161, no. 3, pp. 695-698.

49. Pankov, Zh. "Opticheskie protsessy v poluprovodnikakh" [Optical processes in semiconductors] Moscow: Mir, 1973, 33 pp.

RADIATIVE CHARACTERISTICS OF NONSTOICHIOMETRIC MELT-GROWN AlGaAs/GaAs LASER HETEROSTRUCTURES

P.G. Eliseev, A.A. Zherdev, V.S. Kargapol'tsev, O.N. Talenskiy, G.G. Kharisov

ABSTRACT

Descriptions of different heterostructures for laser diodes are examined. The requirement for growing heterostructures from a thin nonstoichiometric melt layer (~0.05 cm) at low cooling rates is substantiated. It is demonstrated that the threshold lasing current density is independent of the doping element (Sn or Te for *n*-emitters and Zn or Ge for *p*-emitters). The emitter thickness also has a weak influence on this level. Adding aluminum to GaAs shifts the lasing wavelength to 670-690 nm for cooled lasers and to 750 nm for room-temperature lasers. A CW lasing power up to 220 mW (cooled diodes) and up to 65 mW (room temperature) were obtained.

Introduction

The development of highly-efficient injection lasers with broad selection of operating conditions and wavelengths in recent years has led to a significant increase in research and development of heterostructures and other laser structures used in fabricating lasers for a wide range of practical applications. Promising directions involving the fabrication and use of higher power laser radiation can be identified in the applied fields. These include:

1) navigation applications such as laser aircraft navigation and landing systems, maritime navigation systems, etc.;

2) optical data recording on audio and video disks;

3) alphanumeric laser printers;

4) technological applications in electronics and other industrial fields;

5) laser medicine;

6) scientific research equipment, optical frequency standards, photodetector certification, etc.

Visible range lasers are desirable or necessary for most applications (or lasers operating at the shortest wavelengths in the accessible range). The radiation frequency is shifted to the desired range by introducing the necessary component into the semiconductor materials. For gallium arsenide lasers this component is aluminum; adding aluminum shifts radiation to the shortwave frequencies and to the red band in the visible spectrum. In this case the aluminum limit is 37%. The transformation to indirect optical transitions or higher aluminum concentrations significantly degrades laser parameters and causes a sharp jump in the threshold pump current density, in particular. Some additional improvement in laser performance characteristics can be achieved by proper selection of the laser diode geometry: contact width and cavity length. The simplest method of increasing radiated power is expanding the active region. However this leads to degradation of the heat entrainment conditions and increases the threshold currents and also results in multimode operating conditions. Therefore lasers are fabricated for specific operating applications solely by optimizing parameters for specific requirements. Below we will attempt to demonstrate the techniques used to achieve this goal.

One direction based on the application of cooled injection lasers for CW and pulsed operation at a 1 w and higher level has been ongoing at the Lebedev Physics Institute of the Academy of Sciences of the USSR since 1962 [1-4]. This level was first achieved in the IR (0.85-0.87 μm) employing heavily doped laser homostructures fabricated by means of liquid-phase epitaxy (LPE) from a nonstoichiometric melt (1966). Low-threshold heterostructures were subsequently developed to shift the frequency to the visible range. Multiple stripe lasers with ultrathin active layers (quantum dimensional structures) producing CW at room temperature have reached this level most recently. The attempt to use compact, economical semiconductor lasers in acoustic and video disc systems has led to the development of single-mode lasers in the 0.75-0.80 μm range.

Below we describe experiments in the fabrication of heterostructures of different geometries; data are given on the shift of radiation to shorter wavelengths by varying the aluminum impurity; data are also provided from measurements of laser characteristics, primarily radiation power and threshold currents. The multilayered heterostructures were largely fabricated by LPE from nonstoichiometric melts.

1. Heterostructure Types

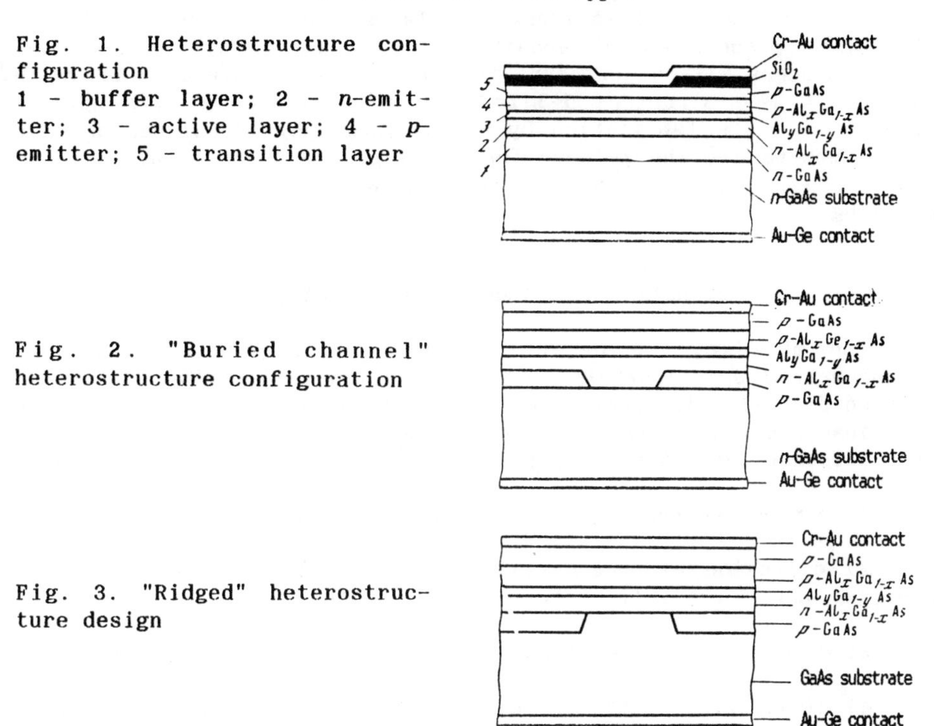

Fig. 1. Heterostructure configuration
1 - buffer layer; 2 - *n*-emitter; 3 - active layer; 4 - *p*-emitter; 5 - transition layer

Fig. 2. "Buried channel" heterostructure configuration

Fig. 3. "Ridged" heterostructure design

Figs. 1-3 show the three types of heterostructures employed in the present study. The heterostructures in Fig. 1 is well-known in the international literature. It was one of the first structures proposed for implementing a narrow stripe contact. Here the contact width is determined by the width of the slots etched by photolithography in the silicon oxide layer deposited onto the upper layer of the heterostructure, and amounts to 15-100 μm. The current is applied through the Cr-Au metallic layer. The width of the radiating region in this structure is determined by the slot width plus the current spreading due to diffusion in the segment between the contact layer and the $Al_yGa_{1-y}As$ active region.

The next structure (see Fig. 2) is called a "buried channel" structure. A *p*-type layer is deposited onto an *n*-type substrate. In this case slots 3-10 μm in width and a depth extending to the substrate are exposed by photolithography. An *n*-type layer is deposited on top to fill in the slots, and the regular sequence of layers continues. A *p*-*n*-junction is thereby formed in the heterostructure; this junction is off with respect to the current flowing from the substrate into the structure and only the slot leaves a path along which current can flow. A drawback of this structure is that after deposition of the first *p*-type layer the substrate containing the layer is exposed and undergoes photolithographic treatment. We developed a similar heterostructure where photolithography is used on the substrate in the fabrication process while deposition of the heterolayers is a continuous process without exposure to air. This structure is shown in Fig. 3. A mesa-structure 5-3 μm in width and 3-5 μm in height is formed during its fabrication by photolithography techniques onto an *n*-type substrate. The *p*-layer forming a *p-n* junction with the mesa-structure is then deposited with the regular sequence of heterolayers following.

2. Heterolaser Fabrication Technique

The multilayered heterostructures were grown from a nonstoichiometric solution/melt in which the crystallization conditions are significantly different from the crystallization conditions from a stoichiometric melt. It is normally assumed that the following factors influence the LPE process [5]:

1) convective transport of the dissolved material;

2) diffusion of the dissolved material to the substrate/melt interface and of the solvent from the interface;

3) deposition of the dissolved material onto the substrate;

4) dissipation of the crystallization heat.

In comparing crystallization from a nonstoichiometric solution/melt and from a stoichiometric solution we can easily see that in the first case the crystallization process is determined by the first three factors, while in the second case it is determined by the fourth factor only. Since the linear rate of growth from a stoichiometric melt is of the order of a few centimeters per hour while the growth rate from a nonstoichiometric solution/melt normally is less than 10^{-2} cm/h we can easily calculate that the quantity of heat expended in crystallization from a nonstoichiometric melt/solution is two orders of magnitude less and this factor can be ignored.

The slow crystallization rate is a determining factor in selecting the heterostructure growing technique. Very thin layers of a few millimeters or less are required for injection lasers. Reproducible fabrication of such layers requires a slow crystallization rate and makes it necessary to use a nonstoichiometric solution/melt. However certain difficulties derive from this method. Concentration supercooling has a significant influence on the first three factors controlling the LPE process from a nonstoichiometric solution/melt; concentration supercooling causes spatial instability at the substrate/-melt interface, produces additional crystallization centers, etc. A number of studies [6-9] have considered the problem of concentration supercooling and have proposed a variety of methods for its elimination: the development of significant temperature gradients at the substrate/melt interface; low solution/melt cooling rates; growing in a pre-supercooled solution/melt and growing from a thin solution/melt. In modern LPE technology there are two primary methods of eliminating concentration supercooling: growing from a thin solution/melt layer and low cooling rates.

A variety of cassette designs limiting the solution/melt thickness have been developed for fabricating a thin solution/melt layer [10-11]. We employed a cassette that makes it possible to fabricate a solution/melt layer of 0.05 cm at a cooling rate of the order of 0.3 deg/min which eliminated concentration supercooling and made it possible to achieve good morphology of the grown epitaxial layers. *n*-type gallium arsenide with a carrier concentration $N \approx (2\text{-}3) \cdot 10^{18}$ cm^{-3} was used in all cases as the substrate. A gallium-based solution/melt was used in the growing process.

Tin or tellurium doping were used to establish the *n*-layers. A series of 15 experiments involving tellurium doping and 19 experiments using tin doping were analyzed by dispersion analysis techniques; this analysis revealed that these series are statistically (level of significance: 95%) nondifferentiable. Data on the current threshold densities were processed. In other words we can state that doping by either element will produce on the average identical thresholds. These data apply to lasers operating at liquid nitrogen temperatures.

Regarding the fabrication of the *p*-type layers a number of studies [12-14] have revealed that the use of zinc as an acceptor impurity is preferable compared to germanium. Zinc is a more difficult material from the viewpoint of technology since it has a higher vapor pressure and a higher diffusion coefficient. However it produces shallower impurity levels as a doping impurity which makes it possible to obtain layers with higher electrical conductivity.

Using germanium it is difficult to obtain high electrical conductivity in the *p*-layer for two reasons:

1) the solubility of germanium in an AlGaAs solid solution drops with growth of the aluminum concentration;

2) The germanium produces deeper (compared to zinc) acceptor levels which causes an increase in resistance particularly at lower temperatures.

The high zinc vapor pressure serves to produce reproducibility of results due to the overall zinc contamination of the growing system. In order to reduce this influence we used a lower temperature of the process: of the order of 700°C and germanium was used as the doping impurity for growing contact layer 4 (see Fig. 1).

However subsequent statistical processing analogous to that done for *n*-emitters has revealed that the threshold current density at liquid nitrogen temperatures for 20 lasers with *p*-layers doped only with zinc and 15 lasers doped by zinc and germanium is statistically non-differentiable. A possible explanation is that the threshold current density is also influenced by other factors that are not accounted for in statistical processing. The discreteness factor $d = r^2$ (r is the correlation coefficient) is 0.43 for the correlation between the threshold current density and the weight of zinc introduced to the melt. This means that a change in the doping level of only 43% determines the change in the threshold current density. The remaining change is caused by other factors.

Figs. 4 and 5 give point diagrams for the dependence of the lasing threshold current density on the thickness of the *n*- and *p*-emitters. Regression line equations were calculated for these point diagrams which were then used to plot the curves. The primary fact of interest is that the thresholds are extraordinarily weakly dependent on the thickness of the *n*-emitters. In this case we can even speak of the absence of a correlation between these characteristics. The correlation coefficient is equal to 0.21 which also indicates a weak correlation.

The dependence of the threshold lasing current density on the thickness of the *p*-emitter is more clearly expressed, although, judging by the correlation coefficient equal to 0.37, the relation can be classified as only averaged. The discreteness factors (0.044 for the *n*-emitters and 0.14 for the *p*-emitters) reveals a relatively insignificant contribution of emitter thickness to the change in threshold lasing current density.

Fig. 6 shows the lasing threshold current density plotted as a function of radiation wavelength for the case of aluminum doping of the laser active medium. These curves are regression curves based on data for 30 lasers operating at room temperature and for 34 lasers operating at liquid nitrogen temperature. For the cooled lasers the curve is analogous to that obtained in study [13] which for room temperature lasers the curve is analogous to those obtained in studies [14, 15]. The curves were derived by the least squares method and therefore they continue the trend towards a continuous increase in the thresholds and towards shorter wavelengths. At the same time the lasing conditions begin to change and the threshold currents jump sharply at a sufficiently large aluminum percentage as indicated at the beginning of the present study. We can state on the basis of our results that the shortest wavelengths at which lasing threshold current densities remain acceptable are 670-690 nm for cooled lasers and 750 nm for lasers operating at room temperature. These data are in good agreement with those given in the literature.

Lasers with broad contacts were fabricated to achieve high radiation power. They are normally mounted in holders to improve heat entrainment from the active region since two-way heat release is provided in such holders. The stripe contacts were fabricated at 15, 20, 50, and 100 μm in width. Silicon dioxide obtained by pyrolysis of

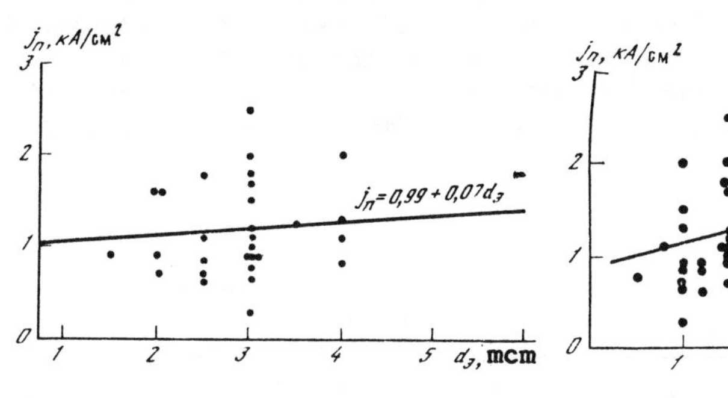

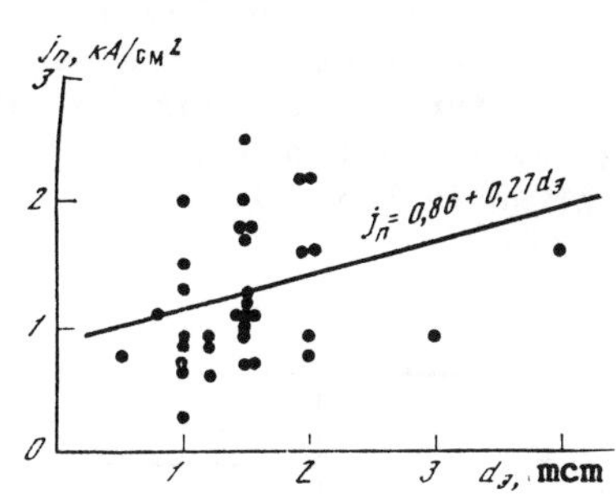

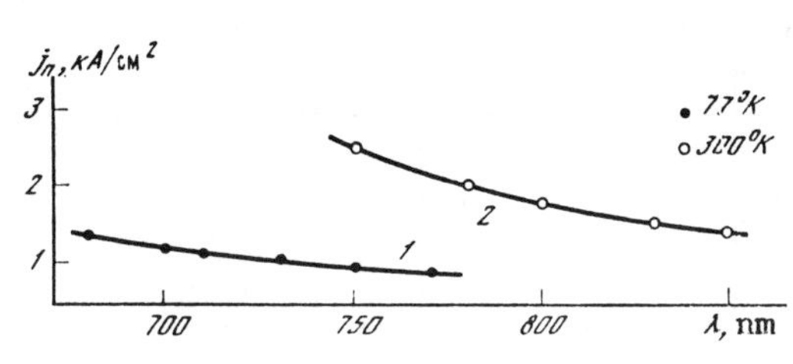

Fig. 4. Point diagram of the dependence of j_T on the n-emitter thickness

Fig. 5. Point diagram of the dependence of j_T on the p-emitter thickness

Fig. 6. The lasing threshold current density as a function of radiation wavelength
1 - T = 300 K, $j_T = 1/[0.39+0.0034\ (\lambda - 750)]$; 2 - T = 77 K, $j_T = 1/[0.741+0.0047\ (\lambda - 680)]$

tetraethoxalyne has been used as the insulator in stripe lasers. A typical thickness of SiO_2 film is 0.2-0.3 μm. A contact is applied to the p-side by vacuum deposition in the form of a two-layered chromium-gold system. A eutectic germanium-gold layer was used as the contact on the n-side. The contacts were fused at 450°C for two to three minutes. Gold up to 5 μm in thickness was then applied to both sides voltaically.

Stripe contact lasers are soldered by means of indium onto copper heat sinks. The p-side of the laser made contact with the heat sink. The series resistivity per unit of area was less than $2 \cdot 10^{-4}$ ohm·cm^2.

3. Laser Characteristics

The E-I and P-I characteristics were measured together with the lasing wavelength in d.c. operation in the fabricated lasers. The threshold characteristics were measured in pulsed operation.

S-shaped characteristics (Fig. 7, curve 1) were identified in measurements of the E-I characteristics of lasers from initial experiments with doped n-emitters. It has been suggested that such characteristics are related to an insufficient doping level. We carried out the following special experiments to test this hypothesis. As we see from Fig. 7 with growth of the doping impurity concentration the section of the E-I characteristic with negative resistance gradually disappears. Therefore the experiment relating to curve 4 was used as the basis for further experiments with insignificant deviations of the doping level. The doping conditions for the different curves of Fig. 7 are given in Table 1.

Table 1

Series No.	Layer doping, at.%			
	n-AlGaAs, doping by Sn	Active layer	*p*-AlGaAs doping by Zn	*p*-GaAs doping by Ge
1	4.2	-	0.5	3.6
2	4.2	-	0.5	3.6
3	6.6	-	0.5	3.6
4	13	-	0.5	6

Table 2

Series No.	Cavity Length L, μm	P_{max}, mW	n_d
1	175	180	0.45
2	200	215	0.51
3	300	220	0.50
4	375	210	0.43
5	475	160	0.36
6	800	120	0.25

Experimental data on the output power of lasers are given in Figs. 8-10. A FD-24K calibrated silicon photodiode was used for power measurements together with calibrated neutral light filters for attenuation when necessary.

Fig. 8 shows the total radiated power as a function of the pump current for wide lasers. The measurements were carried out on an integrating sphere. The lasing wavelength was 7.5 nm with laser dimensions of 400×200 μm^2. The output beam power from the cryostat was 0.5 W for lasers in this series.

The P-I characteristics of lasers with strip contacts are given in Fig. 9. In this case the power was measured by a photodiode in the immediate vicinity of one of the laser mirrors. The measurements were carried out for a series of lasers with different cavity lengths. The width of the stripe contact in all lasers of this series was identical and equal to 100 μm. The radiation wavelength is 700 nm. The cavity length, the maximum power obtained from a single mirror P_{max} and the total external differential quantum efficiency η_d calculated after power addition radiated from both laser sides corresponding to the different curves in Fig. 9 are given in Table 2. It is clear from the curves and the data in Table 2 that the highest radiated power is generated by lasers with a cavity length lying between 200 and 400 μm. For these lasers P_{max} is 220 mW. The maximum power of lasers with 50 μm stripe contact width is 80 mW at 685 nm. P_{max} reached 70 mW (wavelength of 680 nm) for lasers with 20 μm wide stripe contacts.

Fig. 10 give the P-I characteristics of laser diodes operating in CW conditions at room temperature. The radiation power was measured from a single mirror. The width of the radiating section of the diodes was 15 μm with a cavity length of 250 μm. The maximum power per mirror at 840 nm was 65 mw and 40 mw at 780 nm. The external differential quantum efficiency from a single mirror is 17 and 16%, respectively.

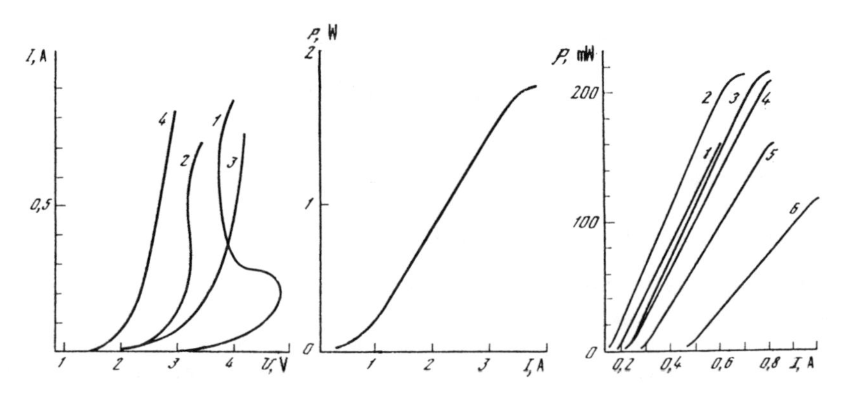

Fig. 7. E-I characteristic of laser diodes as a function of doping level of the n-emitter at T = 77 K
Doping data for the curves in Table 1

Fig. 8. P-I characteristic of broad contact lasers in CW operation at T = 77 K

Fig. 9. P-I characteristics of stripe contact lasers at various cavity lengths at T = 77 K
Data for curves in Table 2

Based on results from a series of lasers operating at liquid nitrogen temperatures the dependence of a quantity inverse to the differential quantum efficiency is plotted as a function of cavity length. As we know the expression for the inverse differential quantum efficiency can be written as

$$1/\eta_d = (1/\eta_{stim.})[1 + \alpha_i L/\ln(1/R),$$

where $\eta_{stim.}$ is the internal quantum efficiency at the lasing threshold; α_i are total internal losses; L is cavity length; R is the reflection coefficient of the mirrors.

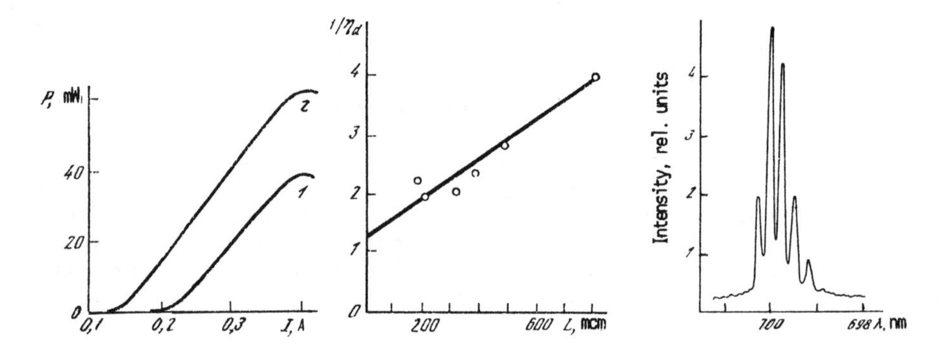

Fig. 10. P-I characteristics of laser diodes operating in CW conditions at T = 300 K
1 - radiation at 780 nm; 2 - radiation at 840 nm

Fig. 11. A quantity inverse to the external differential quantum efficiency plotted as a function of cavity length at T = 77 K

Fig. 12. Lasing spectrum of laser operating at 150 mW at T = 77 K

The derived relation shown in Fig. 11 makes it possible to determine $1/\eta_{stim.}$ and α_i. In extrapolating to L = 0 we obtain $\eta_{stim.}$ = 0.80. Solving the equation for α with any L yields a loss value of α_i = 11 cm^{-1}. However when the cavity length is increased the total internal differential quantum efficiency η_d diminishes.

Most lasers have a monomode radiation spectrum. Fig. 12 shows the lasing spectrum of a laser producing 150 mW. The spectral width at half amplitude is 5 Å.

4. Conclusions

Liquid nitrogen cooled AlGaAs/GaAs laser diodes are the simplest method of obtaining laser radiation of 0.1-1 W in power in CW operation in the visible (red) portion of the spectrum. In spite of well-known successes in the fabrication of laboratory samples of room-temperature diodes by growing layers from hetero-organic compounds, in the near future results presented here will serve as an actual practical solution to this problem, since they are based on existing technical and technological methods. The primary conclusion of the present study is the possibility for obtaining CW laser radiation at upwards of 0.1 W from a single AlGaAs/GaAs laser diode heterostructure at

~700 nm. The actual values of the maximum lasing power from separate samples are greater than 1 W which demonstrates the promise of this technology as well as the possibility for satisfying the stringent requirements on laser radiation sources operating in the visible range (in the applications discussed in the introduction). Further improvements to these heterostructures will make it possible to improve the operating temperature of the laser diodes which will in turn simplify the corresponding auxiliary equipment.

References

1. Eliseev, P.G., Ismailov, I., Krasil'nikov, A.I., et al. "The temperature dependence of the threshold current of injection lasers and CW operation under liquid nitrogen cooling" ZhTF, 1966, vol. 36, no. 12, pp. 2211-2214.

2. Eliseev, P.G., Ismailov, I., Man'ko, M.A., et al. "Spontaneous and coherent radiation from epitaxial *p-n*-junctions" FTP, 1967, vol. 1, no. 9, pp. 1315-1318.

3. Belov, B.I., Eliseev, P.G., Zakharov, Yu.P., et al. "Laboratory diode holders for injection lasers and electroluminescence diodes" PTE, 1968, no. 2, pp. 187-188.

4. Eliseev, P.G., Strakhov, V.P. "CW semiconductor laser with an output power of several watts" ZhTF, 1970, vol. 40, no. 7, pp. 1564-1565.

5. Andreev, V.M., Dolginov, L.M., Tret'yakov, D.N. "Zhidkostnaya epitaksiya v tekhnologii poluprovodnikovykh priborov" [Liquid epitaxy in semiconductor device technology] Moscow: Sov. Radio, 1975, 328 pp.

6. Tiller, W.A. "Theoretical analysis of requirements for crystal growth from solution" J. CRYST. GROWTH, 1968, vol. 2, no. 1, pp. 69-72.

7. Harris, J.S., Snyder, W.L. "Homogeneous solution growth epitaxial GaAs by tin doping" SOLID STATE ELECTRON., 1979, vol. 12, no. 5, pp. 337-341.

8. Crossley, J., Small, M. "Some observations of the surface morphologies of GaAs layers growth by liquid phase epitaxy" J. CRYST. GROWTH, 1973, vol. 19, no. 3, pp. 160-163.

9. Minden, H.T. "Constitutional supercooling in gallium arsenide liquid phase epitaxy" J. CRYST. GROWTH, 1970, vol. 6, no. 3, pp. 228-231.

10. Lockwood, H.F., Ettenberg, M. "Thin soliton multipole layer epitaxy" J. CRYST. GROWTH, 1972, vol;. 15, no. 1, pp. 81-83.

11. Zherdev, A.A., Talenskiy, O.N. "Epitaxial growing of gallium arsenide from a thin solution/melt layer with vertical substrate placement" ELEKTRON. TEKHNIKA. Ser. 11, 1978, vol. 1, no. 4, pp. 76-80.

12. Iton, K., Inoue, M., Teramoto, I. "New heteroisolation stripe geometry visible light-emitting lasers" IEEE J. QUANT. ELECTRON., 1975, vol. 11, no. 7, pp. 421-431.

13. Kressel, H., Hawryto, F.Z. "Red-light-emitting $Al_xGa_{1-x}As$ heterojunction laser diodes" J. APPL. PHYS., 1973, vol. 44, no. 9, pp. 4222-4225.

14. Kressel, H., Hawryto, F.Z. "Red-light-emitting laser diodes operating CW at room temperature" APPL. PHYS. LETT., 1976, vol. 28, no. 10, pp. 598-600.

15. Wada, M., Itoh, K., Shimizu, H., et al. "Very low threshold visible TS lasers" IEEE J. QUANT. ELECTRON., 1981, vol. 17, pp. 776-789.

LUMINESCENCE OF CdS AS A FUNCTION OF CRYSTAL POSITION IN THE GROWTH ZONE

N.A. Martovitskaya, S.A. Pendyur, O.N. Talenskiy

ABSTRACT

This study calculates the vapor condensation rate distribution for an A^2B^6 compound along a growing tube for growing crystals by resublimation in an inert gas flow. The change in photoluminescence intensity of the edge band (λ = 5130 Å) and a group of exciton lines (λ = 4875, 4880, and 4910 Å) of crystals grown at various points in the growth zone was investigated. It was found that the intensity of these luminescence lines changes predictably along the growth tube. This change is compared to the condensation rate distribution and is related to the crystal defects occurring due to the differential in growth rates at different points in the growth zone.

Introduction

Cadmium sulfide is often used as a model substance to investigate many effects observed in semiconductors. A significant fact here is that this material is easy to use: it is possible to fabricate crystals of various type, different levels of structural perfection and with properties that can be controlled over a wide range using sufficiently simple techniques. The latter characteristic is related to the fact that the electrical properties of CdS, like those of other A^2B^6 semiconductor compounds, are highly dependent on the presence of electrically active centers normally associated with local failures of the crystal lattice. The number of defects can be changed using different methods, including technological methods. However most of these methods involve altering prepared crystals (deformations, annealing, etc.) The problem of growing A^2B^6 crystal compounds with predetermined properties is largely an unsolved problem. The capabilities of probably the simplest and most widely employed technique (resublimation in an inert gas flow [1]) have not been considered in sufficient detail. Below we present our observations on this process that will, we believe, make it possible to approach the selection of grown crystals from a new vantage point by predicting the probable nature of their properties in advance.

Description of the Resublimation Process

The process of growing A^2B^6 compound crystals by resublimation in an inert gas flow employs a very simple industrial assembly consisting of a resistance furnace with an internal cylindrical cavity. The furnace contains a silica tube through which the inert gas flow is directed. The front of the tube (with respect to flow) contains a silica boat containing the initial material normally in powder form. A specific temperature distribution is maintained along the tube by developing two or more heating zones in the oven. The temperature is at a maximum in the region containing the boat with the blend; the temperature gradually drops in the direction of gas flow. In the high-temperature region the cadmium sulfide is vaporized and is transported by the gas flow to the lower temperature region where thermodynamic supersaturation conditions have been established, and the vapor condenses, thereby forming a crystal of any specific shape and size which both depend on the degree of supersaturation, the cooling rate and a number of other factors. Crystallization can take place directly on the tube walls. Normally the tube contains devices to decelerate the flow and to prevent multiple crystallization sites [2-3].

This process is subject to analytic description, which is not common in crystal technology [4]. For this process we select two cross-sections separated by a distance Δx in the tube (Fig. 1). (The x-axis lies along tube.) It is assumed that the cross-sectional area s of the tube is constant along its length. In steady-state conditions the material balance in an elementary volume bounded by the selected cross-sections can be written as:

$$\Delta m = wsc\,(x_2) - wsc\,(x_1) = ws\Delta c, \tag{1}$$

where Δm is the quantity of matter reentering the solid phase; w is the rate of gas flow; $c(x_i)$ is the vapor concentration in the corresponding cross-sections. It is obvious that since the temperature in the vaporization zone is higher than in the growth zone the vapor source does not limit vapor flow and is unlimited in the same sense that diffusion is unlimited. We will divide (1) by the elementary volume $\Delta v = s\Delta x$ and with an unlimited reduction in Δv we obtain

$$dN = \frac{dm}{dv} = w\frac{dc}{dx}\,. \tag{2}$$

dN is the quantity of matter reentering the solid phase per unit of volume and per unit of time, i.e., the rate of vapor condensation. Therefore the determination of condensation rate is reduced to finding the positional concentration gradient. We transform (2) to simplify the experimental determination of dN:

$$dN = w \frac{dc}{dT} \frac{dT}{dx}. \tag{3}$$

dT/dx is the positional temperature gradient: a quantity determined by thermograph measurements in experiment. In order to search dc/dT- the temperature concentration gradient - we will use Clapeyron's equation:

$$pv = \frac{m}{M} RT. \tag{4}$$

Dividing both halves of (4) by v and remembering that $m/v = c$, we obtain

$$c = \frac{p}{T} \frac{M}{R}, \tag{5}$$

where p is vapor pressure, T is temperature, M is the molecular weight of the test material and R is the gas constant.

The temperature dependence of vapor pressure has been investigated in sufficient detail for various substances (see, for example, [5]) and hence it is possible to find the changes in concentration c as a function of temperature. Differentiating (5) we obtain

$$\frac{dc}{dT} = \left(\frac{dp}{dT} - \frac{p}{T}\right) \frac{M}{RT}.$$

The rate of gas flow can be expressed through the gas rate Q - a quantity normally measured in experiment. Accounting for the correction to the difference between the temperature T at which the rate is measured and the temperature T in the test tube we can write

$$w = \frac{Q}{\pi r^2} \frac{T}{T_0}, \tag{6}$$

where r is the cross-sectional radius of the tube.

Unifying (3), (5), and (6) we obtain

$$dN = \frac{Q}{\pi r^2} \frac{M}{RT_0} \left(\frac{dp}{dT} - \frac{p}{T}\right) \frac{dT}{dx}. \tag{7}$$

According to (7) the rate of condensation depends on the geometry of the growing assembly and technological parameters: gas rate, pressure and temperature. It is possible to use the selected parameters of the process to find the distribution of the condensation rate of the matter along the reactor.

We carried out the following experiment to verify this relation [4]. A rod with graduated segments 1.8 cm in length was placed along the tube axis. The cadmium selenide crystal growing process was then activated. The segments with grown crystals were weighed and the difference in their weight compared to before the process determined the weight of the crystal sediment in each 1.5 cm section along the tube

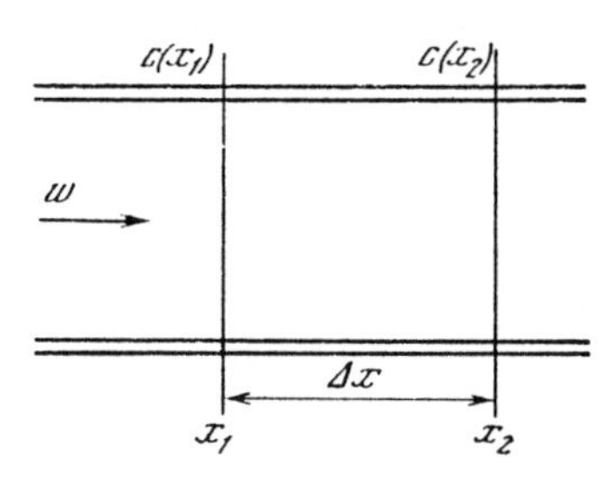

Fig. 1. Configuration for determining condensation rate

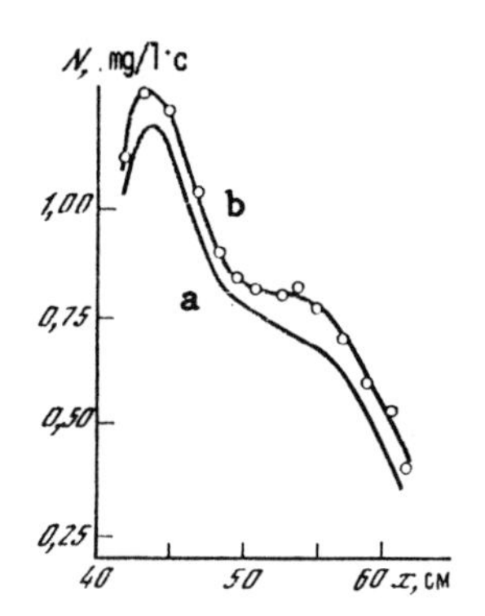

Fig. 2. Comparison of the calculated condensation rate to experiment
a - calculation; b - weight distribution of the CdSe condensate

axis. The curve drafted on the basis of these data was compared to the curve calculated by the parameters of the process. As we see from Fig. 2 the correlation between these curves is quite satisfactory. The knee in the curves results from a specially-generated differential in the temperature profile of the oven. The comparison was purely qualitative since the material condenses onto both the tube walls and partially onto the reactor walls. The scale plotted on the Y axis refers to a theoretical curve. The fact that the vapor condensation process has a maximum can be determined by examining the growth zone. The crystals are thicker in the vicinity of the maximum. The ascending branch of the condensation rate curve begins as early as the vaporization region (over the boat containing the blend) and therefore the curve maximum is at the origin of the crystal growth zone. We can establish visually that needle-shaped crystals with a length-to-thickness ratio of 10^2-10^3 are the primary crystals grown in the section coinciding with the decaying branch.

The number of needle crystals grows as the slope of this section of the curve increases. As has been established [6-7] thin (5-50 μm) crystal platelets grow due to the expansion of one of the needle faces which has a hexahedral shape in cross-section. Hexahedral prisms (length to thickness ratio of 10-15) grow near the condensation rate maximum. It follows from these observations that the rate of vapor condensation in this process determines the configuration and number of crystals in each tube/reactor cross-section.

Photoluminescence of Crystals as a Function of Localization

Many investigators who have studied the properties of CdS platelet crystals grown by this technique have noted that the crystals have significantly different [8] luminescence spectra. For illustrative purposes we will provide the photoluminescence spectra of our crystals (Fig. 2). The first fact that becomes apparent is that there are crystals having edge luminescence without groups of exciton bands (a) and vice versa: crystals having exciton bands with zero edge luminescence (c and d). Bands with different wavelengths dominate in the group of exciton bands in the various crystals. Finally there are crystals in which the intensity of the edge radiation and the exciton radiation are comparable (b).

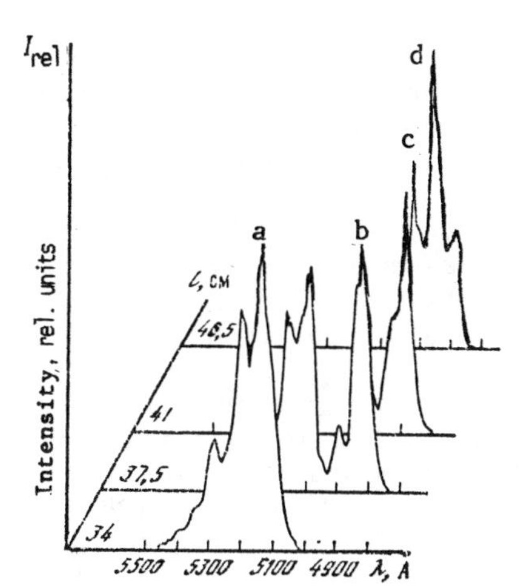

Fig. 3. Photoluminescence spectra of crystals measured at different distances from the reactor end
a - sample 2002-1; b - 200-2; c-2002-8; d - 2001-7

Fig. 4. Change in luminescence intensity of "edge" band (curve 1) and of the neutral donor bound exciton i_2 (curve 2) along the growth zone

Fig. 5. The change in luminescence intensity of the neutral acceptor-bound exciton I_1 (curve 1) and of the free exciton (curve 2) along the growth zone

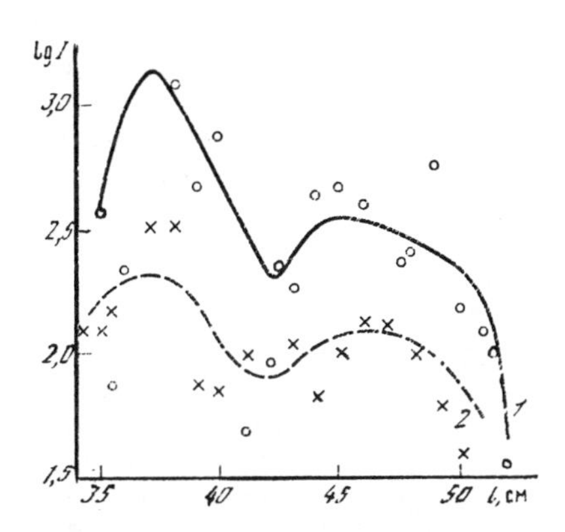

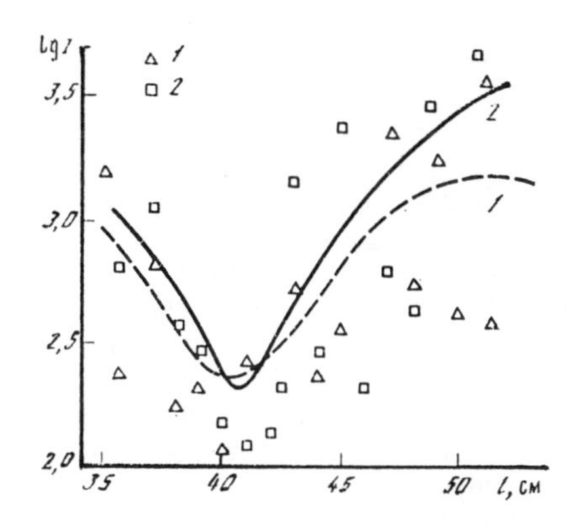

Based on the differential in the condensation rate of CdS vapors described in the preceding section it is natural to attribute the differences in the photoluminescence spectra (as well as in other properties) to the fact that the crystals were grown at different points in the crystallization zone and, therefore, at different condensation rates. The rate of condensation itself directly influences the crystalline perfection of the crystals. The properties of A^2B^6 compound crystals are determined primarily by their defects. A direct confirmation of this arrangement are the spectra shown in Fig. 3 from crystals grown at different points in the growth zone. The spectra show the distances from the end of the growth tube in centimeters.

It would be interesting to determine the regularities of the behavior of crystals properties along the growth zone and to compare these to the vapor condensation rate curve. A comparison was attempted for the photoluminescence of CdS. Processes whose parameters were maintained at an identical level within possible accuracy were implemented. For example, the temperature distribution from process to process varied by less than 0.5°. Ripple-free platelets were selected from the grown crystals. In our assembly a silica platelet was mounted longitudinally in the growth zone; glass moulds serving as crystal growth sites were soldered perpendicular to this platelet. The wafer containing the crystals was removed from the tube after this process. The distance from the moulding from which the crystal was taken to the platelet edge was fixed. Since the position of the silica platelet was fixed in the growing tube, it is possible to determine the distance from the selected crystal to the tube end. This was done to establish a correlation with the thermogram and, consequently, the vapor condensation rate curve which was drafted based on the distances from the tube end.

The spectra were recorded using the standard method (see, for example, [9]) at liquid nitrogen temperatures. Luminescence was excited by means of a helium-neon laser. Crystals obtained in five processes in January of 1981, in three processes in April through May of 1981, in four processes in June of 1983, and in two processes in June of 1985 were investigated. In 1981 and 1985 the photoluminescence spectra were recorded by means of a DFS-12 spectrograph while in 1983 a SF-8 spectrophotometer was used. The RMS error in measuring the relative intensity of the spectral lines in all four series of measurements was statistically equivalent, which made it possible to correlate all values in a single data file. Below we provide results for different luminescence bands.

"Edge" luminescence. The zero-phonon "edge" (green) luminescence band was recorded with a maximum near λ = 4130 Å. There were 77 bands in all. The data were averaged for each coordinate. These averages are given in Fig. 4. Further processing involved use of the averaging technique (three times) [10] where each value of y_i was assigned an average value $\bar{y}_i = (y_{i-1} + y_i + y_{i+1})/3$. The derived points were connected by a curve. In Fig. 4 curve 1 corresponds to such averaged values of the luminescence intensity of the "edge" band.

Radiation from neutral donor-bound exciton. The luminescence band with a maximum in the vicinity of λ = 4880 Å was recorded in 39 samples. The statistical processing routine was the same as for "edge" luminescence. The data are given in Fig. 4 (curve 2).

Radiation from neutral acceptor-bound exciton. Luminescence bands with a maximum in the 4910–4940 Å range were identified with radiation from this exciton. Radiation in this band was recorded for 45 samples. Processing results are given in Fig. 5 (curve 1).

Free Exciton Radiation was measured in 84 samples. The luminescence band had a maximum near 4875 Å. The data are given in Fig. 5 (curve 2).

The intensity of all luminescence bands investigated here have a rather clearly expressed minimum corresponding to distances of 39.5-42.5 cm from the end of the growth tube. The second feature that becomes evident is the similarity of the curves of the "edge" luminescence and the luminescence of the neutral donor-bound exciton, on the one hand, and a like similarity between the radiation from the neutral acceptor-bound exciton and the free exciton, on the other. In order to better clarify this similarity the curves are plotted in pairs in Figs. 4 and 5, respectively.

Discussion of Results

Initially it would be best to note the nigh level of crystalline perfection of the cadmium sulfide platelets. The smooth, ripple-free CdS platelet crystals, after repeated X-ray topographical analyses turned out to be free of dislocations, which coincides with the results from study [6]. Their luminescence features therefore cannot be attributed to "macrodefects", such as crystalline block boundaries, twins or similar dislocations. It has long been known [8] that characteristic luminescence bands are related to "microdefects" of the vacancy type in cadmium and sulfur sublattices or interstitial atoms.

It was noted at the beginning of this chapter that there is a certain relation between the technological growing conditions and the physical properties of the derived crystals. In order to determine the relation to luminescence features Fig. 6 gives the calculated rate of condensation of the cadmium sulfide vapors. The calculation was carried out for the technological parameters used to grow crystals whose luminescence was investigated. In order to carry out the calculation using equation (7) it is necessary to known the variation in vapor pressure as a function of temperature. The relation proposed by Pogoreliy [10] was used for this purpose:

$$\lg p = -7420/T + 6{,}35\lg T - 12{,}91. \quad (8)$$

Subsequent research yielded a virtually identical relation [12].

The first aspect that appears from a comparison of the condensation rate curve to curves of the change in luminescence intensity is the correlation between the maximum of the condensation rate and the minima of the luminescence curves. Therefore the physical properties, particularly the luminescence, are indeed in close correlation with the crystal growing conditions. For further consideration of this relation we note that, as we know [12], cadmium sulfide vaporizes with dissociation by the formula

$$CdS \rightleftarrows Cd + {}^1/_2S_2.$$

Since the components exist separately in the vapor phase we can formally assume that their condensation also occurs separately, while the compound forms on the crystal surface. We can then calculate the condensation rates of the components with this assumption. We carried out this calculation, borrowing the temperature dependencies of the vapor pressure from study [5]:

$$\text{for Cd}: \lg p = 2{,}52874 - 5180{,}713/T - 0{,}00084961T + 2{,}05627 \lg T;$$
$$\text{for S}: \lg p = 109{,}05777 - 8756{,}69/T + 0{,}01105787T - 35{,}68404 \lg T.$$

The calculated component condensation rates are also given in Fig. 6. It is best to avoid a direct comparison of the absolute condensation rates of the components and CdS. The vapor pressures for the elements given by A.N. Nesmeyanov were derived by direct measurement during the direct vaporization of cadmium and sulfur samples. The conditions are completely different in the vaporization of CdS. Specifically, the initial relation between the components is given. Therefore the comparisons can be relative only, and Fig. 6 provides data in a form normalized to the maximum value for each component and CdS. The fact that the maxima of the condensation rates virtually coincide can be attributed to the determinate influence of the temperature distribution on the condensation rate curve.

An important conclusion from a comparison of the curves in Fig. 6 is that the crystal growing conditions in the segment of the growth zone below the condensation rate maximum will be different from the conditions in the region above this maximum. This is based on the fact that the relative condensation rate of sulfur exceeds the relative condensation rate of cadmium in the initial growth region (below the condensation rate maximum). Beyond the maximum the cadmium condenses from the vapor phase at a relatively rapid rate. Consequently the relative condensation rates of the components are quite similar in the vicinity of the maximum. Therefore this segment of the growth zone will produce crystals of either stoichiometric composition or crystals in which the number of luminescence centers, attributable to various defects, is mutually compensated.

We will calculate the ratio of luminescence intensities of the neutral acceptor-bound exciton (I_1) and the neutral donor-bound exciton (I_2). At the front of the growth zone it is ~6 and at the end it is ~21. Near the minimum of the luminescence intensities (35 cm from the origin) I_1/I_2 = 1.58: also a minimum value. This provides basis for the contention that the crystals are virtually completely compensated in this range. Study [13] has demonstrated that the I_F/I_L ratio where I_F and I_L are the intensities of the lines A_F and A_L observed near the exciton ground state n = 1 in the forbidden geometry $E \| C$ (E is the polarization vector of light; C is the optical axis of the crystal) can serve as a quantitative measure for estimating the free exciton lifetime τ. At our request V.V. Travnikov and V.V. Krivolapchuk determined I_F/I_L for crystals selected from three experiments and they found that this ratio is at a minimum near the luminescence intensity minimum. This result confirms the conclusion that compensated crystals grow near the condensation rate maximum. In the case of stoichiometry we should expect maximum exciton lifetimes.

Today the sulfur vacancies V_S are generally regarded as responsible for the "edge" radiation and neutral donor-bound exciton radiation. The growth in intensity of the "edge" luminescence and of the exciton I_2 observed near the condensation rate maximum becomes clear from this viewpoint (see Fig. 4). Since the condensation rate of sulfur here is less than that of cadmium, sulfur vacancies form in conditions of a relative shortage of sulfur atoms in the crystal growing process.

The decay in intensity of these luminescence bands at the end of the growth zone is most likely due to the enhanced crystal structure due to the general slowdown in the condensation rate. However the high intensity of the "edge" band and of the exciton I_2 at the beginning of the growth zone before the condensation rate maximum is not encompassed in this concept, since it is difficult to imagine that with a relative excess of sulfur atoms, sulfur vacancies would form in the lattice. It is more logical to assume that the excess sulfur atoms appear at the interstitial sites. It is possible to find a way out of this difficulty by relying on the viewpoint from study [13]. The authors of this study demonstrated in experiments involving the electron bombardment of CdS that "edge" luminescence can be attributed to the interstitial sulfur. Kingstone, et al. [14] who detected five lines in the "edge" radiation band, assigned two of these lines to excess sulfur, since they are found in crystals grown by adding hydrogen sulfide to the argon flow, producing excess sulfur vapor pressure. However if we accept this viewpoint we must recognize that two different defects are responsible for the "edge" and the i_2 bands: S_1 at the beginning of the growth zone and V_S at the end of the growth zone. Obviously with this assumption we would expect certain, although quite small, differences in the energy characteristics of these centers. Evidently this is the case as demonstrated by Fig. 7 which shows the variation of the maximum of the "edge" band on the wavelength scale as a function of crystal growth position. The curve is drafted based on results from four growing process for 29 crystals. The maximum spread of values is 5.8 meV.

The I_2 exciton is also related to the presence of sulfur vacancies (see, for example, [15]). Our data reveal a single origin of this band and of the "edge" luminescence band since the progression of the luminescence intensity curve of this exciton closely follows the "edge" luminescence curve. The correlation coefficient between the intensities of these two luminescence bands calculated by our data was $r = 0.833$. The high value of r reveals a close relation between these quantities.

The neutral acceptor-bound exciton I_1 is attributable to centers in which the vacancies in the cadmium sublattice are displaced by Li or Na atoms [3]. The behavior of I_1 prior to the condensation rate maximum where cadmium concentration lags behind the sulfur concentration corresponds to this concept; in this range we can expect the formation of a large number of cadmium vacancies.

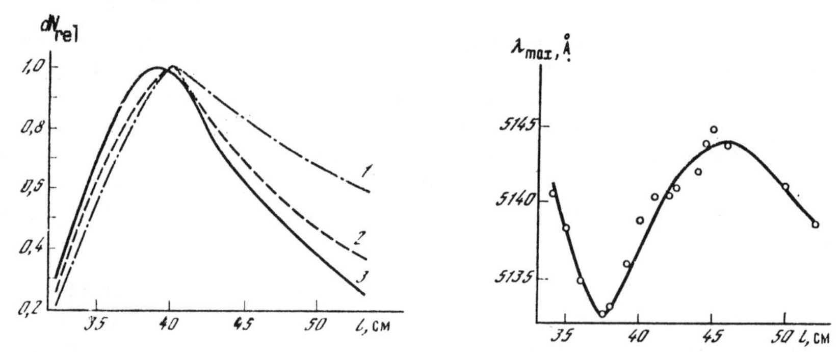

Fig. 6. Condensation rate distribution along the growth zone
1 - Cd; 2 - S; 3 - CdS

Fig. 7. Wavelength variation of the luminescence maximum of the "edge" band along the growth zone

The behavior of the branch of intensity curve I_1 beyond the maximum of the condensation rate does not fit within the framework of this viewpoint, since we would expect in this range an excess of cadmium, and the excess cadmium would function as a donor.

The free exciton bears information on the "ideal nature" of the lattice [18] and in this sense confirms that the crystals have defects near the condensation rate maximum, since the minimum of the luminescence intensity of the free exciton is here. The similarity of its behavior to the progression of the luminescence curve of the I_1 exciton is surprising. The correlation coefficient between the intensities of these excitons is 0.941 and reveals a near functional dependence and that a common cause would be expected behind their behavior.

In conclusion we can summarize by stating that when growing CdS crystals by resublimation in an inert gas flow, the composition of the crystals will vary along the growth zone. The electrical properties (at least the photoluminescence) will also change in connection with the change in composition. Variations in the component condensation conditions along the growth zone are responsible for changes in composition and properties. Evidently similar changes can be expected with other crystal growing techniques with two or more components where transport in the gas phase will occur.

In conclusion the authors wish to express their gratitude to N.I. Timcheniko for the computer-generated condensation rate calculations and to E.V. Alieva and M.V. Luk'yanovich for measurements of the photoluminescence spectra of a number of samples.

References

1. Vil'ke., K.-T. "Vyrashchivanie kristallov" [Crystal growing] Leningrad: Nedra, 1977, pp. 91-92.

2. Corsini-Mena, A., Elli, M., Paorici, C., Pelosini, L. "A simple versatile method to grow cadmium-chalcogenide single crystals" J. CRYST. GROWTH, 1971, vol. 8, no. 3, pp. 297-301.

3. Nassau, K., Shiever, Y.W. "Vapor growth of II-VI compounds and the identification of donors and acceptors" J. CRYST. GROWTH, 1972, vol. 13/14, pp. 375-382.

4. Yakushin, V.K., Talenskiy, O.N. "The fabrication and properties of A^2B^6 and A^4B^6 semiconductor compounds and related solid solutions" In: Tez. dokl. Pervoy Vsesoyuz. nauch.-tekhn. konf. (1-4 Fevral. 1977) [Topic Papers of the First All Union Scientific and Technical Conference (1-4 February 1977)] Moscow: MISnS, 1977, pp. 31-32.

5. Nesmeyanov, N.N. "Davlenie para khimicheskikh elementov" [The vapor pressure of chemical elements] Moscow: Izd-vo AN SSSR, 1961.

6. Chikawa, J., Nakauama, T. "Dislocation structure and growth mechanism of cadmium sulfide crystals" J. APPL. PHYS., 1964, vol. 35, no. 8, pp. 2493-2501.

7. Dierssen, G.H., Gabor, T. "Growth mechanism of CdS platelets" J. CRYST. GROWTH, 1972, vol. 16, pp. 99-109.

8. Thomas, D.G., Hopfield, J.J. "Exciton states and band structure in CdS and CdSe" PHYS. REV., 1962, vol. 128, pp. 2135-2148.

9. Dospekhov, B.A. "Metodika polevogo opyta" [Field testing technique] Moscow: Agronromizdat, 1985, pp. 294-295.

10. Pogorelyy, A.D., "Thermal dissociation of zinc and cadmium sulfides" ZhFKh, 1948, vol. 22, no. 6, pp. 731-745.

11. Goldfinger, P., Jeunehomme, M. "Mass spectrometric and knudsen-cell vaporization studies of group 2B-6B compounds" TRANS. FARADAY SOC., 1963, vol. 59, no. 492, phase transition. 12, pp. 2851-2867.

12. Travnikov, V.V., Krivolapchuk, V.V. "Spectral criteria for estimating the lifetime of free excitons" FTP, 1985, vol. 19, no. 6, pp. 1092-1099.

13. Kulp, B.A., Kelley, R.H. "Displacement of the sulfur atom in CdS by electron bombardment" J. APPL. PHYS., 1960, vol. 31, no. 6, pp. 1057-1061.

14. Kingstone, D.L., Green, L.C., Croft, L.W. "Edge emission bands in cadmium sulfide" J. APPL. PHYS., 1968, vol. 39, no. 13, pp. 5959-5955.

15. Brodin, M.S., Davydova, N.A., Shabliy, I.Yu. "The influence of laser irradiation on the optical spectra of CdS monocrystals" FTP, 1976, vol. 10, no. 4, pp. 625-631.

16. Lider, K.F., Novikov, B.V., Permogorov, S.A. "Application of bound-exciton optical spectra in the study of radiation damage in crystals" PHYS. STATUS SOLIDI, 1966, vol. 18, no. 1, pp. k1-k3.

SUBJECT INDEX